The PRICE of ALTRUISM

The PRICE
of ALTRUISM

George Price and the Search
for the Origins of Kindness

OREN HARMAN

W. W. NORTON
NEW YORK   LONDON

Photograph credits: page x, Library of Congress; page 8 (Kropotkin), Library of
Congress; page 38 (George and Edison Price), courtesy of the Price family; page 38
(George, Alice, and Edison Price), courtesy of the Price family; page 58 (Fisher),
Library of Congress; page 58 (Haldane), Raphael Falk; page 84 (George and Julia Price),
courtesy of the Price family; page 84 (Price family), courtesy of the Price family;
page 108 (von Neumann), Library of Congress; page 108 (Allee), University of Chicago
Library; page 138, *Minneapolis Star Tribune*; page 152 (Smith), University of Sussex;
page 152 (Hamilton), Photo Researchers, Inc.; page 176, courtesy of the Price family; page
194, Oren Harman; page 226, *Nature*; page 256, courtesy of the Price family;
page 282, courtesy of the Price family; page 310, Owen Gilbert; page 332,
courtesy of the Price family; page 348, Oren Harman.

For information about permission to reproduce selections from this book,
write to Permissions, W. W. Norton & Company, Inc.,
500 Fifth Avenue, New York, NY 10110

For information about special discounts for bulk purchases, please contact
W. W. Norton Special Sales at specialsales@wwnorton.com or 800-233-4830

Manufacturing by RR Donnelley, Harrisonburg
Book design by Dana Sloan
Production manager: Julia Druskin

Library of Congress Cataloging-in-Publication Data

Harman, Oren Solomon.
The price of altruism : George Price and the search for the origins of kindness / Oren
Harman.—1st American ed.
p. cm.
Includes bibliographical references and index.
ISBN 978-0-393-06778-1 (hardcover)
1. Price, George Robert, 1922–1975. 2. Geneticists—United States—Biography.
3. Geneticists—Great Britain—Biography. 4. Scientists—United States—Biography.
5. Scientists—Great Britain—Biography. 6. Population genetics.
7. Altruistic behavior in animals. I. Title.
QH429.2.P75H37 2010
576.5092—dc22
[B]
2010011934

W. W. Norton & Company, Inc.
500 Fifth Avenue, New York, N.Y. 10110
www.wwnorton.com

W. W. Norton & Company Ltd.
Castle House, 75/76 Wells Street, London W1T 3QT

1 2 3 4 5 6 7 8 9 0

To Danzi and Mishy with love

Contents

The PRICE of ALTRUISM

Charles Darwin (1809–1882)

Prologue

he men ducked out of the rain into the modest Saint Pancras Cemetery chapel. It was a bleak London day, January 22, 1975. The chapel was spare, its simple pews and white ceiling and walls giving it the feel of a rather uninspiring classroom. Soon they'd follow the hearse down a short path to the burial plot on East Road, where in an unmarked grave the body of the deceased would be laid to rest.[1]

A middle-aged man with a scraggly beard shuffled through the heavy wooden door beneath the ragstone spire, his nose red from whiskey and eyes swollen from fatigue. He'd been in and out of prison, destitute, hard on luck. A big toe jutted from torn sneakers, its nail uncut and covered in grime. Life had not smiled on Smoky. The only person who'd ever truly cared for him was George.

The bearded man was followed into the chapel by four other homeless men, the dead man's final companions, all bundled up in discarded sweaters and scarves found in trash-bins and at the shelters—too small, belonging once to unknown strangers, but welcome protectors from the bitter cold. Some wore belts and socks that George had kindly given them, others pants and overcoats for which he had generously provided the coin. He'd been a true saint, one of them muttered, holding back tears as he passed a few solitary

University of London geneticists sitting uncomfortably in silence. A distinct stench of urine followed the ragtag party as it made its way toward the front of the chapel where the coffin lay. There were ten people in the room, maybe eleven. It was a glum ending to a glum affair.[2]

And there, at the front of the chapel, stood the world's two premier evolutionary biologists, brilliant men and silent rivals. "George took his Christianity too seriously," said Mr. Apps, administering the ceremony on behalf of Garstin Funeral Directors in the absence of any family. "Sort of like Saint Paul," Bill Hamilton whispered audibly under his breath, forcing John Maynard Smith to bite his lip. Then there was a silence. George Price had come over from America to crack the problem of altruism and uncovered something terrible. Now he was dead, the victim of his own hand.[3]

· · ·

From the dawn of time mankind has been contemplating virtue. It began with an act of trickery: ". . . then your eyes shall be opened," the snake whispers to Eve in the Garden of Eden, coaxing her to eat of the fruit, "and ye shall be as gods, knowing good and evil." But if judgment had replaced innocence by way of conniving, it didn't take long before the hard questions arrived. Soon Cain rose up against his brother Abel, killing him to tame his envy. When the Lord came asking for Abel's whereabouts, Cain answered: "Am I my brother's keeper?" It was a question that would reverberate down the paths of history, becoming a haunting companion to humanity.[4]

Then came Darwin.

The devout believed that morality was infused from above on the Sixth Day, religious skeptics that it had been born with philosophy. Now both would need to reexamine their timelines. "He who understands baboon," the sage of evolution scribbled in a notebook, foreshadowing what was to come, "would do more towards metaphysics than Locke."[5]

It was like confessing a murder. If, as the Scottish geologist James Hutton wrote toward the end of the eighteenth century, the earth was so ancient that "we find no vestige of a beginning—no prospect of an end"; if, as Darwin himself argued, life on earth had evolved gradually, over eons, and, far from a ladder was more like a tree; if, just like muscles and feathers and claws and tails, behavior and the mind had been fashioned by natural selection—if all these were true, it would be inconceivable to continue believing that man's defining feature was entirely unique. Whether life had been "originally breathed . . . into a few forms or one" by a Creator, as Darwin suggested, bowing before popular sentiment in the second edition of *The Origin of Species* after leaving him out of the first, virtue was no kind of human invention. More ancient than the Bible, still earlier than philosophy, morality was in fact older than Adam and Eve.[6]

Why do amoebas build stalks from their own bodies, sacrificing themselves in the process, so that some may climb up and be carried away from dearth to plenty on the legs of an innocent insect or the wings of a felicitous wind? Why do vampire bats share blood, mouth to mouth, at the end of a night of prey with members of the colony who were less successful in the hunt? Why do sentry gazelles jump up and down when a lion is spotted, putting themselves precariously between the herd and hungry hunter? And what do all of these have to do with morality in humans: Is there, in fact, a natural origin to our acts of kindness? Does the virtue of amoebas and bats and gazelles and humans come from the very same place?

Altruism was a puzzle. It stood blatantly opposed to the fundamentals of the theory, an anomalous thorn in Darwin's side. If Nature was bloody in tooth and claw, a ruthless battle fiercely fought beneath the waves and through the skies and in the deserts and the jungles, how could a behavior that *lowered* fitness be selected? Survival of the fittest or survival of the nicest: It was a conundrum the Darwinians would need to solve.

And so, starting with Darwin, the quest to solve the mystery of altruism began. It traveled far and wide: From the *Beagle* in the southern seas to the court of the Russian czar Alexander II to the chambers of London's Royal Society; from economics lecture halls at the University of Chicago to Senate hearing rooms on Capitol Hill; from Indiana prairies to Brazilian jungles to Jamaican mountains; from World War I trenches to anti–Vietnam War demonstrations,

Marxist manifestos to Anglican proclamations, Quaker pacifications to Nazi heresies. Some argued that man was all of a part with Nature, bound by his animal beginnings, others that his intelligence rendered him uniquely transcendent. Some championed a return to origins, others the clawing climb of culture away from them, still others an uneasy marriage between the two. Emphatic and zealous, each saw the problem from where they were standing, sometimes translating prior commitments about what's right and wrong directly onto nature. And so the quest remained far from complete. As we shall see, 150 years after Darwin it continues just as passionately.

But if the search for the natural origins of goodness has woven a historical tapestry of unusual complexity and color, of strikingly original science and dramatic personalities and events, one important thread has so far been missing. It is the thread of the unique life and tragic death of the forgotten American genius George Price, atheist-chemist and drifter turned religious evolutionary–mathematician and derelict, the man who rests in an unmarked grave in Saint Pancras Cemetery to this very day. Some of the greatest scientific minds of the nineteenth and twentieth centuries have grappled with the reality of true selflessness, from economics to biology, mathematics to philosophy, ecology to theology to genetics. In the face of the difficult odds posed by egoism, they found it far from an easy problem to crack. Then, like a phantom, came George.

As the thread of George Price's life is woven into the tapestry of the search for the origins of altruism for the very first time,[7] the colors of the pageant suddenly change in radiance and hue. For using Darwin's great insight to penetrate the mystery of kindness, Price came to see what had eluded many before him: Whereas others, in their hunt to fathom goodness, pitted different levels of organization of life against one another—the gene conniving against the individual, the individual subverting the group, one group fighting doggedly against another—this lonely outsider understood that they would all have to be part of a single equation. It was a dramatic flash—a penetration that would forever change our view of the evolution of life. Unknown, untrained, in a foreign country, dejected and alone, he had caught a glimpse of the great canvas of natural selection and seen its splendor and broadness. And, writing the elegant equation, he literally came off the street, anonymous, to present it to the world.

But if George Price's mathematics helped penetrate the origins of altruism deeper than ever before, his life itself was an attempt to answer its most burdensome and mystifying riddle. The level at which selection operates is a technical issue but bears heavily on a fundamental conundrum: If altruism evolved over time in nature, it surely must have served some utilitarian purpose, and if it serves an ulterior purpose it is never what it seems. Part of some natural metric, the purity of selflessness is undermined by the scourge of self-interest: What looks like sacrifice may in fact be the road to personal gain. And so, when it comes to us, a dreaded question arises: Beneath its evolutionary veneer, and despite the refinements of culture, does true selfless altruism exist? It's a question every human since Adam and Eve has sought desperately to answer.

It is a question, too, that modern scientists tackled fearlessly. Whether they ever succeed entirely is doubtful: The problems science is equipped to answer, we shall see, are fundamentally of a different kind. George Price's life, on the other hand, provides a precious and original counterpoint. From the depths of the Great Depression in New York City to Swinging London in the sixties, and from the glamour of secrecy at the Manhattan Project to the humiliation of homelessness off Soho Square, it traces a dramatic trajectory. From militant atheism to religious ecstasy; from comfort and respectability to self-imposed vagrancy; from selfishness to selflessness and finally to the depths of anguish and suicide—George Price's life, like the grander tale of attempts to crack the mystery of altruism, is a powerful reminder of the inescapable duel between biological necessity and the transcendence of the human spirit. But if ultimately, just like science, it too fails to provide a full answer to the mystery of kindness, it illuminates, more clearly than ever before, the meaning of the mystery.

. . .

This is the timeless story of the search for the origins of kindness. It is a tale of animal and man, of nature and politics, of true goodness and deceptive appearances. Its characters are many and colorful: the Russian anarchist Prince Peter Kropotkin; the "Devil's disciple" Thomas Henry Huxley; the Hungarian-born mathematical wizard and father of game theory, John von Neumann; the English polymath and "last man to know all there is to know" J. B. S. Haldane; the "Orwellian" psychologist B. F. Skinner; the father of information theory, Claude Shannon; "the most distinguished Darwinian since Darwin," Bill Hamilton; John D. Rockefeller, Vladimir Ilych Lenin, Adolf Hitler, Franklin Roosevelt, Richard Nixon, the Beatles, and many more.

But more than a personal history of men, this is also a collective chronicle of humanity. From its lowly beginnings in some primordial soup through slime mold to ant to antlered deer to inquisitive monkey, the story of the search for goodness climbs all the way from the oceans and the jungles to titanic twentieth-century battles between systems of government and economics. From the promise of democracy and the free market to communism and the hope of equality, and from the liberations and perils of individualism to the inebriation of nationalism and unity, the quest to crack the altruism code traces an epic voyage. From baboons fighting in trees, to the Russian Revolution, to Nazi Germany, to the atom bomb, to twenty-first-century neurogenetics and brain imaging today, it is mankind's soaring, Sisyphean journey to return to the paradise of the Garden of Eden.

And it is also the tragic tale of the tortured soul of one man, George Robert Price, the hitherto missing thread now woven into the tapestry of the greater story. More than any other person, Price came closest to seeing how sacrifice could be born of the ruthless cull of evolution, even if his penetrating mathematics could never tell him if selflessness was ever truly genuine and pure. Unable to find the answer to this greatest of human conundrums in science, he went looking for it in other places. What he found is a lesson to all those who came before and after him.

And perhaps, in some small sense, it is also an answer to Cain.

Part One

Peter Kropotkin (1842–1921)

Thomas Henry Huxley (1825–1895)

1

War or Peace?

*H*e would wait until dusk. That would be the best time to slip away unnoticed. As he packed a small valise, memories of his animated talk on the glacial formations of Finland and Russia at the Geographical Society the previous evening still lingered in his mind. It had gone well, he thought. The country's leading geologist, Barbot-de-Marny, had spoken up in his favor. It was even proposed that he be nominated president of the physical geography section of the society. Years of journeying to the frozen hearts of faraway places had finally paid off. But now he must concentrate. Now he must flee. "You had better go by the service staircase," one of the servant girls whispered.[1]

A horse-drawn carriage stood at the gate. He jumped in. The cabby whipped the horse and turned onto Nevsky Prospekt the majestic avenue planned by Peter the Great in the city that called itself by his name. It was a short ride to the rail station. and from there, please the spirits, to freedom. Russia was a vast land, and in the remoteness of its eastern expanses it was his intention to start a land league, like the ones that would become so powerful in Ireland in the years just ahead. It was the beginning of spring 1874.

Suddenly a second cab galloped by. To his great surprise there in its carriage, was one of the two weavers who had been arrested the week before, waving his hand at him. Perhaps he has been released, he thought, and has an important communication to make to me. Duly he ordered the cabby to stop, but before he could greet the weaver, a second man, sitting beside him, appeared. Two years of clandestine meetings, disguises, and sleeping in other people's beds had come to an end. Jumping into his carriage, the second man, a detective, cried out: "Mr. Borodin, Prince Kropotkin, I arrest you!" Later that night, in the bowels of the infamous Third Section, a gendarme colonel solemnly read the charge: "You are accused of having belonged to a secret society which has for its object the overthrow of the existing form of government, and of conspiracy against the sacred person of his Imperial Majesty." For Prince Peter Alekseyevich Kropotkin, alias Borodin, the game was finally up.[2]

· · ·

Across the Baltic and North seas at the very same time, Thomas Henry Huxley was fastening his bow tie in preparation to open, together with the president, Sir Joseph Dalton Hooker, the Thursday meeting of the Royal Society in Burlington House, Piccadilly, London.

Pictures of his carriage ride that afternoon flashed like fiery moths in his mind: Magsmen, cracksmen, shofulmen and prostitutes, child fences, religious fakes, and grimy boxers and promoters—this was the Victorian underworld, lying on his route to England's sanctum sanctorum of science. He knew the gutters and fever nests well: He had come from them. As he settled into the velvet-cushioned oak chair, he stole a nervous glance across the room.

Born in 1825 above a butcher shop in Ealing, a small village twelve miles west of London, he was forced at ten to abandon school to earn the pittance his unemployed father could not provide him. At thirteen he was apprenticed to a "beer-swilling, opium-chewing" medical man of a brother-in-law in Coventry, before being fastened to a lowlife mesmeric doctor back in town. At times the young Huxley thought he might drown in what a later biographer would call the "ocean stream of life" that was London—teeming with "whores, pandars, crimps, bullies." He found refuge in the dreary apothecary shop, grinding drugs in solitude. Steadily a rage grew within him. How could the middle class remain so coldly indifferent, he wondered in his diary and in countless letters to friends, to such unabashedly squalid suffering?[3]

With hard work and determination he gained a scholarship to Charing Cross Hospital, and later won the gold medal for anatomy and physiology at the University of London. At twenty, to pay back debts, he joined the Royal Navy as assistant surgeon on board the HMS *Rattlesnake*, surveying the coasts and innards of Papua New Guinea and Australia and dissecting otherworldly invertebrates from the wild southern seas. The specimens and papers he sent back home quickly made a name for him as an authority on the oceanic hydrozoa. At twenty-five he was elected to the Royal Society. Before long he was the professor of natural history at the Royal School of Mines, Fullarian professor at the Royal Institution, Huntarian professor at the Royal College of Surgeons, and president of the British Association for the Advancement of Science. He led important royal commissions, setting out to fix the British world: Trawling for Herrings on the Coast of Scotland, Sea Fisheries of the United Kingdom, the Contagious Diseases Acts, Scientific Instruction and the Advancement of Science.

He shuffled the papers in front of him. The seat of Britain's most learned men of science for more than 300 years, the Royal Society was undergoing dramatic change, mirroring the very face of the nation. Gone were the courtly days of yesteryear, the unchallenged loyalties to Crown and Church. As doctors, capitalists, and even those strange birds, "academics," began ringing at the bell, a fresh spirit was being ushered in. The new patrons were merchants and builders of empires abroad, not "blue blooded dilettantes" and "spider-stuffing clergy." For Britain itself, as for its august Royal Society, the new gods became "utility and service to state; its new priests, the technocrats and specialists." Men, that is, just like Huxley. [4]

Earlier that week he had been hosted by the radical caucus of Birmingham. A statue of the chemist Joseph Priestley was being unveiled, and Huxley seemed just the man to do the talking. With the city fathers hanging on his every word, he painted the townsfolk a vision of "rational freedom" sanctioned by a science-driven state. The post, the telegraph, the railway, vaccination, sanitation, road building—all would be well served if run by government. Improving life for British citizens in this way was the only way to stave off the bloody revolutionary rage scourging the rest of Europe.[5]

Clearing his throat to open the meeting, Huxley had calmed. It was the spring of 1874 and he was secretary of the Royal Society. The imperial botanist, Hooker, sitting beside him, had just declined a knighthood as beneath the dignity of science. Huxley smiled to himself. If a little boy from Ealing could make it, the system must be true and just after all. In a cutthroat world of competition, he had clawed his way from the gutter to the very heights of Victorian living. Brimming with fiery spirit—"cutting up monkeys was his forte, and cutting up men was his foible" the *Pall Mall Gazette* observed of him—he was the unblinkered professional public servant at the service of the modern, benevolent state.[6]

"My fellows, I call this meeting to order."

• • •

When the four-wheeled carriage crossed the Palace Bridge over the Neva some days and interrogations later, notwithstanding the silence of the stout Circassian accompanying officer, Kropotkin knew he was being taken to the terrible fortress of Saint Peter and Saint Paul.

Here Peter the Great had allegedly tortured and killed his son Alexis with his own hands; here Princess Tarakanova was kept in a water-filled cell, "the rats climbing upon her to save themselves from drowning"; here Catherine buried political prisoners alive. And here, too, great men of letters had recently been chained: Ryleyev and Shevchenko, Fyodor Dostoyevsky, Pisarev. The revolutionary Mikhail Bakunin, too, had spent eight hard years there before the czar offered him the choice of banishment to Siberia, which he gladly accepted. Bakunin ended up escaping from Siberia in 1861 to Yokohama and from there to San

Francisco and then New York. When he finally burst into fellow revolutionist and exile Alexander Herzen's apartment in London sometime later, "Can one get oysters here?" was the first thing the great anarchist bellowed. With such thoughts in his mind Kropotkin smiled under his beard and promised himself: "I will not succumb here!"[7]

Immediately he was ordered to strip and handed a green flannel dressing gown and gigantic woolen stockings of "an incredible thickness." Boat-shaped yellow slippers were thrown at his feet, so big that they fell off when he tried to walk. The prince was to be treated like any other inmate. Still, the military commander of the fortress, General Korsakov, a thin and tired old man, betrayed enfeebled vestiges of the tug of stature in czarist Russia, seeming visibly embarrassed by the occasion. "I am a soldier, and only do my duty," he said, not quite looking the prince in the eye. Kropotkin was paraded through a dark passageway guarded by shadowed sentries. A heavy oak door was closed behind him, and a key turned in its lock.

The room was a casemate, "destined for a big gun," Kropotkin later wrote in his memoirs, with an iron bed and a small oak table and stool. The sole window, a long, narrow opening cut in a wall five feet deep and protected by an iron grating and a double iron window frame, was so high that he could hardly reach it with his outstretched fingers. Defiant, he began to sing, "Have I then to say farewell to love for ever?" from his favorite Glinka opera, *Russlan and Ludmila*, but was soon silenced by the basso reproach of an invisible guard.

The cell was half dark and humid. Absolute silence reigned all around. Assessing his surroundings, the prince determined to keep his body fit. There were ten steps from one corner to the other. If he paced them 150 times, he would have walked one verst—two-thirds of a mile. Then and there he decided to walk seven versts every day: two in the morning, two before dinner, two after dinner, and one before going to sleep. And so he did, day in and day out, month in and month out. And he let his mind roam.[8]

. . .

Darwin called him "my good and kind agent for the propagation of the Gospel," though "good" and "kind" were perhaps not quite the words for the slashing rapier of "Darwin's bulldog." For Huxley the alternative was "to lie still and let the Devil have his own way," for the resistance to the logic of materialism and evolution seemed to him nothing short of the workings of Satan. Darwin's nemesis, Britain's leading anatomist, Richard Owen, had called Huxley a pervert with "some, perhaps congenital, defect of mind" for denying divine will in Nature, but this sort of thing only stoked his internal fires. Finally, from the heights of the Royal Society, Huxley, known to his enemies as "the Devil's disciple," could begin to bring about the revolution.[9]

The first blow was struck from Russia. Vladimir Kovalevskii had come to London to work on hippopotamus evolution and was soon befriended by Huxley. Darwin's philosophy of descent with modification by the merciless hand of Nature's blind selector had been fought over with rancor on the pages of popular newspapers and debated with disdain in the halls of museums. But evolution still remained at the margins of true scientific discourse. More than a decade following the publication of the *Origin of Species*, the *Philosophical Transactions* of the Royal Society had yet to print one article related to Darwinism, stubbornly clinging to "facts" and avoiding "theory," and keeping as far from controversy as its blue-bloodedness could afford. But Huxley and his X-Club friends were now the new masters.[10] When the secretary read Kovalevskii's paper to the society, George Gabriel Stokes complained that it was an abomination that a nihilist known to the Russian secret police be allowed to air such folly. Comparing Darwinian speculations with the axioms of Newton was a blow to the very foundations of knowledge. For Stokes, Cambridge's Lucasian professor of mathematics, the "continuous curve" connecting the creative acts was a piece of "divine geometry," the very considered opposite of "Creation by Caprice." Yet Huxley arranged for sympathetic reviewers, and "On the Osteology of the Hyopotamidae" soon appeared in the pages of *Transactions*. It took the hippopotami and a nihilist Russian, but the fiery lad from Ealing had finally traversed the Royal Society's impasse. A flood of "free-thinking" was about to violently burst open the pearly gates of England's scientific holy of holies.[11]

. . .

Kropotkin was born in Moscow in the winter of 1842. His maternal grandfather had been a Cossack army officer—some said of note—but his father's side provided the truly important pedigree. The Kropotkins were scions of the great Rurik dynasty, first rulers of Russia before the Romanovs.[12] At a time when family wealth was measured in numbers of serfs, the family owned nearly twelve hundred souls in three different provinces. There were fifty servants in the Moscow home, and another seventy-five at the Nikolskoye country estate. Four coachmen attended the horses, five cooks prepared the meals, and a dozen men served dinner every evening. It was a world of birch trees, governesses, samovars, sailor suits, and sleigh rides that young Peter was born into, "the taste of tea and jam sharpened and sweetened by the sense of the vast empty steppes beyond the garden and imminent end of it all."[13]

Not all was idyllic. Like other famous sons of Russian landed nobility—Herzen, Bakunin, Tolstoy—Peter would come to despise the particular flavor of Oriental despotism baked in the juices of Prussian militarism and overlaid with a foreign veneer of French culture. Ivan Turgenev's short story "Mumu," describing the misfortunes of the serfs, came as a startling revelation to an apathetic nation: "They love just as we do; is it possible?" was the reaction of sentimental urban ladies who "could not read a French novel without shedding tears over the troubles of the noble heroes and heroines." [14] Images etched themselves on young Peter's mind: the old man who had gone gray in his master's service and chose to hang himself under his master's window, the cruel laying waste of entire villages when a loaf of bread went missing, the young girl who found her only salvation from a landlord-arranged marriage in drowning herself. Increasingly, thinking and caring sons of the ruling elites of imperial Russia witnessed up close the meanness and sterility of the feudal world into which they were born, and fretted over the future of their beloved Russia. Many wondered: What is to be done?[15]

Aleksei Petrovich Kropotkin, a retired army officer who had seen little real action but nevertheless lived entirely according to military custom, thought he knew very well what his son needed to do: Little Peter's artistically inclined mother died of consumption when he was just four, and thereafter he would be groomed for the life of a soldier. When a serendipitous opportunity presented itself to showcase his son at a gala costume party in honor of Czar Nicholas I's twenty-fifth anniversary, eight-year-old Peter's uniform was prepared with particular attention. And there he was, dressed as a Persian page with a belt covered with jewels, and hoisted by his uncle, Prince Gagarin, to the platform, when the czar himself beheld him. Taking the young boy by the arm, Nicholas I led him to Maria Alexandrovna, the pregnant wife of the heir to the throne, saying: "That is the sort of boy you must bring me." [16]

The czar would not live to regret his words, but his heir would. The Corps of Pages in St. Petersburg was the training ground for Russia's future military elite; only 150 boys, mainly sons of the courtly nobility, were admitted to the privileged corps, and upon graduation could join any regiment they chose. The top sixteen would be even luckier: *pages de chambre* to members of the imperial family—*the* card of entry to a life of influence and prestige. When Peter was sent there by his father at fifteen, he already considered it a misfortune. But despite himself he graduated at the top of the class and was made personal liege to Alexander II, Nicholas having died some years earlier. It was 1861, and insurrections were growing more violent and expensive, opposition more damagingly vocal. The new czar was coming under increasing pressure to grant freedom to his serfs. When he finally signed the Edict of Liberty on March 5 (according to the old Russian calendar), Alexander seemed to Peter transcendent. The sentiment was fleeting. The glamour of richly decorated drawing rooms flanked by chamberlains in gold-embroidered uniforms took his breath away at first, but soon he saw that such trifles absorbed the court at the expense of matters of true importance. Power, he was learning, was corrupting.

As he shadowed the czar at a distance, with the requisite combination of "presence with absence," the aureole he once imagined over the imperial ruler's person gradually gloomily eroded. The czar was unreliable, detached, and vindictive, and many of the men around him were worse. With the Corps of Pages, Kropotkin had learned to march and fence, build bridges and fortifications, but his true interests, he already knew, lay elsewhere. Secretly he began to read Herzen's London review, the *Polar Star*, and even to edit a revolutionary paper. When the time came to pick a commission he determined to travel to the far expanses of eastern Siberia, to the recently annexed Afar region. His father and fellow cadets were shocked—after all, as sergeant of the corps the entire army was open to him. "Are you not afraid to go so far?" Czar Alexander II asked him before he was to leave, surprised. "No. I want to work. There must be much to do in Siberia to apply the great reforms which are going to be made." "Well, go; one can be useful everywhere," the czar replied, but with such an expression of fatigue and complete surrender that Kropotkin thought at once, "He is a used up man." [17]

. . .

Thirty years before Kropotkin set out for the Afar, Charles Darwin set sail on the HMS *Beagle*. En route to Buenos Aires in October 1832, Darwin noticed swarms of phosphorescent zoophytes, each smaller than the dot above this *i*. They illuminated the waves surrounding the ship with the glow of a pale green light as it sailed into the dark unknown ocean. Darwin was aware of the prevalent explanation: The tiny marine creatures had been put there by God to help sailors avoid shipwreck on gloomy nights at sea. This was the doctrine of finalism, or teleology, the very backbone of a tradition of natural theology on which Darwin's generation had been reared. [18]

But the young lad from Shrewsbury would not have God's benevolence stand as a proxy for scientific explanation. The glow ordained to direct lost sailors, he was certain, was simply phosphorescence caused by the decomposing bodies of the millions of dead zoophytes caught among the live ones—a process by which the ocean purified itself. This was purpose enough, God's benevolence notwithstanding. The true beauty of Nature could be unmasked only by uncovering her own laws, not God's divinations. The Reverend William Paley figured natural design to be proof of godly design—how else to explain the excellence of the crystalline lens of the eye of the trout, or the aero-dynamic perfection of the wing of the eagle? But the answers in his *Natural Theology* now seemed to Darwin like questions: If God were bracketed, and natural laws sought out in his stead, how could the seamless fit between organisms' forms and functions be explained? How did Nature come to seem so perfect?[19]

One way to look at the problem would be to study Nature's imperfections, long recognized as a puzzle, and unsuccessfully explained away by the argument for design. Why on earth do flightless kiwis have vestiges of wings, snakes relics of leg bones, or moles traces of once-busy eyes? The mysteries of biogeography kept tugging at his mind, too: Why are there fewer endemic species on islands than on the mainland? Where did these species come from? Why are they so similar to mainland species if their natural surroundings are so different? A fixity-of-species man upon embarkation, Darwin returned to England in October 1836 leaning toward a more dynamic view of Nature and her ways. Still unsure of the physical law to explain away all conundrums, he nevertheless arrived after nearly five years at sea with "such facts [that] would undermine the stability of species."[20]

And then something momentous happened.[21] In October 1838 Darwin read *An Essay on the Principle of Population* by the clergyman and former professor of political economy Thomas Malthus. The idea that population increases geometrically while food supply increases arithmetically was meant by Malthus to prove that starvation, wars, death, and suffering were never the consequence of the defects of one political system or another, but rather the necessary results of a natural law. A Whig and a supporter of Poor Law action to ameliorate the condition of the destitute, Darwin was not sympathetic to Malthus's reactionary politics, but applying the clergyman's law to nature was different. Immediately he realized that given the struggle for existence everywhere, "favorable variations would tend to be preserved, and unfavorable ones to be destroyed. The result of this would be the formation of new species. Here, then," he wrote, "I had at last got a theory by which to work." Evolution by natural selection was nothing more and nothing less than "the doctrine of Malthus, applied to the whole animal and vegetable kingdoms."[22]

After all, if one great lesson had been gleaned from the journey, it was the awesome abundance of life on the planet. On the massive vines of "wonderful" kelp off the coast of Tierra del Fuego, plummeting forty-five fathoms into the darkness, Darwin found patelliform shells, troche, mollusks, bivalves, and innumerable crustacea. When he shook, out came "small fish, shells, cuttle-fish, crabs of all orders, sea-eggs, star-fish, beautiful holuthuriae, planariae, and crawling nereidous animals of a multitude of forms." The "great entangled roots" reminded Darwin of tropical forests, swarming with every imaginable species of ant and beetle rustling beneath the feet of giant capybaras and slit-eyed lizards, under the watchful gaze of carrion hawks. The splendor and variation were endless. "The form of the orange tree, the coconut, the palm, the mango, the tree-fern, the banana," Darwin wrote nostalgically, surveying the tropical panorama at Bahia as the *Beagle* pushed for home, "will remain clear and separate: but the thousand beauties which unite these into one perfect scene must fade away; yet they will leave, like a tale heard in childhood, a picture full of indistinct, but most beautiful figures."[23]

In truth, Darwin knew, nature was one grand cacophonous battle—brutal, unyielding, and cruel. For if populations in the wild have such high rates of fertility that their size would increase exponentially if not constrained; if it is known that, excepting seasonal fluctuations, the size of populations remains stable over time; if Malthus was right, as he surely was, that the resources available to a species are limited—then it follows that there must be intense competition, or a *struggle for existence*, among the members of a species. And if no two members of a population are identical, and some of these differences render the life chances, or *fitness*, of some greater than others—*and are inherited*—then it follows that the selection of the fitter over the less fit will lead, over time, to evolution. The consequences were unthinkable, yet Darwin's logic was spotless. From the "war of nature, from famine and death" the most exalted creatures had been created. Malthus had brought about in him a complete "conversion," one which, he wrote to his trusted friend Joseph Hooker in 1844, was "like confessing a murder."[24]

. . .

Prisons to reform, schools to build, tribunals to assemble—the great administrative apparatus of the state was waiting to be marshaled. Wide-eyed, Kropotkin had joined the Cossack regiment, eager to bring justice to faraway districts. Gradually he saw his considered recommendations all dying a silent death on the gallows of bureaucracy and official corruption. When a Polish insurrection broke out in the summer of 1863, Alexander II unleashed a terrible reaction, all reforms and their spirit long forgotten. Disillusioned, Kropotkin gradually turned to nature. Fifty thousand miles he traveled—in carts, on board steamers, in boats, but chiefly on horseback, with a few pounds of bread and a few ounces of tea in a leather bag, a kettle and a hatchet hanging at the side of his saddle. Trekking to Manchuria on a geographical survey he slept under open skies, read Mill's *On Liberty*, and beheld with astonishment "man's oneness with nature."[25]

Kropotkin's primary concern now became working out a theory of mountain chains and high plateaus, but he was keen, too, to find evidence for Darwin's great theory. He had read *The Origin of Species* at the Corps of Pages, and in a way this was his polar voyage of the *Beagle*. What he saw, then, came as a great surprise: Darwin spoke of a fierce struggle between members of the same species, but everywhere he looked Kropotkin found collaboration: horses forming protective rings to guard against predators; wolves coming together to hunt in packs; birds helping each other at the nest; fallow deer marching in unison to cross a river. Mutual aid and cooperation were everywhere.

Like Darwin upon his return from his journey, after five years of adventure Kropotkin had yet to develop a fullblown theory of nature. But if Darwin's belief in the fixity of species had been shaken on the *Beagle*, Kropotkin's assurance of the struggle for existence was completely shattered on the Afar. By the time he arrived in St. Petersburg in April 1867, he wrote, "the poetry of nature" had become the philosophy of his life.[26] At the same time, he had lost all faith in the state: Once a constitutionalist who believed, like Huxley, in the promise of benevolent administration, Kropotkin emerged from the great Russian expanses fully "prepared to become an anarchist."[27]

It was in Switzerland some years later that he became a full-fledged revolutionary. The death of his father finally setting him free, news of the Paris Commune drew Kropotkin to Europe. In Zurich he joined the International, gaining a taste for revolutionary politics. But it was in Sonvilliers, a little valley in the Jura hills, that something really moved him. In the midst of a heavy snowstorm that "blinded us and froze the blood in our veins," fifty isolated watchmakers, most of them old men, braved the weather in order to discuss their no-government philosophy of living. This was not a mass being led and made subservient to the political ends of a few apparachiks. It was a union of independents, a federation of equals, setting standards by fraternal consensus. He was touched and deeply impressed by their wisdom. "When I came away from the mountains," Kropotkin wrote, "my views upon socialism were settled. I was an anarchist."[28]

Back in St. Petersburg he joined the Chaikovsky Circle, an underground group working to spread revolutionary ideas. For two years, between learned debates at the Geographical Society and lavish imperial soirees, Kropotkin became "Borodin." Disguised as this peasant he ducked into shady apartments to lecture on everything from Proudhon to reading and arithmetic, slipping away again like a phantom. Communalism and fraternity were the anarchist response to the state, order without Order. Here was the creed: Left to his own devices, man would cooperate in egalitarian communes, property and coercion replaced by liberty and consent. Progress was being made uniting the workers in revolt against the czar when the police began taking serious counteraction. A group of agitating weavers had been arrested, and a raid on a student apartment produced a revolutionary manifesto authored by one P. A. Kropotkin. It was then that he knew that he would have to leave without delay. Now, pacing in his prison cell, Kropotkin could not help but grimace: if only he had forgone that last talk on glacial formations!

. . .

If competition between individuals was, scandalously, Nature's way, she had forgotten to whisper the news to some of her smaller creatures. Many an ant species, Darwin knew, was divided into fixed, unbreachable castes. The honeypot ant of the American deserts has workers whose sole job is to hang upside down, motionless, like great big pots of sugared water, so that they may be tapped when the queen and her brood are thirsty. Members of another caste in the same species have gigantic heads with which, Cerberus-like, they block the nest entrance before intruders. The leaf-cutter ants of South America sport castes that differ in weight up to three hundredfold, from miniature serene fungus gardeners to giant ferocious soldiers. In the ant world some tend to the queen, others to the nest, others to food, others to battle—each to his caste and each to his fate. What Darwin found amazing was that besides the queen and a few lucky males, all the rest of the ants are effectively neuters. This made no sense if success in the battle for survival was measured by production of offspring.[29]

For Darwin the mystery lay in trying to explain how such different behavior and morphology arose in a single species, for since all the workers had no offspring, natural selection could hardly be fashioning their traits through their own direct kin. What this meant was that the queen and her mate were somehow passing on qualities through their own progeny—massive heads, gardening scissor teeth, and mysterious altruistic behavior—that they themselves did not possess, an obscurity that Darwin found "by far the most serious special difficulty, which my theory has encountered."[30] This was a problem of heredity: How could traits, both of form and of behavior, perform such Houdini acts in their journey from generation to generation?

It was also a glaring exception to "nature, red in tooth and claw."[31] If evolution by natural selection was the doctrine of Malthus applied to the whole of living creation, little ants and bees and termites were islands of chivalry in a sea of conflagration. Why—how—this anomalous sanctuary of "goodness"?[32] To solve the mystery Darwin asked a simple question: Who benefits? The answer, he thought, was the "community,"[33] for those who could forage or fight would surely free others to partake in procreation, on the very same principle that rendered the division of labor "useful to civilised man." If selection sometimes worked at a level higher than the individual, even the ultimate sacrifice of the stinging bee or ant centurion could evolve. This was quite an idea, for the very essence of Darwin's theory, as he declared in

The Origin of Species, was that "every complex structure and instinct" should be "useful to the possessor." Natural selection could "never produce in a being anything injurious to itself, for natural selection acts solely by and for the good of each."[34] And yet it did.

Darwin was impressed. It was the strongest evidence yet, he thought, for the incredible power of natural selection. In truth, he would come to believe, it was actually entirely much bigger. "The social instincts," he wrote in *The Descent of Man*, "which no doubt were acquired by man, as by the lower animals, for the good of the community, will from the first have given him some wish to aid his fellows, and some feeling of sympathy."[35] Evolution was the key to the beginnings of morality in humans.

. . .

Rheumatism had almost killed Kropotkin. With the help of family connections and a friendly doctor's note, he was transferred after twenty-one months to the detention house and from there to the military hospital. Finally, even though he was sickly and frail, there was a glimmer of hope: The hospital was not nearly as well guarded as the fortress. The day of the escape was fixed. It was to be June 29 (old style), the day of Saints Peter and Paul—his friends having decided to throw, he later wrote, "a touch of sentimentalism into their enterprise."[36] A red balloon climbing into the blue sky would be the signal to make a dash for the gate, where a carriage would whisk him to freedom. But the impossible happened that day: No red balloons could be found in all of St. Petersburg, and when one was finally discovered and snatched from the hand of a howling boy, it would not fly, nor would the apparatus for making hydrogen, hurriedly bought from an optician's shop, revive it. The woman who finally strung the flaccid balloon to an umbrella, walking up and down behind the hospital wall, did not help either: The wall was too high and the woman too short, and the signal never reached poor Kropotkin.

The next morning a relative came to visit at the hospital carrying a watch that she asked that he be given. Unsuspected, it was passed to him, though the timepiece was far from innocent. Hidden inside was a cipher, detailing the new plans for escape that very day At 4 p.m.

Kropotkin went out to the garden for his afternoon stroll. When he heard the cue of an excited violin mazurka, he made a desperate dash for the gate. "He runs! Stop him! Catch him!"—a sentry and three soldiers were in hot pursuit, so close that he could feel the wind of the bayonet thrust toward him.[37]

That evening they clinked glasses at Donon's, St. Petersburg's finest restaurant: The secret police would never think of looking there. The escape was a feat of true altruism: Untold accomplices had selflessly braved grave danger—one signaling with handkerchiefs, a second by means of synchronized cherry eating, a third distracting the guard, a fourth playing the violin, a fifth commanding the carriage—Kropotkin was aglow with pride. But he would have to leave. Soon he crossed the Finnish border and was on a steamer headed for London.

. . .

Mady Huxley died of pneumonia on November 20, 1887. The great neurologist Jean-Martin Charcot had come to England to examine her—the loss of vision and voice having led her father to fear "the worst of all ends—dementia." She was to him a "brilliant creature," his fair and beloved third child. A specialist on "hysteria" and teacher to the young Sigmund Freud, Charcot determined that Mady suffered from a grave mental illness and invited her to Paris for hypnosis. It was too late. Arriving at the Salpêtrière Hospital, exhausted, she succumbed before the treatment could remove her emotional "conflicts."[38]

Staggering, in pain, Huxley traveled to Manchester for a talk he felt honor bound to give. As the train sped north through the West Midlands—Coventry, Birmingham, Wolverhampton, Stoke-on-Trent—he glanced at England passing by. For more than four years now he was president of the Royal Society, the winner of medals, and very sun of the scientific *orbis terranum*.[39] But if Huxley had come a long way from above the butcher's shop in Ealing, so too had England from its affluent "Age of Equipoise." Dissenters and Nonconformists had waged a battle for meritocracy against Church and Crown in the 1850s, '60s, and '70s, but this was long yesterday's triumph. Great boring machines were now miraculously digging the Channel tunnel deep beneath the sea, yet millions in the cities and countryside took to bed hungry. The "interminable Depression" had coincided with "a specialist age"; at its finest hour technology was failing the masses.[40]

And the masses were swelling. Britain's population had reached 36 million and was adding nearly 350,000 hungry mouths every year. As growth rates had surged, a new phrase made its way from France and Germany. English Darwinians had by tradition few qualms about folding the social into the biological: For them animals and man bowed just as humbly before Nature and her laws. But as political socialism took a bite at the Malthusian core of survival of the fittest; as suffrage, labor unrest, and the "Woman Question" ushered in a new age of extremes—a currency was needed to remind civilization of its beastly beginnings. Were the teeming congestion and competitive strife not confirmation enough of Malthus's prediction? Huxley called industrial competition with Germany and the United States a form of international "warfare," and *Nature* and the *Times* applauded.[41] But if mercantalism had morphed into an all-out image of battle, if Darwin's Malthusian struggle had been writ large on the world as a whole—there were those who were prepared to fight it. It was against such men and women that "social Darwinism" would now be wielded.[42]

Leading the way was Herbert Spencer, Huxley's X-Club companion, a "bumptious" man with a "breathless vision" of evolution galloping ahead to perfection. No one had swallowed Darwin so wholly, even if some (including Darwin himself) thought it had gone down the wrong pipe. For the "Prince of Progress" the physical, biological, social, and ethical all danced to the tune of evolution.[43] Historical destiny was like the womb and the jungle, the growth of civilization "all of a piece with the development of the embryo or the unfolding of a flower."[44] An eminent Victorian who had dabbled in phrenology, he had coined the "survival of the fittest."[45] Friend to Mill, follower of Comte, and a lover of George Eliot, Spencer honed to perfection the belief in human perfectibility. But it was the strong, not the meek for him, who carried the future in their bones, their struggles and triumphs the true holy of holies, their might—the right and just. To let it be, government would need to step aside, even when its actions seemed "progressive." Intervention, after all, was really a curse disguised as a blessing, the conquest by maudlin sensibility of the necessity of natural law. Unfettered competition alone would lead to the advance of civilization in the long run—endowments and free education be damned and myopia forlorn.

From the Left other voices came buzzing. Henry George's *Progress and Poverty,* a popular appeal for land nationalization, was rapidly gaining readers.[46] Touting Rousseau's noble savage, George led a frontal attack on property and competition. Even the codiscoverer of natural selection, Alfred Russel Wallace, from his retirement nest in Dorset, took a jab at "Darwinism"—a term no one had done more than he to establish.[47] Women, he now claimed, when liberated economically by socialism, would freely choose the righteous among men. As such they would be humankind's great redeemers, breeders of goodness into future generations.[48] This was a different woman from Darwin's in his *Descent of Man*, to say the least.[49] But even if Wallace's utopia seemed far-fetched and would need to be nudged along by higher forces (a spiritualist, he had removed man from the arena of natural selection), it hardly mattered anymore. Beaten over the head by natural rights, ancient communes, and the promise of equality, the Darwinian establishment was reeling. Perhaps competition was not the natural law they said it was. Perhaps their "religion of Science" was an illusion, nothing but a false "religion of despair."[50]

Huxley was taking the fire. After all, he had fashioned himself the very embodiment of "science as panacea." Spencer was a "long-winded pedant," he thought, a "hippopotamus," as misguided in his sacrifice of the masses on the altar of Darwinism as he was in his belief in the inheritance of acquired characteristics. And yet nature really *was* brutal, like "a surface of ten thousand wedges" each representing a species being "driven inward by incessant blows."[51] Success always came at the expense of another's failure. How then to escape the trap into which the patrician Spencer had willfully fallen? How to wrest morality for the masses from the bloody talons of Nature?

These were his mind's torments as the train pulled into London Road Station, Manchester. At Town Hall, before his crowd, the darkness in his soul poured itself onto the natural world. Glassy-eyed and imagining his daughter, Huxley unmasked the vision of Nature's butchery: "You see a meadow rich in flower & foliage and your memory rests upon it as an image of peaceful beauty. It is a delusion. . . . Not a bird that twitters but is either slayer or [slain and] . . . not a moment passes in that a holocaust, in every hedge & every copse battle murder & sudden death are the order of the day."[52]

As "melancholy as a pelican in the wilderness," as he wrote to a friend, Huxley was sinking into depression.[53] The Manchester address was printed in February's *Nineteenth Century*, and soon became a disputed cause célèbre. In "The Struggle for Existence in Human Society: A Programme" Huxley asked readers to imagine the chase of a deer by a wolf. Had a man intervened to aid the deer we would call him "brave and compassionate," as we would judge an abetter of the wolf "base and cruel." But this was a hoax, the spoiled fruit of man's translation of his own world into nature. Under the "dry light of science," none could be more admirable than the other, "the goodness of the right hand which helps the deer, and the wickedness of the left hand which eggs on the wolf" neutralizing each other. Nature was "neither moral nor immoral, but non-moral," the ghost of the deer no more likely to reach a heaven of "perennial existence in clover" than the ghost of the wolf a boneless kennel in hell. "From the point of view of the moralist the animal world is on about the same level as a gladiator's show," Huxley wrote, "the strongest, the swiftest, and the cunningest . . . living to fight another day." There was no need for the spectator to turn his thumbs down, "as no quarter is given," but "he must shut his eyes if he would not see that more or less enduring suffering is the meed of both vanquished and victor."[54]

Darwin and Spencer believed that the struggle for existence "tends to final good," the suffering of the ancestor paid for by the increased perfection of its future offspring. But this was nonsense unless, "in Chinese fashion, the present generation could pay its debts to its ancestors." Otherwise, it was unclear to Huxley "what compensation the *Eohippus* gets for his sorrows in the fact that, some millions of years afterwards, one of his descendants wins the Derby." Besides, life was constantly adapting to its environment. If a "universal winter" came upon the world, as the "physical philosophers" watching the cooling sun and earth now warned, arctic diatoms and protococci of the red snow would be all that was left on the planet. Christians, perhaps, imagined God's fingerprint on nature, but it was Ishtar, the Babylonian goddess, whose meddling seemed to Huxley more true. A blend of Aphrodite and Ares, Ishtar knew neither good nor evil, nor, like the Beneficent Deity, did she promise any rewards. She demanded only that which came to her: the sacrifice of the weak. Nature-Ishtar was the heartless executioner of necessity.[55]

But what, then, of man—was he, too, to bow in deference to the indifferent god of inevitability? As for all other creatures "beyond the limited and temporary relations of the family," for man too the "Hobbesian war of each against all" had been the normal state of existence. Like them he had "plashed and floundered amid the general stream of evolution," keeping his head above water and "thinking neither of whence nor whither." Then came the first men who for whatever reason "substituted the state of mutual peace for that of mutual war," and civilization was born. Self-restraint became the negation of the struggle for existence, man's glorious rebellion against the tyranny of need. But no matter how historic his achievement, ethical man could not abolish "the deep-seated organic impulses which impel the natural man to follow his non-moral course." Chief of these was procreation, the greatest cause of the struggle for existence.[56]

Of all the commandments, "Be fruitful and multiply" was the oldest and the only one generally heeded. Despite his best intentions, then, ethical man was locked once more in the nonmoral "survival of the fittest." Population was driving him to war. For the dark moment Huxley could see no tonic, though, contra the socialists, he was certain that no "fiddle-faddling with the distribution of wealth" could deliver society from its tendency toward self-destruction. Industrial warfare having replaced natural combat, the corporatist Huxley made a plea for state-sponsored technical education. But this was as much medicine, he knew, as an eye doctor's recommending an operation for cataract on a man who is going blind, "without being supposed to undertake that it will cure him of gout."[57]

For Darwin morality had come from the evolution of the social instincts, but for Huxley they were a vestige of amoral beginnings.

Instincts were *antisocial*; their primeval lure the bane of man's precarious existence. Desperately seeking the cure for social ills, Huxley nevertheless would not search for it in nature. Nor, weeping over the loss of Mady on the train ride back from Manchester, could he find any solace there.

. . .

Malthus was already dead when *Russian Nights* became a best seller in the 1840s. The novel's author, Prince Vladimir Odoesky, had created an economist antihero, driven to suicide by his gloomy prophecies of reproduction run amok. The suicide was cheered on by the Russian reading masses: After all, in a land as vast and underpopulated as theirs, Malthusianism was a joke. England was a cramped furnace on the verge of explosion; Russia, an expanse of bounty almost entirely unfilled. But it was more than that. "The country that wallowed in the moral bookkeeping of the past century," Odoesky explained, "was destined to create a man who focused in himself the crimes, all the fallacies of his epoch, and squeezed strict and mathematical formulated laws of society out of them." Malthus was no hero in Russia.[58]

And so when the *Origin of Species* was translated in 1864, Russian evolutionists found themselves in something of a quandary. Darwin was the champion of science, the father of a great theory, but also an adherent to Malthus, that "malicious mediocrity," according to Tolstoy.[59] How to divorce the kind and portly naturalist Whig from Downe from the cleft-palated, fire-breathing, reactionary reverend from Surrey?[60] Both ends of the political spectrum had good grounds for annulment. Radicals like Herzen reviled Malthus for his morals: Unlike bourgeois political economy, the cherished peasant commune allowed "everyone without exception to take his place at the table." Monarchists and conservatives, on the other hand, like the Slavophile biologist Nikolai Danilevsky, contrasted czarist Russia's nobility to Britain's "nation of shopkeepers," pettily counting their coins. Danilevsky saw Darwin's dependence on Malthus as proof of the inseparability of science from cultural values. "The English national type," he wrote, "accepts [struggle] with all its consequences, demands it as his right, tolerates no limits upon it. . . . He boxes one on one, not in a group as we Russians like to spar." Darwinism for Danilevsky was "a purely English doctrine," its pedigree still unfolding: "On usefulness and utilitarianism is founded Benthamite ethics, and essentially Spencer's also; on the war of all against all, now termed the struggle for existence—Hobbes's theory of politics; on competition—the economic theory of Adam Smith. . . . Malthus applied the very same principle to the problem of population. . . . Darwin extended both Malthus's partial theory and the general theory of the political economists to the organic world." Russian values were of a different timber.[61]

But so was Russian nature. Darwin and Wallace had eavesdropped on life in the shrieking hullabaloo of the tropics. But the winds of the arctic tundra whistled an altogether different melody. And so, wanting to stay loyal to Darwin, Russian evolutionists now turned to their sage, training a torch on those expressions Huxley and the Malthusians had swept aside. "I use this term in a large and metaphorical sense," Darwin wrote of the struggle for existence in *Origin*. "Two canine animals, in a time of dearth, may be truly said to struggle with each other which shall get food and live. *But a plant on the edge of a desert is said to struggle for life against the drought*, though more properly it should be said to be dependent on the moisture."[62] Here was the merciful getaway from *bellum omnium contra omnes*, even if Darwin had not underscored it. For if the struggle could mean both competition with other members of the same species *and* a battle against the elements, it was a matter of evidence which of the two was more important in nature. And if harsh surroundings were the enemy rather than rivals from one's own species, animals might seek other ways than conflict to manage such struggle. Here, in Russia, the fight against the elements could actually lead to cooperation.

• • •

London did not keep Kropotkin for long. The Jura Federation that had turned him anarchist during the blizzard in Sonvilliers beckoned once again, and within a few months he was in Switzerland knee deep in revolutionary activity. On March 18, 1877, he organized a demonstration in Bern to commemorate the Paris Commune. Other leaders of the Jura feared police reaction, but Kropotkin was certain that in this instance violence would serve the cause. He was right. Police brutality galvanized the workers, and membership in the federation doubled after the demonstration. The peacefulness of the Sonvilliers watchmakers notwithstanding, Peter was developing a political program: collectivism, negation of state, and "propaganda of the deed"—violence—as the means to the former through the latter.

It was the young people who would bring about change. "All of you who possess knowledge, talent, capacity, industry," Kropotkin wrote in 1880 in his paper *Le Révolté*, "if you have a spark of sympathy in your nature, come, you and your companions, come and place your services at the disposal of those who most need them. And remember, if you do come, that you come not as masters, but as comrades in the struggle; that you come not to govern but to gain strength for yourselves in a new life which sweeps upward to the conquest of the future; that you come less to teach than to grasp the aspirations of the many; to divine them, to give them shape, and then to work, without rest and without haste, with all the fire of youth and all the judgment of age, to realize them in actual life."[63]

On March 1 (old style), 1881, Alexander II was assassinated in Russia. Once his trusted liege, Kropotkin welcomed the news of his death as a harbinger of the coming revolution. But he would have to watch his back now. The successor, Alexander III, had formed the Holy Brotherhood, a secret counteroffensive that soon issued a death warrant against Kropotkin. Luckily Peter had been expelled from Switzerland for his support for the assassination, and now, back in London, he was given warning of Alexander's plot. Undeterred, he exposed it in the London *Times* and Manchester *Chronicle*, and a deeply embarrassed czar was made to recall his agents. Still, if Kropotkin had escaped with his life, he was less lucky with his freedom. Despairing of the workers' movement in England, he traveled to France, where his reputation as an anarchist preceded him. Within a few months he was apprehended and sentenced, and spent the next three years in prison. It was soon after his release following international pressure that the news arrived: his brother Alexander, exiled for political offences, had committed suicide in Siberia.[64]

It was a terrible blow. Alexander had been his lifelong friend, perhaps his only true one. But his suicide also made Peter all the more determined to find confidence in his revolutionary activities. Increasingly he turned to science: the science of anarchy and the science of nature. They had evolved apart from each other, but the two sciences were now converging, even becoming uncannily interchangeable. When Darwin died in the spring of 1882, Kropotkin penned an obituary in *Le Révolté*. Celebrating, in true Russian fashion, the sage of evolution entirely divorced from Malthus, the prince judged Darwin's ideas "an excellent argument that animal societies are best organized in the community-anarchist manner."[65] In "The Scientific basis of Anarchy," some years later, he made clear that the river ran in both directions. "The anarchist thinker," Kropotkin wrote, "does not resort to metaphysical conceptions (like the 'natural rights,' the 'duties of the state' and so on) for establishing what are, in his opinion, the best conditions for realising the greatest happiness for humanity. He follows, on the contrary, the course traced by the modern philosophy of evolution." [66] Finding the answers to society's woes "was no longer a matter of faith; it [is] a matter for scientific discussion."[67]

· · ·

Meanwhile, navigating anxiously between Spencer's ultraselfish ethics and George and Wallace's socialist Nature, Huxley had found an uneasy path to allay his heart's torments. If instincts were bloody, morality would be bought by casting away their yoke. This was the task of civilization—its very raison d'etre: to combat, with full force, man's evolutionary heritage. It might seem "an audacious proposal" to create thus "an artificial world within the cosmos," but of course this was man's "nature within nature," sanctioned by his evolution, a "strange microcosm spinning counter-clockwise." Huxley was hopeful, but this was optimism born of necessity: For a believing Darwinist any other course would mean utter bleakness and despair.[68]

Like Darwin, Huxley saw ants and bees partake in social behavior and altruism. But this was simply "the perfection of an automatic mechanism, hammered out by the blows of the struggle for existence." Here was no principle to help explain the natural origins of mankind's morals; after all, a drone was born a drone, and could never "aspire" to be a queen or even a worker. Man, on the other hand, had an "innate desire" to enjoy the pleasures and escape the pains of life—his *aviditas vitae*—an essential condition of success in the war of nature outside, "and yet the sure agent of the destruction of society if allowed free play within." Far from trying to emulate nature, man would need to combat it. If he was to show any kindness at all outside the family (to Huxley the only stable haven of "goodness"), it would be through an "artificial personality," a conscience, what Adam Smith called "the man within," the precarious exception to Nature-Ishtar. Were it not for his regard for the opinion of others, his shame before disapproval and condemnation, man would be as ruthless as the animals. No, there could be no "sanction for morality in the ways of the cosmos" for Huxley. Nature's injustice had "burned itself deeply" into his soul.[69]

. . .

Years in the Afar, in prisons, and in revolutionary politics had coalesced Kropotkin's thoughts, too, into a single, unwavering philosophy. Quite the opposite of Huxley's tortured plea to wrest civilized man away from his savage beginnings, it was rather the *return* to animal origins that promised to save morality for mankind. And so, when in a dank library in Harrow, perusing the latest issue of the *Nineteenth Century*, Kropotkin's eyes fell on Huxley's "The Struggle for Existence," anger swelled within him. He would need to rescue Darwin from the "infidels," men like Huxley who had "raised the 'pitiless' struggle for personal advantage to the height of a biological principle."[70] Moved to action, the "shepherd from the Delectable Mountains" wrote to James Knowles, the *Nineteenth*'s editor, asking that he extend his hospitality for "an elaborate reply." Knowles complied willingly, writing to Huxley that the result was "one of the most refreshing & reviving aspects of Nature that ever I came across."[71]

"Mutual Aid Among Animals" was the first of a series of five articles, written between 1890 and 1896, that would become famously known in 1902 as the book *Mutual Aid*. Here Kropotkin finally sank his talons into "nature, red in tooth and claw." For if the bees and ants and termites had "renounced the Hobbesian war" and were "the better for it" so had shoaling fish, burying beetles, herding deer, lizards, birds, and squirrels. Remembering his years in the great expanses of the Afar, Kropotkin now wrote: "wherever I saw animal life in abundance, I saw Mutual Aid and Mutual Support."[72]

This was a general principle, not a Siberian exception, as countless examples made clear. There was the common crab, as Darwin's own grandfather Erasmus had noticed, stationing sentinels when its friends are molting. There were the pelicans forming a wide half circle and paddling toward the shore to entrap fish. There was the house sparrow who "shares any food" and the white-tailed eagles spreading apart high in the sky to get a full view before crying to one another when a meal is spotted. There were the little titis, whose childish faces had so struck Alexander von Humboldt, embracing and protecting one another when it rains, "rolling their tails over the necks of their shivering comrades." And, of course, there were the great hordes of mammals: deer, antelope, elephants, wild donkeys, camels, sheep, jackals, wolves, wild boar—for all of whom "mutual aid [is] the rule." Despite the prevalent picture of "lions and hyenas plunging their bleeding teeth into the flesh of their victims," the hordes were of astonishingly greater numbers than the carnivores. If the altruism of the hymenoptera (ants, bees, and wasps) was imposed by their physiological structure, in these "higher" animals it was cultivated for the benefits of mutual aid. There was no greater weapon in the struggle of existence. Life *was* a struggle, and in that struggle the fittest *did* survive. But the answer to the questions, "By which arms is this struggle chiefly carried on?" and "Who are the fittest in the struggle?" made abundantly clear that "natural selection continually seeks out the ways precisely for avoiding competition." Putting limits on physical struggle, sociability left room "for the development of better moral feelings." Intelligence, compassion and "higher moral sentiments" were where progressive evolution was heading, not bloody competition between the fiercest and the strong.[73]

But where, then, had mutual aid come from? Some thought from "love" that had grown within the family; but Kropotkin was at once more hardened and more expansive.[74] To reduce animal sociability to familial love and sympathy meant to reduce its generality and importance. Communities in the wild were not predicated on family ties, nor was mutualism a result of mere "friendship." Despite Huxley's belief in the family as the only refuge from nature's battles, for Kropotkin the savage tribe, the barbarian village, the primitive community, the guilds, the medieval city—all taught the very same lesson: For mankind, too, mutualism beyond the family had been the natural state of existence.[75] "It is not love to my neighbor—whom I often do not know at all," Kropotkin wrote, "which induces me to seize a pail of water and to rush towards his house when I see it on fire; it is a far wider, even though more vague feeling or instinct of human solidarity and sociability which moves me. So it is also with animals."[76]

The message was clear: "Don't compete! Competition is always injurious to the species, and you have plenty of resources to avoid it." Kropotkin had a powerful ally on his side. "That is the watchword," he wrote, "which comes to us from the bush, the forest, the river, the ocean." Nature herself would be man's guide. "Therefore combine—practice mutual aid! That is the surest means of giving to each other and to all the greatest safety, the best guarantee of existence and progress, bodily, intellectual, and moral."[77]

If capitalism had allowed the industrial "war" to corrupt man's natural beginnings; if overpopulation and starvation were the necessary evils of progress—Kropotkin was having none of it. Darwin's Malthusian "bulldog" had gotten it precisely the wrong way around. Far from having to combat his natural instincts in order to gain a modicum of morality, all man needed to find goodness was to train his gaze within.

. . .

War or Peace, Nature or Culture: Where had true "goodness" come from? Should mankind seek solace in the ethics of evolution or perhaps in the evolution of ethics? Should he turn to the individual, the family, the community, the tribe? The terms of the debate had been set by its two great gladiators, and theirs would be the everlasting questions.

Huxley died at 3:30 p.m. on April 29, 1895. He was buried, as was his wish, in the quiet family plot in Finchley rather than beside Darwin in the nave of Westminster Abbey. No government representative came to the funeral; there was no "pageantry" or eulogy either. But there were many friends—the greatest of England's scientists, doctors, and engineers; museum directors, presidents and councils of the learned societies; and the countless "faceless" men from the institutes who had taken the train down from the Midlands and the North—all bowing their heads in silence. His had been a life of pain and duty: from Ealing to the Royal Society, from rugged individualism to corporatism, Unitarianism to agnosticism, and finally back again to the merciful extraction of human morality from the pyre of Nature-Ishtar. He was placed in the ground in a grave that, the *Telegraph* noted, had been "deeply excavated."[78] In line with his view of the exclusive role of family in nature, it was above his firstborn son, Noel, who had died aged four in 1860, that Huxley would come to rest.

When the revolution in Russia finally broke out in February 1917, Kropotkin was already old and famous. On May 30, thousands flocked to the Petrograd train station to welcome him home after forty-one years in exile.[79] Czarless and reborn, Russia had revived his optimism in the future. But then came October and the Bolsheviks, and like years before in the Afar, the spirit of promise soon wasted into disappointment. "We oppose bureaucrats everywhere all the time," Vladimir Ilyich Lenin said to Kropotkin when he received him in the Kremlin soon after. "We oppose bureaucrats and bureaucracy, and we must tear out those remnants by the roots if they are still nestled in our own new system." Then he smiled. "But after all, Peter Alekseevich, you understand perfectly well that it is very difficult to make people over, that, as Marx said, the most terrible and most impregnable fortress is the human skull!"[80]

Kropotkin moved from Moscow to the small village of Dimitrov, where a cooperative was being constructed. Increasingly frail, and working against the clock on his magnum opus, *Ethics*, he still found time to help the workers.[81] "I consider it a duty to testify," he wrote to Lenin on March 4, 1920, "that the situation of these employees is truly desperate. The majority are literally starving. . . . At present, it is the party committees, not the soviets, who rule in Russia. . . . If the present situation continues, the very word 'socialism' will turn into a curse."[82]

Lenin never replied. But he did give his personal consent when Peter Kropotkin died on February 8, 1921, that the anarchists arrange his funeral. It would be the last mass gathering of anarchists in Russia.

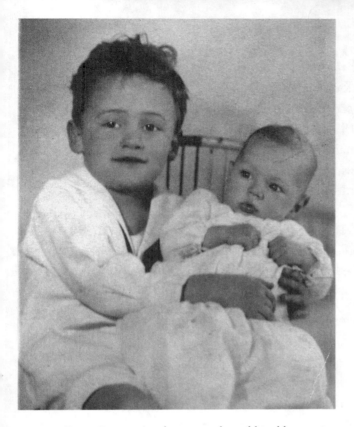

George Price at age thirteen weeks and his older brother, Edison, 1923

Teenagers George and Edison with their mother, Alice

2

New York

William ran up the backstairs into the Belasco Theatre. Showtime was at 8:00 p.m. sharp, and none of the fixtures were in place. The director was going to kill him. He was sweating, out of breath. As he ran past the stage, spinning excuses in his mind, he glimpsed, just for a split second, a pair of sparkling eyes behind a curtain in the dark. His heart stopped: He'd never seen anything more beautiful.[1]

William Edison Price was a man on a mission. "A light for every purpose" was the slogan; to be "Pioneers of Progress"—the intent. Baby Hercules Flood Lights, Cyclorama Reflector Strips, Display Reflector Borders—all were in great demand. There were fashion shows, pageants, expositions, exhibits. There were traveling attractions, which usually used Portable Switchboard Dimmer Boxes. And, since a mere pile of merchandise was no longer sufficient to attract passersby, there were show windows, of course, calling for instant adjustment of color and the direction of light where desired. But most of all there was the stage. "The theatre itself is old as the dimmest page of history," the company's catalog explained, "but scientific lighting of the theatre is in its infancy." Together with his business partners John Higham and Michael Kelly, William Edison Price was the Display Stage Lighting Company, Incorporated, at 334 West Forty-fourth Street, New York City, and business, thank goodness, was booming.[2]

It was "the Bishop of Broadway" who really cranked up the contracts. David Belasco was born in San Francisco in 1853 to Portuguese Jewish parents whose real name was Velasco; his father had been a mime in London before seeking greater fortunes in America. Escaping from a monastery to a circus at the age of twelve, David soon landed his first real job: callboy at the Metropolitan Theatre. By twenty-nine he left for New York City, having acted in 170 plays and written or adapted at least a hundred. A stage manager at Madison Square Garden and then for Daniel Frohman at the Lyceum, he was moving in New York's flashy show-business circles. But it was the Civil War romance *The Heart of Maryland* in 1895 that really made Belasco's name as playwright, producer, and director. Then came the French adaptation *Zaza*, the farce *Naughty Anthony*, and, at the turn of the century, *Madame Butterfly*, the Japanese-set tearjerker destined for operatic fame in the hands of the great Giacomo Puccini.[3]

Dressed in the clerical collar that gained him his moniker, Belasco was known for his tantrums. One favorite trick was to stamp on his watch, smashing it to smithereens. (Only very close associates knew that he kept a stock of cheap, secondhand watches for just such occasions.) But if Belasco was a mix of calculated melodrama, so were his plays—and audiences flocked to the theaters to enjoy them. Maudlin and sensationalist, they were a far cry from Ibsen, Chekhov, or Strindberg. Still, if the Europeans had brought emotional realism to their characters, Belasco would bring technical realism to his stage. His settings were famous for accuracy down to the most minute detail: a functioning laundromat, a reproduction of a Childs Restaurant with actors brewing coffee and cooking pancakes onstage. Once he even purchased a room in a flophouse, removed it from the building, brought it to his theater, cut out one wall, and presented it as the set for a production.

Lighting was his passion. From under his hands colored silks and gelatin slides were ushering in a revolution. He was the master of mood and of tension, his "real" sunsets a spectacle to behold. When he bought the Stuyvesant Theatre in 1907 he was particular about the fly space and hydraulic systems, about the Tiffany lighting and ceiling panels, rich woodwork and murals. By 1910 it was renamed the Belasco, and the great actors of the age—David Warfield, Lenore Ulric, Frances Starr, Blanche Bates—were all on board and working. The address was West Forty-fourth Street between Sixth and Seventh avenues, just a few blocks from the "Pioneers of Progress" at Display.

Thank goodness, the fixtures were finally in place. Shaking Belsaco's hand and taking a moment to catch his breath, William Edison remembered the curtain. *The Auctioneer*, a comedy in three acts, was being restaged after its successful run in 1901. "An Old Friend Back at Belasco," the *Times* exclaimed, touting the former vaudeville actor David Warfield's Simon Levy, a bittersweetly comical Lower East Side peddler down on his luck. It was a great hit, Simon's cry, "Monkey on a stick, 5 cents!" having lost "none of its plaintive melancholy."[4] Broad faced, rather short, and with the sparkling eyes that had captivated him, Alice Avery was cast in a small part, playing a Misses Compton alongside Warfield's woeful salesman. Summoning up his courage, William walked over to introduce himself.

. . .

In fact Alice was not her real name. She had been born Clara Ermine Avery in Bellevue, Michigan, in 1883, the daughter of Emma Addale Gage and her husband, Frank Avery, a well known Bellevue jeweler. Emma's father was Dr. Gage, a prominent physician who rode horseback through that part of the country, with old-fashioned saddlebags.[5] A devout Christian and faithful church worker, at the age of twenty-two Emma had joined the Methodist church just opposite the old Gage family homestead. When her husband, Frank, died young in 1890, she was left to bring up her four children on her own; they were Clara, Mary, Gage, and George. But Emma had the good fortune of a progressive education, graduating from the private Christian coeducational Olivet College thirty miles south of Lansing, one of the first in America to admit women.[6] She became a beloved mainstay of her community—a third- and fifth-grade teacher—and when her daughter Clara graduated from Big Rapids High School, so did she.

In 1905, after Olivet, Clara became the principal of Hersey High School, but she already knew inside that the theater was her calling.[7] New York City's Broadway lights shone brighter than anything in Hersey. Armed with a stage name and inborn dramatic panache, she would seek new fortunes there.

William Edison Price, to be perfectly honest, was not his real name either. Like his wife-to-be, he had been born in 1883. While she was a daughter of the Christian foundational classes of the Midwest, he was the son of Henry and Etta, Yiddish-speaking Russian Jews, perhaps of the name Preis, who had arrived on boat to America's shores, and from there to Chicago.[8] Brother to Anna, Lena, Robert, Sadie, and Rosie, all born in America, his own given name was Isak. By 1910 the Chicago census listed him as an "electrician" with a positively new appellation. Isak must have figured that the blend of Americanism and professional pedigree in his new name might be a harbinger of future success, and, more or less severing all ties with his family, he got on a train for New York City to invent himself anew.[9]

And so, in a theater named Belasco after a man whose real name was Velasco, William Edison, whose real name was Isak, and Alice, whose real name was Clara, fell in love. The wedding took place on June 6, 1917, in Stamford, Connecticut, a town whose real name had been Rippowam before it was bought from the Indians in 1640.[10] This was America, where dreamers could dream.

· · ·

Dreams didn't come easy, though, at least not for Alice. "Dear managers —" she had written, in a poem printed in the *Morning Telegraph*,

> *I'd like to ask a favor please of you:*
> *Give me your definition of a Broadway ingénue.*

We are too slim or else too fat, or have no shape at all;
Too short—we have our heels built up and then we are too tall.
Our hair's too light or else too dark, or has a tinge of red
That really wouldn't go too well with the leading lady's head.
Our accent's flat or else too broad; we never could "get by"
Our eyes are just a bit too small; our brows a bit too high."[11]

As her prospects grew dimmer in the theater, Alice increasingly helped with the lighting at Display. Soon a first child was born to the unlikely couple, Edison, in the summer of 1918. Life was lived from order to order. The theaters, the moving pictures, schools, clubs, universities, hotels, dance halls, and restaurants—every one, it seemed, needed a lighting fixture. And then there were the suppliers: the Cutler-Hammer Co., Wilmington Fibre Co., Corning Glass Works, Asco Supply, Rome Wire Co., and scores more—each with their trucks and shipments and schedules and "unbeatable" deals.[12] It was into this world of gelatins, clamps, screws, and electric fibers wholesales and showtimes; vaudeville actresses, quick-tongued salesmen and fix-it men; glitzy clubs, Broadway theaters and their melodramatic impresarios that on October 16, 1922, weighing eight pounds twelve ounces, George Robert Price was born in Scarsdale, New York. "Don't you care," a family friend wrote upon his arrival, half in consolation. "This world needs boys for lonesome girls."[13]

. . .

It was an age of invention, and William Edison Price embodied it. Connection Plug patent 1,454,858, Lamp Fixture patent 1,351,681, Clamping Device patent 1,562,052——all these and more were his creations.[14] The theaters loved him, and one in particular would leave its mark.

The Manhattan Opera House had been built by Oscar Hammerstein in 1906, just west of Madison Square Garden on Thirty-fourth Street. Hammerstein had the bold intention of taking on the Metropolitan Opera by offering New Yorkers cheap seats to opera productions; culture, he thought, belonged to everyone. Sensing that it would be out of business in no time, the Met offered

$1.2 million to the upstart establishment to stop producing opera for a decade, a deal even "champion-of-the-masses" Hammerstein could not judiciously refuse. Hammerstein eventually sold to the Shubert brothers, who turned the place into a "combination" house featuring vaudeville on weekdays and cheap opera concerts on Sundays, once again undercutting the Metropolitan. In 1926, when George was not yet two years old, Warner Brothers chose the venue to premiere *Don Juan*, the first commercially released film featuring a recorded musical soundtrack.[15]

By the time Display Stage Lighting Incorporated began contracting at the Manhattan Opera House, the building was owned by the Ancient Accepted Scottish Rite of Free Masonry, which built a grand ballroom on the seventh floor. It may well have been this connection that led William to join the order, all the more so to disguise from his boys their immigrant father's true origins.[16]

Business was booming, and there was no time to look back. On the occasions when the past could not be escaped, it was changed: In the 1920 New York census, the place of origin of William's parents had miraculously migrated from Russia to England. Alice's devoutly Christian side of the family didn't mind William Edison's bouts of religious forgetfulness and invention. "How is the old outfit that broke away from your business," Emma Gage Avery wrote to her daughter regarding some Display internal affairs, adding, rather tellingly, "The Jews."[17]

And so, when in February 1927 William Edison Price died suddenly of pneumonia, the man whose "name was known wherever there was a theatre" was accompanied to his grave by priests of the Masonic Temple. It is unclear how many of the 150 people attending the funeral knew about his true beginnings. Thanks to Alice's acquiescence, his two little sons, Edison and George, positively did not. Isak Preis, forty-five, took his secret to the grave.[18]

· · ·

"Stock Prices Crash in Frantic Selling," the *Washington Post* reported on October 3, 1929, just as the Price family was beginning to grasp the loss of its husband and father. Following a decade of unprecedented growth, the market predicted recovery. "Brokers Believe Worst Is Over," a sanguine *New York Herald Tribune* headlined toward the end of the month, adding that investors should look for "real bargains." Even after "Black Thursday" and "Black Tuesday" optimism abounded: "Very Prosperous Year Is Forecast," predicted the *World*, in December 1929.[19]

Whether the crash had been the result of "extraordinary speculation" as the economist of the welfare state, John Maynard Keynes, thought (England's Chancellor of the Exchequer, Philip Snowden, called it more bluntly "a perfect orgy of speculation"), no longer seemed to matter much.[20] A market that had begun at a high above 380 points in September had by Thanksgiving fallen below 200. As surely as William Edison Price would not rise from his grave, there was not going to be a "recovery." By 1932 stocks would lose nearly 90 percent of their value.

An immigrant haven, New York was one of the cities in America hit hardest by the crash. By the spring of 1930 there were more than fifty bread lines on the Lower East Side providing fifty thousand meals a day to the hungry. Before he embarked on a second campaign for the presidency of the United States, Governor Franklin Delano Roosevelt was made to face a cold reality: Half of the city's manufacturing plants were closed, one in three citizens was unemployed, and roughly 1.6 million were on some form of relief.[21] Having incorporated before the crash at one hundred dollars a share, Alice Avery Price and her company were buried deep in the Great Depression, holding on for dear life, and the two little boys along with them.

They had moved from Hartsdale to the city, to a tenement on West Ninety-fourth Street. Rain came through the ceilings in the front hall, the bathroom, and the kitchen. The living room light fixture was falling down. The toilet did not flush properly, nor did the refrigerator work or sit evenly on the crooked linoleum. Alice wasn't sure which was worse: the plaster coming apart in the bathroom, the three-pane windows ready to fall out in the bedrooms, or the water taps that ran as if with a mind entirely of their own. The more she wrote to her landlord to complain, the more he wrote back threatening legal action for her lack of payment, as did the Laundromat owner Isidore Stern; the boys' friendly eye doctor, Ralph Singer; and the *New York Times* for overdue payment for an advertisement.[22]

Money had to come from somewhere, anywhere, quick. Alice decided to send her older boy, Edison, to a farm upstate owned by a family called the van Akins. With any luck his room in the city would fetch seven dollars a week. It made sense; after all, she must have offered to her aching conscience: Little George could be lugged along to the office, but Edison was often at home until late hours by himself, and Alice was afraid of "prowlers."[23]

Display had gone bankrupt in September 1931. William Edison's former partner John Higham had secretly formed a corporation for night baseball lighting, stealing thousands of dollars of equipment and running up a huge debt in the company name. Meanwhile the foreman of the shop, the diminutive Joseph Levy, and his accountant brother Saul joined hands in conspiracy with the secretary, Mr. Kook. When the lease on the company building expired, Alice was told that the only thing to do was to remodel a bunch of old stables at 410 West Forty-fourth Street. Unbeknownst to her it was Kook's father-in-law who received the bid for the remodeling, and after it was pushed through against her wishes for a considerable sum, the Levys and Kook resigned and started a competitive business.[24]

At the same time the Irving Trust Receivership Company began selling off William Edison's patents. Since he had been more a dreamer than a businessman, many of his inventions were assigned to the company rather than to his own name. The buyers, like Meyer Harris of the Columbia Stage Lighting Company, were invariably competitors, and Alice and Display were going under.[25]

The social fabric of America's greatest city was coming apart at the seams. Privately funded mutual aid societies in the city were collapsing: From six thousand in 1920, only two thousand now remained. More than four hundred private social service institutions—one-third of the agencies in New York—closed their doors by 1933, and abandonment of women and children by husbands and fathers rose more than 135 percent. When hard times hit, it was now apparent, people had to fend for themselves.[26]

. . .

To Alice trust seemed just about as rare as a lucky dime.

First there were the Display men. Then a friendly lodger skipped town without payment, leaving Alice doubly despondent over sending away her boy for nothing but a swindle. A modicum of belief in the goodness of people was restored when a Mr. Walter began coming into the office from the garage to express his heartfelt sympathy. When she saw that he was hungry, Alice even gave him a daily food allowance. They were becoming friends. Walters started borrowing money for his hotel room (he was going to be thrown out, he said) on the promise that he would pay back with forthcoming moneys. It was only after he went off with little George's glasses, provided by Alice after Walters broke his own in a supposed taxi accident, that she discovered that the checks he wrote her were fakes and that his real name was Ward. "These I should like to recover," she now wrote stiffly to Chief Police Inspector G. G. Henry, "as my son starts school on the 26th." George had a squint and his eyesight was growing bleaker, as, they were all learning, were their fortunes. It was a lonely time. "The children and I," Alice wrote to the Transfer Tax Commission, "have nothing left in the world except each other."[27]

A leading officer at the Scully Steel and Iron Company in Chicago, Gage H. Avery stepped in to save his sister. "There is no easy road," he wrote to Clara in the winter of 1932. "It's up to you to stop dreaming and go after the business."[28]

Taking a deep breath and a loan from her generous brother, Alice bought back the name of Display from the receivers and decided to make a go of it. It was a hell of a mess. Creditors were knocking loudly and invariably filing suit. "I have always had a reputation for honesty," she wrote to one of them, between court case and arraignment, angrily begging for his patience.

> *It stands to reason that I have no money or I should be able to have my children together in a home instead of living in a room with one of them while another little fellow is away on a farm. . . . You will*

be wasting your time and mine to start any proceedings against me. Better let me have my time to conduct this little business and try to work out of my present difficulties. The theatrical business is nothing to brag of right now. We are in the lighting end and competition is very keen. It is a case of just existing.[29]

But Alice proved tough, and soon began to pick herself, as well as Display, up from melancholy doldrums. One by one theaters were won over, creditors paid back, suppliers assuaged, new contracts delivered. Even burlesques were no longer out of the picture. To the contrary, in venues like the Kelly Club and the Apollo Theater, burlesque had become the Great Depression's signature entertainment. "Gorgeous glorified maids and models in a landslide of loveliness," the signs said, lit to perfection with fixtures from Display.

There were different versions about how it had evolved. Some said that a chorus girl at the famous Minsky Brothers' Winter Garden Theater on the Lower East Side had gone onstage accidentally without her starched collar and cuffs in what passed for public nudity in 1916. Others swore that the burlesque actress Hinda Wassau broke a strap of her chemise while singing the grand finale and, rather than dash offstage in midsong, bravely finished her number to enthusiastic applause. Whichever was true, burlesque now become the perfect form of entertainment for a depressed city and one of its only growth industries. Slapstick comedy "bits" and scads of scantily clad chorines didn't exactly cost an arm and a leg to produce. Hundreds of out-of-work actors and actresses, comedians, and dancers were again on stage, and the audiences flocked to the theaters.[30]

Citing prostitution and public immorality, Mayor Fiorello H. La Guardia promised to put an end to the "incorporated filth," and ordered City License Commissioner Paul Moss to shut down the venues. This time, at least, he had everyone's God on his side: "Information that has come to me of the spread, evil influence and destructive results from these disgraceful and pernicious performances," the Catholic archbishop of the city wrote to Moss, citing "the cause of great concern." The chairman of the board of Jewish ministers and a rather prudish figure himself, Rabbi Samuel H. Goldenson, agreed:

"These houses cater to the lowest appetites and passions of men and women and altogether are a menace to the moral life of the community."[31] But as long as 'these houses" brought in business, it didn't matter one bit to Alice. One dollar was as green as another, whether it lit Shakespeare, Belasco, or tassels.[32]

However prudish, La Guardia was a master at choosing deputies. One in particular made a name. They called him the Baron Haussmann of Second Empire Paris, and Robert Moses deserved it. Some said he loved automobiles more than humans, but unlike city planners in other parts of the country, Moses came prepared to take advantage of a federal buck. Carving new shorelines, building bridges across the boroughs and highways in the sky—transforming neighborhoods forever—the "master builder" of New York was riding high on the New Deal. At one point one-quarter of federal construction dollars were being spent in New York, and there were eighty thousand men and women working under him.

He was a controversial figure. Some said he later caused the departure of the Brooklyn Dodgers and New York Giants, destroyed Coney Island's amusement parks, ruined the South Bronx and abandoned the blacks. But others could see that however heavy-handed, with the moneys from the Works Progress Administration and Civilian Conservation Corps, he was sculpting New York into a paradigm of modernism. From low-income housing to Rockefeller Center to European Modernist mansions designed by William Lescaze, New York was rising from the ashes of the Great Depression.[33]

. . .

The thirties were an exciting time to be a boy in New York City. The Yankees had released Babe Ruth in 1935, but the following year the rookie Joe DiMaggio led the team to an amazing World Series victory against the rival New York Giants. Cowboy movies were in vogue; a quarter got you through the turnstile to watch the latest adventure of Hopalong Cassidy. In music Artie Shaw and Fletcher Henderson had sparked a new era—the Big Band; it was so big that Benny Goodman gave the first nonclassical concert at Carnegie Hall in 1938.[34] That very same summer the writer Jerome Siegel and the artist Joe Shuster mailed their creation to National Periodical Publications, and Super-man was born.[35] George Price was a quiet and reserved little boy, but around him New York was hopping.

On one occasion he went down to the pier with his mother to see her lodger, Mr. Kafuco, off on the *Europa*. George and Alice presented him with a Waterman pencil, black with gold band trim-ming, and marveled over the "great boat." Mr. Kafuco was some kind of Japanese dignitary, and George was overcome by the line of "distinguished looking Japs" who filed into his beautiful stateroom, each making a "very profoundly deep bow."[36]

The world, he was learning, was a big place after all—bigger, to be sure, than motor-driven dimmers, shouting foremen, even the Display Stage Lighting Company. His brother, Edison, toughened by his months at the van Akins ("Hope you are warm in the shack these cold days," Alice wrote to him), would be the one to enter the family business. The Prices had moved to a new and bigger apartment at 311 West Ninety-fifth Street, and Edison was spending more time at home and at the office. By fifteen he was already instructing his mother on how to deal with her creditors and debtors, not to men-tion coming up with lighting inventions of his own.[37]

In 1935 Edison graduated from the Birch Wathen School on West Ninety-third, where George was enrolled on a full scholarship behind him. The senior play that year was Sir James Barrie's Edwardian-style *Alice Sit-By-the-Fire*, which seemed to fit the school just right. Estab-lished in 1921 as a private coeducational institution, it was run ador-ingly by its pair of staid—and perfectly coiffeured—headmistresses, Miss Louise Birch and Miss Edith Wathen. It was widely known that Miss Birch had been the tutor to the young Nelson D. Rockefeller, and even spent the summer of 1907 teaching Teddy Roosevelt's grand-children at the White House. DeWitt Davidson, Margot Lindsay, Warren Milius, Virginia Plaut: These were the sons and daughters of New York City's socialite Protestant classes. At Birch Wathen they would be instilled with the appropriate aspirations. Music and the arts came before academics, socials and drama trumping science and math. Edison was reclusive and didn't quite fit in. His yearbook page spoke of brilliance and profundity, of a "rare quality of intensity" and a fascination with Indian mysticism. His technical bent didn't quite seem to correspond. Nor, in particular, did his dark Jewish looks.[38]

But George was different. Reddish blond and lanky, with a tie on he looked like an arrival from a Scandinavian Eton. His eyes were an intense hazel, his ears rather large, and he combed his hair straight back. He was well kempt. His manner was somewhat haughty. Athletic but not particularly strong in team sports (he was a substitute on the basketball and soccer teams), he joined the chess team and student council. His academic talents were obvious. Inspired by the name and scenes at the Apollo, at fourteen he produced a seventy-page paper on Greek temple architecture that would have satisfied any university professor. He loved to write, especially about affairs of state, and even more to think about physics and numbers. "We would like to repeat our thanks to those members of the faculty who spent much of their time in aiding the staff to publish this magazine," the editorial of the *Birch Log* read in the spring of 1936. "To the person that made this magazine possible, Miss Birch, we owe an infinity of thanks." Then it added: "According to Price the infinite is finite but our thanks are not." George knew he was bright and was sharp at mathematics, and was more than happy to let other people know it.[39]

Business at Display was still rocky. The fiscal year 1937–38 had been tough; "we are right down to rock bottom at the bank," Alice reported. George had been helping out at the shop all summer long, and Edison's little new light, the Mite-Lite, was launched and promising to be quite a success among display men. "But alas!" Alice bemoaned in the fall in a letter to the school's Miss Birch "we have no working capital to give it the send-off it deserves." The fifty-dollar school registration fee was simply too steep. If the year would be a good one at Display, Alice would send George back to graduate. In the meantime he had successfully passed the citywide exam, and would be starting the eleventh grade gratis at Stuyvesant.[40]

*   *   *

There could not have been a more opposite school in the city to Birch Wathen. Down on 345 East Fifteenth Street, just a few blocks from Union Square, Stuyvesant had been built on the old *bowerij*, or farm, of its seventeenth-century namesake, Peter Stuyvesant, the Dutch colonial governor of New Amsterdam. At the laying of the cornerstone on a crisp fall afternoon in late September 1905, William Henry Maxwell, the Irish-born, monocle-wearing, mustached super-intendent of schools, declared a new era in the history of education. Millions were to be expended "in order that every boy, no matter how poor his parents may be, may have opportunities equal to those given to the sons of the rich."[41]

Stuyvesant would be a school combining technical training with an academic program, servicing the bright boys of the lower classes. Whereas there had been fewer than five hundred thousand immigrants arriving annually in America in the last fifteen years of the nineteenth century, the number more than doubled in each of the years in the decade after 1904.[42] Brooklyn, the Bronx, Staten Island, Queens, and Manhattan had been consolidated in 1898, and Greater New York was now the largest city in America. At the dedication earlier in the winter, Professor Thomas M. Balliet, dean of the School of Pedagogy at NYU, explained that the classical system of education was outmoded and unfit for industrial society. "Now we are a nation of manufacturers and traders," he exclaimed. Schools like Stuyvesant, Bronx Science, and Brooklyn Tech, Maxwell thought, were "the educational hope of democracy."[43]

Its reputation grew quickly. By October 1907 the *New York Times* called it "a school that excels anything of a similar nature in the country."[44] The face of the city was changing. In 1890 the Irish and Germans counted for more than a half of New York City's 1.5 million; by 1920 the population had grown to more than 5.5 million, and its largest groups were now Jews (1.5 million) and Italians (eight hundred thousand).

And so when he arrived in September 1938 at the massive five-story structure topped by a sixty-foot flagstaff, George took his place among a new type of classmate. There were Emanuel Schmerzler, Morton Rosenbluth, Jerry Lachman, and the Bader twins—Mortimer and Richard. There were Remo Bramanti, Rosario Pipolo, and the senior president, Anthony Gandolfo. There were Photiadis the Greek, Boyarsky the Pole, and Gallagher and O'Connor and McDougall.

Jews were the majority. It was from schools like Stuyvesant that they emerged in the 1930s "on a ladder from the gutter" to become more than half of the city's doctors, lawyers, dentists, and public school teachers. Perhaps not quite understanding why—after all, his father's origins were unknown to him—George felt comfortable in his new environs, and most of his friends were Jewish.[45]

The boys would commute from Bensonhurst, Brighton Beach, Sheepshead Bay, Canarsie, Flushing, Elmhurst, Washington Heights, and Inwood. The ones from Manhattan, like him, would arrive by train from uptown or walk north from the Lower East Side. At the exit to the subway station on Fourteenth Street there was an Automat where you could buy a couple of slices of bread for a dime, and make a sandwich with free ketchup and mustard. Passing by the girls' school, Washington Irving, on Fifteenth Street and Second Avenue, "sparks would fly"; the Stuyvesant boys may have been nerdy, but they had hormones like everyone else. Private school kids like the ones in Birch Wathen looked positively like "creatures from outer space" to them, but this, most admitted to themselves privately, was undoubtedly a projection.[46]

They were grinders, geeks. Even though by 1934 the citywide exam had been instituted to stem the influx of numbers, classes still had to divide into two sessions: juniors and seniors between 7:40 and 12:35 and freshmen and sophomores from 12:40 to 5:20—and everyone took studies seriously. Every week at assembly they'd sing the school song—"Our Strong Band Can Ne'er Be Broken"—and end with "America the Beautiful." And once a week, for a whole period, they were made to take a hot shower. Many of the boys came from cold water flats, and Stuyvesant wanted them clean.[47]

The school was built on an H-plan, with academic classes on one leg, shops on the other, and labs and lecture rooms on the crossbar. The shops were a proper technical universe: There were pattern making, drafting, wood turning, blacksmithing, forging, and machine work. There was even a 1.5 ton copula installed in the foundry, with a brass melting furnace, a core oven, a melting pot, a molding machine, a dry grinder, and a polish and buffer. In the basement, steel racks for storage of lumber and metals were packed alongside machines for cutting up stock. There were twenty-four-inch surfacers, twelve-inch hand jointers, a double-arbor sawing machine, and a water tool grinder and grindstone. And then there were the labs. Lewis Mumford, the American historian of technology and literary critic, remembered the excitement of it all. His physics teacher had once held up his pencil and said: "If we knew how to unlock the energy in this carbon, a few pencils would be enough to run the subways of New York."[48]

. . .

After school George would stay around for his electives. There was the Experimental Physics Lab, the Arista Club, the American Rocket Society and, most important, the Chess Team. His year was quite a class—with the highest scholastic record in the school's history—not an easy one to shine in. There was Joseph File, who would become a Princeton nuclear engineer, and a pioneer of MRI. There were the Bader twins, legendary future professors of medicine. There was Joshua Lederberg, a year behind, who would win a Nobel Prize for his work in bacterial genetics. And there were Nat Militzok, the six-foot-three-inch future New York Knicks forward, and Kai Winding, the composer and bebop jazz trombonist.

George had a squint in his eye, and a strange affect no one could finger, but even among this crowd he was distinctive. In solid geometry class he would sit quietly in the back of the room and just once in a while offer a comment that no one could quite understand (including the bespectacled teacher, Mr. Solomon Greenfield). His voice was high-pitched and squeaky, though somehow also soft. Most thought him strange, mechanical, even perhaps slightly autistic. Richard Bader, the class valedictorian, could sense that he was leagues above everyone. His genius was baffling, even a little unsettling. Awkward, he signaled an almost defiant self-confidence in his intelligence. First Board, he led the Chess Team to the city championship in 1940, and no one, least of all George, was surprised.[49]

Above a faint cartoon of a Clark Kent type with a fedora, the *Indicator* of the class of June 1940 provided "a picture of the typical senior—the results of a class poll":

Age: 17
Weight: 155
Height: 5'9"
Ambition: Engineer
Favorite Subject: Mathematics
Hobby: Model Airplanes, Photography
Favorite Magazine: Reader's Digest
Favorite Comedian: Jack Benny
Favorite Radio Star: Bob Hope
Favorite Actor: Paul Muni, Errol Flynn
Favorite Actress: Hedy Lamarr, Bette Davis
Favorite Orchestra: Glenn Miller, Kay Kyser
Favorite Athlete: Lou Gehrig, Glenn Cunningham
Favorite Author: John Steinbeck, Sinclair Lewis
Favorite Ball Club: N.Y. Yankees, Brooklyn Dodgers
Favorite Columnist: Jimmy Powers, Walter Winchell
Favorite Morning Newspaper: New York Times, Daily News
Favorite Evening Newspaper: N.Y. Post, N.Y. World-Telegram
Favorite Radio Program: Bob Hope Show, Fred Allen Show
Favorite Motion Picture: "Gone With the Wind"

For his part George hoped to be a "research physicist," and his year-book photograph agreed. Serious and cocky, his gaze was at once both dreamy and focused, and the paradox fit him perfectly.

Parting from the class, Mr. Bradshaw, the track coach, compared success in life to the storming of a citadel, reminding everyone that the average boy earned $8,500 fifteen years upon graduating and wishing them luck in life's "battle."[50] The principal, Sinclair J. Wilson, took a more spiritual approach to education. As the graduating class and their families settled into their seats at Carnegie Hall on a warm summer day on June 25, 1940, the Lincolnesque New England headmaster started with a soft but stern tone, pointing his elongated finger at the hushed crowd:

In an age which weighs the good in the scales of the profitable, men all too ready discard ideals and moral principles. . . . You were ushered into life in a world which had just lately witnessed the fall of mighty empires. Your early years opened upon vistas and dreams of material wealth and a world of comfort unknown to former generations. Peace and plenty seemed about to be assured for all times. Instantly that secure world crashed. Misery and want beset it all around. Now fear and suspicion and hatred would join this evil pair to make a shambles of the garden which you would cultivate. . . . We lose our horizons because our vision becomes dimmed through greed or lust or want of charity. And nations also lose their horizons for the very same reasons. If you would know where to find the lost horizon, then hold fast to the ideals of your youth and no matter what the cost, choose to do the right and follow the better way, for Shangri-La lies wholly within you.[51]

George headed upstate to Walton to the farm he visited in the summers. His home was a tent in a maple grove, and besides cooking carrots with onions and butter and discovering new ways to kill flies, there was really not much to do but sleep and read and go for walks and wander. He had brought twenty books with him, and in particular enjoyed Bertrand Russell's *Introduction to Mathematical Philosophy*. To members of his old chess team he sent opening gambits, challenging them to reply: P to K4 ????, N to KB3 ??? It didn't seem worthwhile to waste a whole postcard on just one game, so he initiated four separate games in each letter. His hands were callused from chopping wood, and his back was "burnt to cinders." But he had graduated second in a class of 708 and was as free as a lark now in the woods just behind him.[52]

. . .

Meanwhile war was looming on the horizon. Hitler had invaded the Low Countries in May and with the help of Mussolini had now battered France into submission. Radio broadcasts were full of news of freighters on the Atlantic being sunk by U-boats, and crackly transmissions by Edward R. Murrow from a London under siege.

America had not yet entered the war, but everyone felt that it wouldn't be long now. President Roosevelt had just instituted the first peacetime draft in September, and down at the old Custom House on Battery Park thousands lined up for their physicals.[52] Packing up his books and pans and chess set, George left the grove and traveled to Sea Bright, New Jersey, for a week at the beach with Alice. Back in New York, he had nine nontransferable tickets to the World's Fair at Flushing Meadows, and a few more weeks before college.

The New York World's Fair invited its visitors to take a look at "the world of tomorrow"; forty-four million had already visited since opening day in the spring of 1939. Walking among the exhibits, George marveled at a futuristic car-based city by General Motors, at Bell Labs' first keyboard-operated speech synthesizer—the "Voder"—and at an amazing new invention, television. Passing by the West-inghouse Time Capsule that was not to be opened until A.D. 6939, he might have thought about the words of Principal Wilson; about marking his horizon and finding Shangri-La within—"no matter what the cost." His grades had won him a coveted Freshman Schol-arship to Harvard, even if the partly bemused, partly awestruck interviewers thought him "an arresting rara avis" and "a baffling kid" as he described to them his experiments to render glass invisible. "I am interested in following "Knowledge like a sinking star, Beyond the utmost bound of human thought,' he cuddled Tennyson in his application, and come September 13, wide-eyed as much as he was determined, set out for Cambridge. "Might go hay-wire but will never be humdrum," the Harvard interviewers concluded. George was on his way.[54]

Ronald A. Fisher (1890–1962)

J. B. S. Haldane (1892–1964)

3

Selections

atie was on her back, legs spread in the air, sweating and panting. Her husband, George, the second half of the successful art-auctioning house Robinson & Fisher, was at her side in the spacious Hampstead bedroom. It was February 17, 1890, and the baby was on its way.

A silence fell over the room. She was deeply religious, which made comprehension of what had happened more difficult. Minutes passed, accompanied by anguished moans. George looked to the side. Katie closed her eyes. The baby had been stillborn.

Just as the midwife was getting ready to clean her up there was a quiver. And a kick. And another. Katie kindled. Something was still alive inside her! Moments later a second baby was pulled out into the world, as tiny as a pink grain of rice, and entirely unexpected. His mother named him Ronald Aylmer Fisher, his friends called him "Piggy," and he would grow to become the man who built the mathematical foundations of evolution.[1]

"I attempted mathematics," Darwin wrote toward the end of his days, ". . . but I got on very slowly. . . . I have deeply regretted that I did not proceed far enough at least to understand something of the great leading principles of mathematics; for men thus endowed seem to have an extra sense."[2] Darwin had not been entombed fifteen years when, at the age of three, little Ronnie asked his nanny what a half of a half was. And a half of that. And a half of that. "Then I suppose a half of a sixteenth must be a thirtytoof," he said.[3]

When he was fourteen his beloved and devout mother died prematurely of peritonitis and numbers became his world. Hampered by poor eyesight, he developed a geometrical imagination, learning to figure out problems in his head rather than writing them down on paper. Many of England's brightest had come up through Harrow, but his tutor, Arthur Vassall, would later divide his untold brilliants into two categories: Fisher and all the rest. Mathematically he was unparalleled. Still, when a school prize offered the winner a choice of books, Piggy's selection divulged a second love that would draw him just as powerfully: Coming to Cambridge on a mathematics scholarship, he darted up to his quarters at Gonville and Caius lugging a suitcase packed with thirteen volumes: the complete works of Charles Darwin.

· · ·

It was 1909, the Darwin centenary and the jubilee year of the publication of *The Origin of Species*. An anonymous donor had endowed a chair "to be devoted to that branch of biology now entitled Genetics," a term that had been invented three years earlier by the mustachioed Mendelian, William Bateson. Still, the electric excitement could not conceal the brewing tension: Bateson's *Mendel's Principles of Heredity* and Francis Darwin's edition of his father's unpublished essays of 1842 and 1844 were both being printed by Cambridge University Press at the other end of King's Parade from Fisher, literal bookends of a fractious biological world. Mendelism and Darwinism were at war.[4]

How had this happened?

Darwin's theory of evolution only worked if there was enough new constant variation for natural selection to play with. But as the eccentric Scottish engineer (and dramatist, linguist, actor, and critic) Fleeming Jenkin had pointed out at the time, since the heredities of mother and father *blend*, red and white resulting in pink, populations would be eternally regressing to the mean. Evolution, that is, would be stuck in the mud before it ever got started, for any new variations would be quickly "swamped" into mediocrity. Darwin stuttered. What was needed, he knew, was precisely what he and his generation lacked: a proper theory of heredity.[5]

When the Moravian monk Gregor Mendel's laws of heredity were rediscovered at the turn of the century they might have been Darwin's saving grace but for the fact that so many took them to be the last nail in his metaphorical coffin. After all, the kinds of dramatic "mutations" the Mendelians were beholding, those that turned a fly's eyes from red to white, or its wings from straight to wrinkly, were a far cry from the supposed tiny variations Darwin's natural selection was "daily and hourly scrutinizing." They were spontaneous disruptive sports, not gradual continuous fluctuations.[6]

Darwinians (calling themselves "biometricians") were doubtful about the existence of invisible discrete "genes"—figments, they thought, of the Mendelians' overwrought imagination. Enough to score phenotypes, without falling into flights of fancy; if there was such a thing as genetics, it might be good for lab flies but not much else. Still, Bateson and his gang were not about to relinquish Nature's playing field, and they rejected the notion of the laboratory artifact. Even if they couldn't be seen, genes were as real as the dew-covered jacket of a *Pisum*. Internally generated, mutations rendered external selection superfluous: Darwin's great theory was wrong.

. . .

Fisher was the kind of chap who would always walk through a revolving door first. His legs were short, his shoulders narrow, and with a pointy reddish beard adorning a massive head, he could look like a keen, deliberate terrier. His father might have lost his fortune by now, but Ron was encouraged by a different kind of nobility, unbeholden to possessions. Together with a group of Cambridge friends he formed a select society, the We Frees, immersing itself in whiskey, the songs of Catholic reactionary Hilaire Belloc, bicycle excursions to the countryside, picnics, Icelandic sagas, a secret language, and Nietzsche.

Anglican by birth and even more by maternal devotion, Fisher held on to God while aspiring to become one. "What man is to the ape," he quoted Zarathustra in a talk he delivered as a student to the newly formed Eugenics Education Society in 1913, "a joke and a sore shame: so shall man be to beyond-man, a joke and a sore shame," adding, in his own words: "We can set no limit to human potentialities; all that is best in man can be bettered."[7]

War would reveal man's truest and purest mettle. Between shrieking bullets and thudding cannons echoes of "beyond-man" would be heard. In the eugenic vein he suggested that winners of the Military Cross be encouraged to procreate beyond their fraction, fortifying civilization, inching it closer to its "potentialities." Whether the "submerged tenth," the dregs of society, were reproducing faster than everybody else was of little concern to him. It was the noble, not the base, which counted in his universe. Strength, gallantry, intelligence, moral fiber: All were of a constitution; like prime beef, and top-grade milk, they could be bred in the right bodies to produce more of their kind.[8]

Except that Fisher himself would never step on the battlefield: On account of impaired eyesight he'd been rejected for military service. A First from Cambridge could hardly blunt this slight, though his chosen, the naïve seventeen-year-old Eileen Guinness, whom he called Nicolette, after the heroine in a medieval troubadour's tale, might dampen it through some serious procreation. Meanwhile, it would be his younger brother Alwyn who would represent the family in the Great War.

Beset by temporary feelings of inadequacy, Fisher followed Belloc's cry, "one man, three acres and a cow," moving his new family from London to the Berkshire countryside and leasing a gamekeeper's cottage and some land. Perhaps a man's mettle could be shown by subsisting off small farming short of charging ahead courageously in battle. Still, "going natural" would have to be bridled by civility; he was a Hampstead boy, after all. While he was off in the day teaching math at Bradford College, it was Nicolette and her Wee Free sister Geraldine (dubbed Gudruna) who tended to the chickens, pigs, and cow. At night they'd convene to read out loud from Frazer's *Golden Bough* and Gibbon's *Decline and Fall of the Roman Empire*, debating the fate of civilization. Then the news arrived one pleasant morning: Alwyn had been killed in action.[9]

. . .

It was around that time in 1918 that Fisher wrote a mathematical paper for the *Transactions of the Royal Society of Edinburgh* that would change the history of science.[10] Fifty years after that city's son had set his teeth into Darwin, Ron penned his reply-by-proxy to the long-deceased Fleeming Jenkin. The problem of variation could be overcome. With just a little bit of explaining, biometrician arms outstretched to choke Mendelian throats could be transformed into embracing appendages.

Here was the problem and its solution: The biometricians rejected genes because no one had ever seen them. But they also snubbed them on theoretical grounds. Scoring a population trait like height, for example, revealed a smooth bell curve, not isolated bars, and this was true for the overwhelming traits in nature. How could genes be responsible for such variation, if they produced large, discontinuous "mutations"? Fisher's reply was that the smooth curve could be explained by imagining small mutations working on many underlying genes. If traits were the result of lots of genes, each affecting them just slightly, the angular could be made to flow. Hereditary factors were discrete, this much was obvious, but their products would look effortlessly continuous![11]

Here, finally, was a mind-blowing reply to the incredulous Scot with the unlikely name: Genes and their chromosome abodes were immortal. Unlike paint, they did not blend. Whether they were dominant and expressed or recessive and latent, whether they interacted ("epistasis") or were simply additive, they passed from generation to generation more faithfully than endowments and even surnames. Mendelians thought they'd buried Darwin's mechanism with their own, but not only did genetics solve the conundrum of blending, it provided the missing piece to the puzzle of evolution. Since "swamping" was not a problem, variation would be preserved and natural selection's diet securely protected. Evolution moved along by the shuffling and selection of adaptive genetic mutations.[12]

Fleeming's fulmination could be put to everlasting rest. From their Westminster Abbey and Brno monastery graves, Darwin and Mendel were finally being wed.

· · ·

"Oh, it's you!" said His Royal Highness the Prince of Wales when the reeling lieutenant with the banged-up head plumped heavily into his military vehicle. It was May 1915, the Battle of Aubers Ridge, and J. B. S. Haldane needed a ride to the infirmary.[13]

The Haldanes were descendants of a Scottish military clan employed by Lowland farmers to protect their cattle from Highlanders' raids. By the time Jack was born in Oxford on Guy Fawkes Night, 1892, they were an intellectual and political powerhouse. His maternal namesake, great-uncle John Burdon Sanderson, was the first Waynflete Professor of Physiology at Oxford. His father's brother Richard Burden Haldane had been the secretary of state for war and lord chancellor for successive Liberal governments. And his father, John Scott Haldane, known to loved ones as "Uffer," had gained a reputation as the most fearless scientist in the history of Britain.[14]

A world expert on mines and what gases and explosions do to the men who work in them, Uffer was famous for sealing himself into chambers to breathe in lethal gases and record their effects on body and mind; for descending into shafts, the underground, and submarines. It was something of a joke in the Haldane home on the banks of the Cherwell River that a succession of identical telegrams would arrive after Uffer had been summoned to investigate a mine disaster: It wasn't that he felt overly obliged to calm every one's nerves back home (such sentiment would be very un-Haldane-like), it was simply that the carbon monoxide had gotten to his memory. Like his wife's uncle, he was an Oxford professor of physiology, but he also loved philosophy, especially the idealism of Kant and the dialectics of Hegel.[15] In the tradition of Darwin and Huxley, he did his science from home assisted by his children. Beside the lab and dairy farm, there were three hundred guinea pigs on the lawn, two of them named Bateson and Punnett after England's leading geneticists. In a loving but unvarnished way, Jack and his baby sister Naomi were their human counterparts.

"Friends, Romans, countrymen," Jack began in the bowels of a North Staffordshire mine. He was eight years old, the story went, and Uffer had instructed him to stand up and recite from *Julius Caesar*. He collapsed just as he had arrived at "the noble Brutus." When he later came to on the ground above, he had learned the salutary lesson that methane is lighter than air and innocuous. That was the nature of the bond between Uffer and "Boy"; a correct reply to "What's the formula to soda-lime?" winning him a trip to a submarine, a mastering of the pressure tables—a job testing decompression suits for the navy. When he was thrown off the backseat of Uffer's bike one day, speeding down South Park Road, the doctors at the Radcliffe Infirmary didn't think the little assistant would make it. But despite a cracked skull he recovered; by some unknown miasma, friends would later say, the accident had turned a bright boy into a genius.

It wasn't all that surprising, then, after the Dragon School and Eton, that JBS enlisted at Oxford, just down Banbury Road. Nor, as the story goes, that he went into the wrong entrance-examination door by mistake, gaining a scholarship in maths instead of classics. In the end he would graduate from New College with a First in both and be offered a fellowship in physiology; all depended on returning in one piece from the war.

Which would not be easy. His men in the elite Scottish Black Watch Brigade, known for its fierceness and tartan kilts, dubbed the obviously somewhat touched lieutenant in charge of grenades and mortars the "Rajah of Bomb." Self-initiated solo raids into no-man's land, a makeshift bomb workshop, and the time he drove a bicycle across a gap in front of the unbelieving eyes of the Germans (having calculated that the enemy's incredulity would assure his safety), all made plain what "Bombo" Haldane himself admitted: He was having a ball in war.[16]

And so when he clambered into the Prince of Wales's car after having been blasted at Aubers it would not be the last time he was wounded. JBS was cocksure, and entitled. His zeal for battle was so great that he was once pushed midaction into a ditch by his own gunners, unenthused about the retaliatory fire his immanent mortar would draw to their battery. Still, his men respected him, revered him, even. With his imposing frame, premature balding head, massive forehead and Celtic-warrior-like mustache, he looked like an "alert walrus" in a skirt. It was alternately comical and bloodcurdling. What other officer in the trenches was writing a scientific paper with his sister describing one of the first-ever examples of genetic linkage in mammals? What other officer, as a confidence-building measure, made smoking compulsory in his bomb-making workshop (he was selecting out the fainthearted)? And what officer had been hurriedly called away at the behest of his uncle, the lord chancellor, in order to put his mind to that of his father's at a makeshift lab in Saint-Omer, the quicker to meet the challenge of German gas attacks on Allied forces in Belgium?[17]

Haldane's men had heard many Oxford stories. How, in 1913, blacklegs had been hired by the municipality to replace the striking horse-tram drivers. With their services provided, the drivers' demands went unmet, and successive attempts to unharness the stand-in horses were defeated by baton-wielding bobbies. Until JBS came along, that is. Marching up and down Cornmarket Street, solo, in solidarity with the workers, chanting the Athanasian Creed and the Latin psalm *Eructavit cor meum*, Jack drew a crowd, blocking the blacklegs and allowing the strikers to set loose their horses: "A scene-stealing cameo role, a feat of memory, a lengthy canonical Latin quotation, a snook cocked at authority and an Oxonian irony: with this piece of street theatre, JBS established his modus operandi."[18] The university fined him two guineas. It was "the first case for over three centuries," Haldane boasted, "when a man was punished in Oxford for publicly professing the principles of the Church of England."

Often dismissive and always demanding, Jack nevertheless had a soft spot for the underdog, and his men and commanding officers could feel it.[19] He was also fiercely loyal: He was wounded twice trying to get back to his brigade from Saint-Omer after he and Uffer gassed themselves silly with chlorine. Haldane's prodigious talents and his love for genetics and the everyman would in due course propel him to fame. In the meantime he was back in action on the Mesopotamian front, and injured again in a valiant attempt to gain control over a fire in a bomb depot at the end of 1916. Presently he was lying bandaged in a military hospital. If a war hero had to be selected for increased procreation, J. B. S. Haldane would have won out over anyone.

．　．　．

We can no more accept the principle of arbitrary and casual variation and natural selection as a sufficient account, per se, of the past and present organic world, than we can receive the Laputan method of composing books (pushed á l'outrance) as a sufficient one of Shake-speare and the Principia.[20]

That was England's leading astronomer and natural philosopher, John Herschel, reviewing the *Origin* in 1861. If variation was due to chance, and evolution depended on variation, then Darwin's entire edifice rested on shockingly haphazard foundations. Herschel called it "the law of higgledy piggledy," and many nodded with approval. After all, were men really to believe that all the vaunted glories of nature were created in the preposterous manner of imaginary dwarves toiling blindly over *Romeo and Juliet* and Newton?

Fisher was now the chief statistician at the Agricultural Research Station at Rothamsted. Already his *Statistical Methods for Research Workers* fashioned him the modern inheritor of Gauss and Laplace.[21] Soon he would be hailed Darwin's greatest successor, too, the unchallenged master of not one but two disciplines. His nonbloodline pedigree might have played a part: Maj. Leonard Darwin, childless eugenicist and Darwin's third son, had become his filial mentor, making Fisher Darwin's spiritual grandson. For nearly a decade now Major Leonard had been prodding him to ground his father's theory in mathematics. Wary of his wont for folding a plenitude of insights into a grain of words, he warned him: "One idea one sentence is, I think, a good rule."[22]

The Genetical Theory of Natural Selection was finally published in 1930, and while Major Leonard's warning wasn't heeded, it immediately became a classic. It had been dictated the previous year in the evenings to Nicolette, a matrimonial duty alongside reading him the *Times* each morning while he ate an unrushed breakfast. In his dense, frustratingly opaque prose Fisher replied, once again with authorial assurance, to the foible of a long-deceased contrarian. "Yes," came his retort to Herschel's rhetorical Swiftian allusion, the Laputans' method seemed just about right: The very power of natural selection stemmed from the fact that it is "a mechanism for generating an exceedingly high degree of improbability."[23]

Everything hinged on the existence of large populations, low mutation rates, and a sea of single genes. Like a Laputan lunging at a random letter to compose the *Principia*, a mutation was a "leap in the dark." But in large populations, where sex would mix things up, mutations could be tested across an enormous range of genetic combinations. Natural selection would guarantee that their presence would almost never invite indifference: Either they made their possessor fitter and were selected, or they reduced fitness and were summarily expelled.

Working out the partial-differential equations for gene frequency change in a population, Fisher could show the precise probability for the survival of a mutant gene. A mutation that conferred a 1 percent advantage would have close to a 2 percent chance of establishing itself in each generation. If the population was large enough, it would take no time at all for such a mutation to spread to every one of its members. In fact, even minute selective pressures, fractions of a percent, could propel mutations to take over a population: The larger the population, the faster the conquest. Had it not been proven lawful, natural selection might have been mistaken for a miracle: The proxy of divine Creation, it transformed unintended accident into the semblance of deliberate, most wonderful design.

The centerpiece of the *Genetical Theory* he called "the fundamental theorem of natural selection": "The rate of increase in fitness of any organism at any time is equal to its genetic variance in fitness at that time." A brew of Anglicanism and Nietzsche, it encapsulated all that he believed in. But what did these mysterious words actually mean? Like an Icelandic spell or a secret We Free enchantment they defied simple understanding. No one but Fisher seemed to know.[24]

What they appeared to mean was this: The second law of thermodynamics spoke of a world settling constantly into disorder: Heat spreads in a cold room, gas diffuses in an empty flask, matter always strives to become homogenous. Fisher knew this well; he had spent a year studying quantum physics with James Jeans at Cambridge after graduation. From the very start, then, God had set himself the task of combating decay.

Thankfully he had created natural selection to help him wage his battle. The fundamental theorem was the biological analog to the second law, only climbing in the opposite direction. Far from matter descending into chaos, life was persistently ordering itself guided by the greatest of impresarios. His proxy, natural selection, was the proud reply to the winding down of the universe. The more variation in fitness there was to work with, the faster was its progress, the fitter did the population grow. Fisher could hardly have asked for a more perfect foundation for his faith: Two scientific laws described the ups and downs of the universe; but it was his, the fundamental theorem, which held all the promise for mankind.[25]

Zarathustra had gotten to him. Already in his 1913 talk to the Eugenics Education Society, Fisher had fathomed the true scope of nature's miracle:

> *From the moment we grasp, firmly and completely, Darwin's theory of evolution, we begin to realize that we have obtained not merely a description of the past, or an explanation of the present, but a veritable key to the future; and this consideration becomes the more forcibly impressed on us the more thoroughly we apply the doctrine; the more clearly we see that not only the organization and structure*

of the body, and the cruder physical impulses, but that the whole constitution of our ethical and aesthetic nature, all the refinements of beauty, all the delicacy of our sense of beauty, our moral instincts of obedience and compassion, pity or indignation, our moments of religious awe, or mystical penetration—all have their biological significance, all (from the biological point of view) exist in virtue of their biological significance.[26]

Huxley had cautioned man to take a good look at Nature and then run, morally, in the opposite direction. But Nietzsche saw that ethics were threatened by evolution only if nature was considered improper. What Fisher was showing was that there were no grounds for such dirty thoughts. On a visit to the London Zoo in 1838, Darwin had scribbled in his notebook: "He who understands the baboon would do more towards metaphysics than Locke." Fisher agreed but thought he'd taken it one step further. In his mathematical Mendelian-like appendix to the *Origin* he had proved what his spiritual grandfather could only imagine: that natural selection had planted and would continue to water the very seeds of human kindness.[27]

If only man didn't get in the way. In the grand evolutionary tale, morality made man fitter, which was precisely why Fisher was growing worried. In his folly man could ruin in generations what Nature had accomplished over eons. The demon was wealth, for the more possessions man held, the fewer incentives he had to procreate. Abundant children, after all, meant a thinner sliver of the pie for each; why toil only to see the hard-earned fruits of your labor waste away in the next generation? In modern times the infanticidal impulse could be replaced by contraception, but it was only the well educated who ever deigned to use a sheath. Since he was certain that the better-quality genes were clustered in the upper strata of society, the multiplying of the dregs was a problem.

God had met entropy by introducing natural selection, but avarice might frustrate his design. In the long run, Fisher was certain, Nature would utter the final, compassionate word: The more contraception was used the less it would appear in future generations, for those who harbored covetous sentiments would leave fewer of their kind, while those who didn't would breed the practice out of the population. Ultimately civilization would learn evolution's lessons and adapt. The question was: Did humanity have the time to linger?[28]

More and more it seemed as though it didn't. Luckily, however, in his benevolence, God offered the dual gift of responsibility and occasion, arming his creatures with the capacity for free will. The fundamental theorem showed that natural selection had done almost all the work for us, inching man closer to his genetic optimum. Still, Fisher told a BBC Radio audience, "in the language of Genesis, we are living in the sixth day, probably rather early in the morning." God was beckoning to complete his Creation: As sure as animals, men could navigate their destiny; nothing was yet determined or prearranged. If only family allowances were provided to propel the desirable (meaning the upper classes) to procreate, mankind could resume its natural course. Far from an instrument of decay, heredity would be its confederate. In the end, after all was said and done, genetics was its greatest and only hope.[29]

· · ·

Before he could react to the whistle of the bomb, the elderly lady on the bench right beside him had been torn to shreds. Blown to the ground, miraculously unhurt, Haldane could hardly believe his luck. Damn those bloody fascists! He was in a park, in the middle of an air raid, Madrid, the summer of 1936.

It was only a few days earlier that he fitted his large frame into leather pants and jacket, and, with the ears of his hat flapping in the wind and goggles fastened tight, made his way to Spain on his motorcycle from London. Haldane was now professor of genetics at University College, a fellow of the Royal Society, and, following the recent attempted coup d'état against the government of the Second Spanish Republic, speeding fast in the direction of his convictions. When Uffer had died earlier that March, Jack had been trusted with scattering his ashes over the family lands in Cloan. He would only join the Communist Party officially in 1942, but already on the train to Scotland insisted on traveling with the urn third class. JBS had been stung by Marx.

Back from the Great War he had been appointed reader in biochemistry at Cambridge where, in 1924, he had given an interview to a journalist named Charlotte Burghes and fallen in love. The old guard at the university weren't happy about the affair; Charlotte was a married woman, and England's leading Mendelian, William Bateson, didn't appreciate Haldane running the streets "like a dog."[30] Still, after he won his appeal against a committee of six senior university members who had sought his dismissal, Charlotte divorced and the Haldanes were married. JBS took pleasure in pronouncing the committee's Latin name, *Sex Viri*, with the requisite soft *W* for a *V*, and a seventh member was added to escape his ridicule. "We must learn not to take traditional morals too seriously," he would say, and there was no one who questioned his sincerity.[31]

Meanwhile he was wowing the world with his talents, emerging as a top-flight biochemist (identifying cytochrome oxidase in different taxa), geneticist (finding more instances of linkage), physiologist (inventing a method to relieve divers of the bends), enzymologist (deriving the Briggs-Haldane equation), popular science writer (writing *Daedalus* and *Callinicus*), and intrepid self-experimenter (drinking ammonium chloride—more than had killed a dog—to prove that breathing is controlled by carbon dioxide). Even by Oxbridge standards he was a character extraordinaire. Alongside these achievements, as the following note to his secretary shows, he'd begun searching for his ideological home:

Wanted:
1) Insignia of the Order of the Holy Cross
2) Mr Ford's bank balance
3) Clara Bow
4) Info on all trains bet. 11 A.M.–5 P.M. from London-Paris
5) £5 (cheque cashed)
6) safety razor blades
7) If you can manage, a max of botches
8) Karl Mark's Kapital[32]

Fascism was ascendant in Europe, economic depression every-where. With Britain failing miserably either to show political resolve or to stage a recovery, disaffected liberals had fallen into lethargy and ennui. The barren, self-pitying "spiritual muddle" drove Haldane, and an entire "social relations of science movement," to rage. It was morally bankrupt and even worse, scientifically illiterate. After all, "the progress of society depends . . . on the progressive applica-tion of science," but England still cuddled "essay-writers" instead of scientists.[33]

Not so in Russia. There, only the ballerina could compete with the scientist for cachet. Haldane, like many other left-leaning West-ern scientists, was completely enamored. In 1928 he and Charlotte had visited Russia on the invitation of the famously handsome and fearless plant geneticist Nikolai Vavilov. Lavished with a gala din-ner and carted around to busy institutes, JBS took little notice of the shabby apartments and empty canteens. Even as they stood in the long line waiting for their exit visas he pontificated to Charlotte about how impressed he was by the Soviets. Russia was where the future lay because in Russia science had been endowed with the authority of religion. "Our present rulers and those who support them," he pronounced upon return, "will be well advised explicitly to imitate the extremely capable Bolshevik leaders, and adopt an experimental method."[34]

Marxism could combat fascism where liberal democracy had failed. It divined the winning attitude for using science and technol-ogy to save civilization. It honored the everyman and gave respect to the underdog. But Marxism provided something even more funda-mental: the tools to grasp reality, to see the world the way it is rather than how people want it to be. Before his conversion Haldane joked that his favorite Marx was Groucho, but he was quickly coming into the fold. Whether this was lip service or the real thing, he'd been using the method of dialectical materialism to get a grip on evolution. Or so at least he claimed.[35]

The Causes of Evolution, published in 1932, was the popular exposition of nine pathbreaking technical papers (with a tenth still on its way) detailing "A Mathematical Theory of Natural and Artificial

Selection."[36] Haldane's equations were not pretty, he was a "plodding algebraist," not, like Fisher, a "mathematician's mathematician."[37] And yet, just like Fisher, he spoke to Mendelians and biometricians both, showing convincingly how natural selection, operating on genes in a population slowly and gradually could account for dramatic evolutionary change. More like a greasy-hands repairman/engineer than a pristine apostle of Nietzsche, he felt no need to posit a fundamental theorem. To Haldane evolution was a process, not an edict.

Partly this was related to dialectics. Fisher worshiped the purifying hand of natural selection, but Haldane thought it had to be "negated" somehow by the chanciness of genetic mutation.[38] This didn't mean that natural selection wasn't a force to be reckoned with: Thirty years before the Oxford geneticist E. B. Ford sent his student Bernard Kettlewell to the forests outside Birmingham to prove it, JBS predicted that the dark peppered moth, the melanic *Biston betularia*, could hold up to a 50 percent selective advantage over its white counterpart. In the span of only a few generations, its ratio in the population compared with the white morph had jumped from 2 percent to over 90, the Industrial Revolution having provided blackened birch trunks to hide from preying birds, and left the rest to natural selection.[39]

But if natural selection was driving populations to higher fitness, it was also being frustrated by mutation. Fisher argued that fitness depended on variation, but this led to a paradox: After all, the more natural selection weeded, the less variation was left. This meant that the mutation rate had to be higher than he allowed for. But since most mutations were harmful, if their rate was too high the fitness of the population would drop. On the other hand, if natural selection scrutinized too harshly, the population could be wiped out entirely. There had to be a dialectic, to JBS this was clear. From the mouth of Marx via Hegel,[40] whatever the fundamental theorem promised, there was a cost to natural selection.[41]

Evolution had accomplished amazing feats but was no panacea. Where Fisher had turned to a divine Nature and the upper classes, Haldane chose radical politics. He'd been back from the Spanish civil war for some years now, and had recently been elected to the executive board of the Communist Party of Great Britain. Sure, he had heard of the rise in the Soviet Union of the Ukrainian farmer Trofim Lysenko, who marshaled Marxist doctrine to lambast "Western, bourgeois, exploitative" genetics, calling its practitioners "fly lovers and people haters." He had heard that the "barefoot professor" was promising a "genuine," "proletarian" agricultural revolution based on false Lamarckian claims, and whispering into Stalin's ear that "determinist" and "racist" genetics needed to be shut down. He had heard the rumors of colleagues losing their jobs, of the disappearance of his gracious host Vavilov, even of executions and purges. But he waved them all aside. There was no hard evidence. Besides, even if Lysenko's claims were "seriously exaggerated" there might be a grain of truth to them.[42]

Haldane was a hereditarian. Back in 1932 he had remarked: "The test of the Union of Soviet Socialist Republics to science, will, I think, come when the accumulation of the results of human genetics, demonstrating what I believe to be the fact of human inequality, becomes important." He enjoyed quoting Engels's assertion that "the real content of the proletarian demand for equality is the demand for the abolition of classes"; any demand that goes beyond that, he said, "of necessity passes into absurdity." Though not a rabid eugenicist like Fisher, he judged the logic behind its tenets sound. He was convinced that the mean IQ was falling due to the abstinence of the educated classes. Genetically speaking, capitalism was frustrating the survival of the fittest. By abolishing classes and getting rid of the differential birth rate, socialism would be the perfect corrective.[43]

This was hardly the reactionary Fisher's plan, but it was not exactly Lysenko's either. Pointing his finger at the Nazi perversion of genetics, the "madman" from the Ukraine had convinced Stalin that Mendelism was a noxious sham. It was by changing the environment, not manipulating imagined genes, that man's destiny would be forged and provided. A French colleague said of Haldane: "Ce n'est pas un homme, c'est une force de la nature!" By all accounts he was a Goliath, a bastion of reason and logic. And yet somehow it didn't seem to bother him—one of the few modern prophets of evolution—to be apologizing for a man who denied the existence of the gene and called survival of the fittest a bourgeois plot. What mattered was the cause. As long as capitalism had yet to show that it could keep people from want and communism that it could not, Haldane was going to side with the Party.[44]

．　．　．

Sewall Wright was no kind of Marxist. Nor, in truth, was God usually on his mind. A Harvard trained theoretical population geneticist, he looked a bit like a very handsome mouse. And as colorful as Haldane and Fisher were, Wright was the son of a small college professor from the Midwest whose "idea of a fun night out was a discussion with his university chums at the local faculty club."[45]

Still, just like his English counterparts, he'd applied equations to evolution, and even though he was sedate and mousy, the fundamental theorem of natural selection just blew him away.[46] Emboldened by its generality, he offered a theory of evolution of his own, grander than Haldane's engineerlike approach, more visual than Fisher's obscure but soaring mathematics.[47]

Life, he explained, was like a landscape of valleys and mountains. Organisms could be imagined in the process of climbing in an upward direction, improving their fitness as they scaled a crest. As the fundamental theorem showed, once an ascent had begun there was no turning back, for natural selection could push in only one direction. This was well and fine. But what if the summit was in fact a foothill, not all that towering: Would organisms find themselves stranded with nowhere else to climb?

Wright's answer was that in addition to natural selection another force sometimes operated, called "random drift." Imagine a crowd of a million angry demonstrators walking down Main Street in the direction of Town Hall and one falling accidentally into a pothole. However tragic to the particular individual, the torrent, and the demonstration, will not have been affected. But if only three friends are walking down an alleyway toward a demonstration and one falls randomly into a pothole, it's almost certain they'll never make it to the rally; their direction and cause will have changed dramatically. As with demonstrators, so with genes in a population: Chance events will influence small groups much more than large ones. And just as the pothole may have swallowed the friend least interested in the demonstration, genetic drift will be divorced from the "goals" of evolution. Unlike its deliberate older brother, drift would not always push organisms to higher genetic peaks: If natural selection was the taskmaster of fitness, drift was the shifting sands underfoot, changing the course of weak and strong alike.

Still, there was a bright side. If populations were not large and homogeneous, if they didn't all walk together through Main Street but rather advanced in parallel alleys in small, partially isolated groups, drift could do what natural selection could never accomplish: alter the genetic structure enough to allow a dramatic change in direction. In Wright's model this was tantamount to blowing a group from a peak into a valley. Working against the fundamental theorem, in the short run it meant a decrease in fitness. But just as a grieving letter to the mayor from one of the two remaining friends might do more for the cause, in the end, than the mass rally, it could ultimately provide a surprising opportunity. For a blow off a foothill might lead to the foot of a new and higher mountain. What looked like a dive would actually mean liberation—the start of a glorious, transcendent new ascent.[48]

He called it the "shifting-balance" theory, since the population shifted between the poles of directed selection and random drift, respectively. Wright thought of it as an enrichment of the fundamental theorem, chance blowing new tunes into the sometimes limited winds of progress. But the "adaptive landscape" left Fisher cold. Mutations were both necessary and random, but it was the guiding hand of selection that ultimately shaped life. Without it there could be no purpose or end. Wright's "landscape" was "picturesque" but biologically nonsensical.[49] Small isolated groups where random events could leave their mark were either on the way to extinction or becoming new species. The sweeping majority of creatures in a population were continually connected genetically, since each had a fair chance of mating with the other. They were all marching steadily together down Main Street. Drift might shift a handful of genes in a vast, never-ending sea; from the point of view of the population it was negligible.

Nature was uniform, not a landscape of troughs and peaks. And the great ocean of life was constantly rising.

• • •

Fisher, Haldane, and Wright would become known as the architects of the evolutionary synthesis, each in his own way wedding Darwin to Mendel.[50] However different their models, all three celebrated the power of natural selection to bring about evolution. For that reason, too, just like Darwin before them, they were puzzled by the conundrum of the ants: How could traits that reduced fitness be selected, going against the interest of those who bear them? Traits, that is, like altruism.

Fisher's reply came via butterflies and caterpillars. Naturalists were familiar with butterflies (and other insects) that are terribly distasteful to the birds that try to eat them. This would be a wonderful defense if only the birds knew it before taking a bite. Since such understanding usually came too late for the butterfly, Fisher wondered how "distaste-fulness" could have evolved.[51]

When his entomologist friend E. B. Poulton told him that most butterflies, like the monarch, have bodies tough enough to survive a bird's first bite, it looked as if the problem went away: If the monarch could survive the initial munch, its distastefulness would mercifully save it. But what about monarch caterpillars? They had "nauseous flavors," too, and yet were as soft as silk. If a bird took a chomp out of one of them, no matter how distasteful, the defense would have been too little too late.

The answer resided in the family. Before caterpillars grow wings they are less mobile and tend to stick together, usually in groups of related kin. If a hungry bird looking for a meal chanced on such a clump it might take a bite out of one but spare the rest, disgusted. The unlucky caterpillar would have sacrificed itself for its brothers and sisters, losing its life in a final act of gallant altruism.

It made sense, Fisher thought, because kin are genetically related: Natural selection could produce kamikaze distastefulness in individuals to ensure that more of their shared genes lived on in the bodies of the brothers and sisters who were spared. The less related the caterpillars in the clump, the more the selective effect "will be diluted," and the less chances of altruism to evolve.[52]

The importance of relatedness had been clear to Fisher as far back as 1918, when he presented in his famous paper a table calculating genetic distances between kin. But it was Sewall Wright, four years later, who provided the variable that was adopted to measure kinship.[53] He called it the "coefficient of relationship," or r, and proved that for an individual and any of its direct descendants it was equal to 0.5^n where n is the number of generations that separate the two. Between mother and son it equals 0.5 (0.5^1), between grandparent and grandchild 0.25 (0.5^2) and so on. r was the measurer of genetic relatedness—the probability that two individuals shared a common gene. More generally, then, Wright could now calculate the r of any two related individuals, not just those who were direct ancestors and descendants.

Figuring out the r of two maternal cousins, for example, worked like this: Calculate the probability that both possess a gene, A, that they inherited from their maternal grandmother (for each cousin separately this equals 0.5^2, meaning that together the probability is $0.5^2 \times 0.5^2 = 0.0625$). Then employ the same procedure to calculate the probability that both cousins got gene A via their maternal grandfather, which likewise equals 0.0625. Finally, add the two sums to get the probability that the cousins share gene A from either of their maternal grandparents ($0.0625 + 0.0625 = 0.125$). Two maternal cousins, that is, have an r of 1/8, or a one in eight chance of sharing a gene inherited through their maternal grandparents.[54]

Even though it would one day prove a key to solving the conundrum, Wright never applied r to the problem of altruism. Instead, more than twenty years later, in 1945, he offered a theoretical model of the evolution of altruism that paid little heed to relatedness. Once again random drift was at the crux.[55]

By definition altruistic traits reduce fitness, and are therefore bad for the individual. But imagine a mixed population of altruists and selfish folk marching together down Main Street: If, just by chance, a bunch of altruists exclusively got sidetracked into an alley, they would make it to Town Hall quicker than the rest of the group. The reason would be that they'd all be helping one another rather than some of them competing to get there first, as would be the case on Main Street. The remaining altruists on Main Street, on the other hand, would not fare as well, since while they were taking time to help others walk down to the destination, the selfish folk would be speeding down to Town Hall.

Within a mixed group, that is, an altruist would suffer from *intergroup* competition at the hands of those out for themselves, but between, or *intragroup,* competition would favor groups with a greater proportion of altruists. In terms of the adaptive landscape, if altruists were blown off different peaks and somehow found themselves all grouped at the foot of the same mountain, they might start an ascent that would place them on the highest peak of all.

Wright's model depended on "group selection," on natural selection sometimes weeding out or favoring whole groups rather than individuals. But while it made sense that groups with more altruists would do better than groups with fewer, the math proved a problem: It would work only if the alleyways were almost completely isolated from one another, which in turn made it very difficult for altruists to stumble into them by chance. Naturalists, as opposed to pen-and-paper men like Wright, knew that such stringent conditions would be tough to find in nature.[56]

Where Fisher turned to caterpillars, and Wright to isolated groups and drift, Haldane looked to the ruminants for his stab at the natural origins of kindness. "Of two female deer," he wrote in an essay from 1928, "the one which habitually abandons its young on the approach of a beast of prey is likely to outlive the one which defends them; but as the latter will leave more offspring, her type survives, even if she loses her life."[57] The logic of selection was often cruel but always plain.

Still, what did distasteful caterpillars and protective mother deer have to do with humans? Could the very same animal logic pertain to man?

However much he believed in worker solidarity, Haldane was a lover of mankind who trusted no one. He agreed with Fisher that evolutionarily speaking, altruism was a family affair. "I doubt if man contains many genes making for altruism of a general kind," he wrote, "though we probably do possess an innate predisposition to family life. . . . For insofar as it makes for the survival of one's descendants and near relations, altruistic behavior is a kind of Darwinian fitness, and may be expected to spread as a result of natural selection."[58]

Remembering his brother, Fisher agreed. After all, the hero might lose his life in battle, but his heroism would survive. The reason was that it was coded in his genes. Since relatives share genes and since, "by repute and prestige," the act of valor confers an advantage on the family, the grieving but proud surviving kin will pass "hero genes" along in greater numbers.[59] Alwyn's sacrifice in Flanders had not been in vain: Fisher's eight children with Nicolette would make sure of that.

Whether he had carried his aristocratic sensibility into the realm of determinist genetics or the other way around, a man was the sum of his genes. It was a sentiment Haldane could warm to, however far he lay from Fisher on the political spectrum. "Let us suppose," JBS wrote, fleshing out the argument,

> that we carry a rare gene that affects your behavior so that you jump into a flooded river and save a child, but you have one chance in ten of being drowned, while I do not possess the gene, and stand on the bank and watch the child drown. If the child's your own child or your brother or sister, there is an even chance that the child will also have this gene, so five genes will be saved in children for one lost in an adult. If you save a grandchild or a nephew, the advantage is only two and a half to one. If you only save a first cousin, the effect is very slight. If you try to save your first cousin once removed the population is more likely to lose this valuable gene than to gain it. . . . It is clear that the genes making for conduct of this kind would only have a chance of spreading in rather small populations when most of the children were fairly near relatives of the man who risked his life.[60]

He then added, with his usual mix of bravado and wit: "But on the two occasions when I have pulled possibly drowning people out of the water (at an infinitesimal risk to myself) I had no time to make such calculations."[61]

Haldane's wit aside, a generalized mathematical model was still lacking. To truly assert itself—Fisher, Haldane, and Wright's prodigious talents notwithstanding—the evolution of kindness would have to wait for a new champion.[62] Still, relatedness had met altruism, for the benefit of the drowning, scaling the political gamut where ideology had failed. Huxley aside, not to mention organized religion, altruism was an offspring of evolution. It was a breathtaking notion. Legend had it that Haldane had first described its logic stooped drunk over a beer in London: "I'll jump into a river for two brothers and eight cousins," he was said to have blurted out just before collapsing on the bar.[63]

• • •

Fisher was knighted in 1952, and sought a quiet retirement in Australia. After the war friends and colleagues at Cambridge wondered about his friendship with Mussolini's former demographic adviser, Corrado Gini, and with Otmar von Verschuer, a medical geneticist and "one of the most dangerous Nazi activists of the Third Reich."[64] He was a strange bird, Fisher—aloof, exacting, imperious, oblivious; too reactionary, people said, to be either a Nazi or a fascist. Instead, he went to his grave an apostle of Nature. It was she who knew what was best for mankind.

Haldane, meanwhile, sank deeper into the Lysenkoist bog.[65] It was official now: genetics had been banned in the Soviet Union, Vavilov brutally murdered in the Gulag; science and justice were not marching hand in hand. Vain and obstinate, JBS continued to offer disclaimers while quietly stopping to pay his Party dues. When all was said and done, he was a scientist.

He was looking for a gracious exit. When Britain joined France and Israel in attacking Egypt as it nationalized the Suez Canal, he found his pretext at long last. To the provost of University College he announced that he no longer wished to be a subject of a state "guilty of aggression by the overwhelming verdict of the human race." It was 1957, and the old soldier was moving to India, where peace and kindness were a philosophy.[66]

To those who knew both men well the irony must have been biting. They had both perceived the relation of kinship to valor, that being able to pass on goodness depends on having children. Fisher, unbattleworthy, had sired a veritable clan. But waving good-bye to England as he got on the plane, Haldane was leaving no one behind. Nor, despite a desire more burning than any other passion, would he ever become a father.

*George and Julia Price in Washington Square,
New York City, 1947*

*George, Julia, Annamarie, and Kathleen—
Morristown, New Jersey, 1949*

4

Roaming

*A*t 7:30 p.m. September 20, 1940, the new freshmen shuffled into the Union at Memorial Hall. With Dean Chase presiding, the Speaker of the Massachusetts House of Representatives, followed by the chairman of the Committee on Admission, followed by John H. Finley, Jr., associate professor of Greek and Latin, each extended their blessings and articulated their thoughts. It was a long evening, but the boys listened in silence. They were the Harvard class of 1944, America's "best and brightest," and by every account of those present at matriculation, its very future and hope.[1]

The college had been named for John Harvard in 1636, but in truth he was merely a generous fellow who had donated his library and a sum of money. There had been no single founder when the Massachusetts legislature, long before there was a nation, created it of the community, by the community and for the community. Its "duty, or in Protestant terms, its task," was to serve the ambitions of American society, capitalism, and democracy. Now, in the fall of 1940, as the winds of war began to blow westward across the Atlantic, the freshman class had for the first time in history been augmented above the regular one thousand, to "take care of any possible vacancies" in case of a draft or voluntary enlistment. To accommodate the increased number of freshmen, the college opened a new dormitory at 24 Quincy Street, just across the street from the Yard. It was there, at Farlow House, that George Price settled into his new digs.[2]

Figuring that he was a Stuyvesant boy and therefore cleverer than the rest, George enrolled in advanced graduate courses in chemistry and biology, as well as in German A even though his talent for languages was appalling. The gamble backfired, and he was struggling. To well-groomed boys in his year, like future economist Lloyd Shapley from Exeter and future philosopher of science Thomas Kuhn from Taft, everything seemed to come easy. Cocksure upon arrival, he increasingly became reclusive, and, like many another freshman from the public schools, began to wonder if Harvard had made a mistake.[3]

Wartime tumult on the campus made keeping his mind on his work an even greater task. At graduation earlier in the spring, the Ivy Orator was booed by the seniors when he said something about America's not being "too proud to fight." Returning after the summer with France surrendered and London bombed, faculty and students were changing their tone, increasingly calling for intervention. They were led by Harvard's president, James Bryant Conant, the chemist, who on every possible occasion would remind his company that "fear of war is no basis for a national policy." The "American Defense, Harvard Group," headed by Ralph Barton Perry, professor of philosophy, operated from the top floor of Widener Library. With more than six hundred faculty on board it carried much weight both in the college and in the national press, helping to house British refugee children in America and calling to fight fascism and tyranny in Europe.

Others of opposite disposition organized themselves in the Committee for Democratic Action and echoed the isolationist call to "Stay out of the war!" Already President Roosevelt had called for fifty thousand planes to make America the "arsenal of democracy," and the

Civil Aeronautics Authority, a branch of the federal government, had announced that it would train fifty to one hundred Harvard students to be pilots at the cost of only forty dollars a year. Courses in navy and military science at the college were overflowing. And, starting on October 16, undergraduates twenty-one and over began to pour through the Widener Reading Room, signing cards and answering questions in registration for the draft.[4]

In December isolationists and pacifists led by Professor William Yandell Elliott and Lloyd's famous astronomer father, Harlow Shapley, met at Emerson Hall to protest the "imperialist" war in Europe. Outside, hundreds of students formed a picket line, shooting red flares into the sky and singing "There'll Always Be an England." A war of signs now raged in the Yard: "Books! Not Guns!" and "Why Fight in Europe? Strike for Peace" intermingling uneasily with "Liberty before Peace" and "War is Hell, Hitler is Worse!" Activists offered passersby a tongue-in-cheek choice between an American flag and a swastika. At the Harvard Students Union, bitter debates raged over Congress's Lend-Lease Bill, which interventionists called "cowardly" for representing "aid short of war."[5]

President Roosevelt signed the bill on March 12, and, one student remembered, "Harvard took a deep breath and waited." As George was vying for the 165 pound spot on the wrestling team and beginning to fathom that his grades might not satisfy his scholarship conditions, the world outside was reeling. A wave of strikes in defense industries swept the nation, British forces were defeated in Egypt and Greece, and America hesitated on the brink of sending out U.S. convoys across a U-boat-infested Atlantic.

We all felt a certain lack of cohesion in American society; somehow twentieth-century democracy had not created sufficient loyalties, had not built up a sufficiently organic social system, to resist the inroads of fear and despair. Perhaps we had become, in the most fundamental sense, complacent and self-satisfied, too convinced, under the tutelage of John Dewey, that man was a good fellow at heart and could make a better world by his own hands and his own wits. . . . Look where we would, we found no harmony, no order in our society. . . . Our universe appeared to be nothing more than a series of events connected in an unknown and unknowable manner.[6]

On account of his poor eyesight George had been classified 4F, and would not be joining the armed forces. Nor would he be continuing at Harvard; a few poor grades prevented the scholarship from being renewed. Its unofficial poet, David McCord, might have been giving his thoughts expression:

Is that you,
John Harvard?
I said to his statue.
"Aye—that's me," said John
"And after you're gone."[7]

In May, disappointed and cowed, George left Farlow House and went back home to New York City. That summer Alice traveled to visit her family in the Midwest, and the Price brothers were left to run business at Display. Since he was "supposed to be the more dependable," George was managing the money, and having a hell of a time fighting off creditors.[8] Still, on account of his strong Stuyvesant performance and Harvard credits, he was accepted with a scholarship as a Home Study sophomore at the University of Chicago. Things were looking up. Then fall came, and winter, and on the morning of Sunday, December 7, 1941, the surprise attack by the Japanese on the U. S. naval base at Pearl Harbor. Finally America was at war.

. . .

The year 1887 wasn't a good one for John D. Rockefeller. The founder of Standard Oil, and America's most powerful magnate, he looked around him and sighed. Congress had just passed the Interstate Commerce Act, clamping regulation on the big boys by making secret rebates and price discrimination illegal. Capitalism just wasn't what it used to be.

A major philanthropic enterprise seemed a welcome diversion. Baptist congregations had been growing like wildfire in the Midwest, and business, too, was growing. What was missing was a house of learning to rival the great Ivys of the East, to fashion the Midwest a true alternative to Washington, New York, and Boston. And so, when Chicago clergy approached him, the wealthiest Baptist layman in the land agreed to contribute to the founding of a university. Chicago's leading merchant, Marshall Field, donated the site on the Midway. The "Wunderkind" William Rainey Harper, Hebrew scholar and workhorse extraordinaire, was secured as first president. By 1892 the gates were open. Rome was not built in one day, people said. The University of Chicago almost was.[9]

Harper sought to bring together the attributes of Oxbridge and the great German universities: to combine undergraduate and graduate, teaching and research, all at the very highest level. "The times are asking not merely for men to harness electricity and sound," he declared at commencement, "but for men to guide us in complex economic and social duties." Scholarship would be a "matter of the gravest moment," mere fact-finding frowned upon in favor of knowledge applied. If not Rome, then like Athens at the zenith of its glory, the University of Chicago would be a republic.[10]

Many experts were sure the experiment would fail. It would be a "veritable monstrosity," they said, a Midwestern "Harper's Bazaar." Not if he could help it. Born to a Scottish-Irish family that ran a general store in New Concord, Ohio, the first president was a genius salesman. With Rockefeller's dollars in hand, he went to work shamelessly robbing Yale, Harvard, Cornell, and Johns Hopkins of their very best men. Chicago was looking for "home runs." By 1910 Rockefeller had spent $35 million building the university. Senator Jonathan Dolliver of Iowa didn't like it. "The University of Chicago," he complained, "smelled of oil like a Kansas town." His objection went unheeded. Intellectual gravitas had soon become the distinctive mark of the university. A new titan of education had been born.[11]

. . .

"There is a right and wrong way," the *University of Chicago Magazine* declared,

> *to crawl through bushes, to fall with a rifle and pack, to throw a hand grenade, to bandage a leg, to read a map, to bayonet a man. Next quarter approximately one thousand young men, unused to military life, will spend three hours a week in the fieldhouse learning these right ways. In the Basic Military Training course they will also learn elements of army organization, military law, first aid, and other phases of modern warfare.*[12]

By the time George arrived on campus in January 1942, the University of Chicago was like a military anthill. Cadets from the Army Signal Corps bumped into U.S. Public Health Service nurses. In all colors of uniform—army, navy, air force, Red Cross—and at any given time, five thousand men and women could be seen scurrying across the Quadrangles. The Institute for Military Study had already been running since October 1940, preparing more than two thousand students and civilians. At the Institute of Meteorology all the army, navy and civilian meteorologists in the country were now being trained and sent to service. Bartlett Gymnasium, Sunny Gymnasium, and the men's residence halls had been turned over to the navy; its radiomen and signalmen were now in training at the university and would need the accommodation.[13] More than one hundred contracts with the federal government were signed, involving thousands of men and millions of dollars. Even Norman Maclean, assistant professor of English (and former forest ranger) had traded in Shakespeare for shooting, and was teaching an advanced course in rifle marksmanship.[14]

For some the onslaught of the armed forces was not all bad: The meteorology cadet Charles Hlad just happened to hold the world record in the high hurdle and led a traditionally weak Chicago side to three surprising victories at the start of the track season. But even though the old-timers at Stagg Field got a kick out of their opponents' bewilderment, for most the "extinction of the civil ans" seemed a glum, uninvited affair.[15]

And so, when President Robert Maynard Hutchins stood before those present at the twenty-second annual trustees' dinner, he stated what to students and faculty was already obvious: "We are now engaged in total war," the commanding six-foot-three-inch former dean of the Yale Law School declared. "Total war may mean the total extinction, for the time being, at least, of the characteristic functions of the University of Chicago. I say this as flatly and crudely as I can. . . . We are now an instrumentality of total war."[16]

More than developing new theories in ecology or economics, the university, President Hutchins explained, was now in "service of the nation and the world." With the combination of the year-round quarter system and the scheme to award bachelor's degrees in the sophomore year to students who had started their studies as juniors in high school, the wartime necessity of "speeding up" education presented no problem at Chicago. It did mean, however, that in addition to the military presence there were hundreds of underclassmen scurrying about the halls.[17]

Many of the freshers must have been left speechless by the annual senior Mustache Race, which began that winter. A school tradition for more than twenty-five years, it consisted of the shaving on February 20 of the aspiring candidates, and judgment on March 5, by Brad, proprietor of the Reynolds Club Barber Shop, of the man (or woman) who had sprouted the most magnificent mustache. The annual Botany Pond Brawl, "theoretically" featuring a tug of war in the shallow algae-infested body of water lying behind the Botany Building—but always degenerating into a wild free-for-all—followed soon after the Mustache Race. Underclassmen might have been careful, too, where they rested their bottoms. "Dire penalties await the cynic," the school newspaper warned regarding the C Bench in front of Cobb Hall, "who defies tradition to park his worthless carcass on the Bench." It was into this blend of custom and mayhem that George now arrived, trying to get serious once again about his studies.[18]

· · ·

Inorganic Chemistry: 231, Quantitative Analysis: 240, Vertebrate Zoology: 205, Chemical Thermodynamics: 361—he was packing in his science courses. In truth this fitted in well with the general war atmosphere at the university: Outside the mechanical, physical, and natural sciences, entire divisions in the humanities and social sciences were feeling sterile and superfluous. Technology and science were going to win the war; this seemed as clear as day to almost everyone. Luckily President Hutchins was there to remind them of the broader purpose.

"The disorders of our civilization," he opined that fall, "result in part from our conviction that the advance of technology will solve all our problems." Not even sure that it might help to win the war, Hutchins was certain that it could never buy moral understanding. "We must have something more," he declared, and instituted two new graduate programs. The Committee on Communications was to "undertake to unite public opinion on the war, to develop an intelligent national morale, and to prepare the nation for a rational approach to post-war problems." The Committee on Social Thought was to concern itself with the issues at the base of social and political organization. And so, in the spirit of his president, George added a general course in social sciences alongside the advanced chemistry and mathematics that he was now taking—and acing—in the fall.[19]

It was shortly after he joined the wrestling team that year that he met Al Somit in the weight room. Al was an angular-jawed, wise-cracking Jewish kid from Chicago whose Ukrainian-born father, just like George's, had died when he was four. He had arrived that summer from college in Kentucky, where he had belatedly learned, two days after his arrival on a music scholarship, that something called "chapel" was mandatory. Promptly back in town, and with a growing sense of both his musical and technical limitations, he enrolled at the University of Chicago as a political science and history major.

In the weight room, Al could see that George had a wiry body and worked harder and sweated more than anyone. Intense hazel eyes planted beneath a broad forehead made contact in unsettling bursts. Even before he spoke, you'd pick him out as different, but when he did his voice was high and squeaky and started and stopped in spurts. He had a baby face, which was confusing. He flexed his muscles weirdly in front of the mirror. Eccentric, unscripted, he was a one-off.

Some felt uncomfortable in his presence. There was something brutally honest about his gaze. You couldn't escape it. It made you search for words, look at your feet, wonder whether he saw completely through you. Toting around his Harvard bag he seemed smug, above it all, juvenile. But there was a truth in him you walked away from at your peril: George Price knew things. He never said anything familiar. He was strange, and the way in which it was unclear whether he knew this or not somehow lent weight to his glances and opinions. Al was intrigued by his defiant obliviousness and boldness. George's genius for numbers provided a confidence he lacked in himself. Both loved to argue and, even more, to win an argument. They were lanky and opinionated and had a quirky sense of humor. Before long they were rooming together in a dank basement off Fifty-fifth Street.[20]

. . .

Sensitivities on campus were at an all time high. The secret constitutions of Chi Rho Sigma and other fraternities barred Jews and Negroes from the organizations. "Such discrimination pulls a Pearl Harbor on campus democracy every day that it continues to exist," the *Daily Maroon* attacked with language that seemed most salient. Many co-ops had sprung up, fighting back against "restrictive covenants." Only in 1948 would a Supreme Court decision render such covenants illegal. Meanwhile, race and religion were on everyone's mind.[21]

On warm days George would make the thirty-minute walk from the university down to Lake Michigan and offer to apply girls' sun lotion as they lay bathing on its shore. He was an oddball, Al thought, an oddball among oddballs, never quite grasping his relation to the world. In the fall he joined the Ellis Eating Co-Op, made up mostly of Jewish students. Al would soon be leaving for the Engineer Corps, and besides, $4.83 was a good deal for twenty meals ($.03 extra charge for additional beverages).[22]

"What do you expect of Jews?" George would say loudly while waiting in the queue, somehow missing the point that anti-Semitic jibes might actually offend anyone, not to mention Al. At first it seemed meaningless enough, just kind of strange, really. After a while, though, it was pissing people off. The more irritated the Ellis gang became, the more George pursued them. In a move that baffled every one, he declared a "battle against the non-Aryans," and then proceeded directly to join the Jewish Ellis Housing Co-Op.[23]

He was taking dancing lessons, alone, at Arthur Murray's. Angular and off rhythm, he seemed lost in a weird and mechanical internal world. Some put it down to geekiness compounded by the trauma of losing a father, and let it go, but most found it difficult to "get" him. Whether oblivious or spiteful, detached or goading, to George it all seemed like one big game. He was positively titillated by attention, even when it was negative. In the spirit of their offbeat humor, Al counseled his friend to plead freedom of speech when the constitution-touting co-op finally put up a vote to oust him. It didn't help. George was a contrarian; he got his kicks from pushing the limits just as hard as he possibly could.[24]

Sometimes Al wondered whether his friend was slightly nuts. Then he'd shrug the thought away with a forgiving smile. Between manic bench-pressing, supercilious racism, and all-nighters crutched on Benzedrine, George, he knew, was "just George" and that was that. When he'd question him about his offensive remarks George would offer an innocent glance and shrug his shoulders. He knew that people thought him awkward, but it didn't bother him, he claimed. Sometimes deeper feelings would escape, like slivery rays of sunlight penetrating a cloud. "It helps my morale," he wrote to a friend, "to learn that everyone doesn't dislike me."[25]

• • •

In the spring the "Mecca of the caffeine addicts," the Coffee Shop, was appropriated by the army, and with it the "last vestige" of peacetime campus life.[26] In the short time that George had been at Chicago, the university had undergone momentous changes. It had shifted from support from endowment to tuition fees, from individuals to corporations, from private donors to government funding. By 1944, 198 federal contracts had been executed, all on a not-for-profit basis. The university budget had tripled. Civilian enrollment was down 30 percent from the prewar level, and the normal ratio of three men to two women had reversed itself, and worse.[27]

But things were beginning to change. With America already well mobilized, university training programs across the country began to close one by one. As the G.I. Bill came in, the army units went out, and regular university life at Chicago, it seemed, was actually returning to normal.

Or so people thought. In truth enormous resources were now being spent on a top-secret underground project. In 1939 Albert Einstein had sent a letter to President Roosevelt, urging him to call upon the nation's resources to develop atomic weapons to fight the Nazis. Roosevelt complied, and within a year scientists at Cal Tech and Columbia had theoretically demonstrated the awesome explosive potentials of the isotope uranium 235 and an element just recently discovered called plutonium. Soon after Pearl Harbor a group led by the Nobel laureate Arthur H. Compton was set up by the government for consolidating plutonium research at Chicago. The outfit, called the Metallurgical, or "Met," Lab, was the cover given to Compton's facility, and it was tucked away behind the ivy-covered sandstone facade and glistening rectangular windows of unassuming Eckhart Hall.[28]

The Manhattan Project charged Compton with producing chain-reacting "piles" of uranium to convert to plutonium, with finding ways to separate the two, and, ultimately, with building an atomic bomb. As George had been innocently taking his undergraduate Organic Combustion Analysis and Differential Equation examinations, Glenn T. Seaborg and his team at the university secretly isolated the first weighable amount of plutonium from uranium, irradiated in cyclotrons.

But there was still the business of building uranium-and-graphite piles (later called reactors) that could be brought to critical mass in a controlled, self-sustaining nuclear reaction. When a labor strike prevented such work at a designated laboratory at the Argonne Forest thirty miles southwest of Chicago, the famed Italian physicist Enrico Fermi, together with Martin Whittaker and Walter Zinn, set out to build a pile in a squash court under abandoned stands in the west wing of Stagg Field. The pile was a crude construction, made of black bricks and wooden timber, but, miraculously, it worked. On the bitterly cold day of December 2, 1942—unknown to George, his fellow students and professors at Chicago, and practically everyone else in the world—the pile went critical at 3:53 in the afternoon. The nuclear age had dawned.[29]

· · ·

George graduated in September 1943, Phi Beta Kappa. A star in the Department of Chemistry, he was awarded the Eli Lilly Fellowship and invited to continue for his doctorate. Starting off on enzyme chemistry, his project quickly changed. He'd be working now under the supervision of a Dr. Samuel Schwartz, he was told, and joining the Manhattan Project.

A medical doctor and research scientist from the University of Minnesota, Schwartz was the son of Russian Jewish immigrants who had grown up poor in north Minneapolis. One of seven children, he sold candy and ice cream as a boy to help his family, and after high school snagged a job washing beakers in a lab at the university. The head of the lab, C. J. Watson, was a bigwig at the medical school, and, noticing his quick mind, took Schwartz under his wing. Watson sent him to medical school, and the investment paid off. Now at the Manhattan Project at the Met Lab in Chicago, the bearded, broad-smiling Schwartz headed a twenty-five-man team studying the biological effects of atomic radiation and metals. Day in and day out, George would walk up the sandstone steps of the George Herbert Jones Chemistry Building, nodding hello to the namesake's bronze bust as he whisked through the entrance hall and down to the lab.[30]

His new project was technical and had nothing to do with enzymes. If an atomic bomb was built and ended up being used, small traces of toxic uranium would find their way into human bodies. In order to be able to treat such people, a sensitive method had to be devised to help detect the traces of uranium. A generation earlier a pair of researchers at the Carnegie Institute had shown that the intensity of uranium fluorescence is greatly increased by fusion with sodium fluoride, and that its light, therefore, could be measured. The method was all right but needed to be made better. It was a needle-in-a-haystack problem, and George would try to solve it.[31]

He went to work under a veil of secrecy. How to best detect these tiny amounts of harmful substance? Besides the chemistry of the analytical procedure there was also the construction of suitable photoelectric fluorophotometers. It was a task that brought back childhood memories of Display and challenged his talents as a physicist and tinkerer. Various Manhattan Project teams across the country were working on the problem, and the pressure was on. Identical sets of thirty urine samples containing known amounts of uranium were prepared at the Rochester Project, and sent for analysis to seven of the strongest competitors. In March 1945 a uranium analysis conference was held at the University of Rochester, New York, to present the results. Six of the groups had completed the analysis. Of these, one employed spectroscopy and the rest fluorescence. Statistical analysis showed that the method devised by the Chicago group was the most sensitive and accurate, and George returned to Eckhart Hall with a winning smile. Aloof and often indifferent to people, he had emerged the champion of sensitivity with molecules and photons.[32]

But science and life, by some mysterious osmosis, were beginning to leak into each other. It was around this time that he met Julia Madigan, a dark-haired zoology graduate of the University of Michigan who had arrived at the Project after two years in medical school. On her mother's side, it was thought, she was the daughter of immigrant German Jews who had converted on the boat, and on her father's of Roman Catholic Irish who'd fled the great potato famine in the 1850s. Raised in the small paper mill town of Munising, Michigan, Julia's mother, Barbara Kinde, was the telegraph operator in the Upper Peninsula town of Marquette, and her father, James, the train station manager. With the help of two of his younger brothers, Frank and Michael, James bought the local hardware store, and before long the Madigans had secured profitable Forest Service contracts revolving around the logging trade with Canada. The Madigans were hardworking community people, and gave thanks for their prosperity by generously donating to the local church. When she came of age they sent little Julia to attend Sacred Heart with the nuns.

When Al Somit heard from his best friend that he was beginning to fall for her, he rubbed his eyes in disbelief. Julia was sexually conservative, George naturally prurient; Julia was a devout believer, George a militant atheist; Julia was mercurial, George a prankster; Julia respected convention, but George reveled in extremism. Working together at the Manhattan Project, the two couldn't even agree on the bomb (George was for, Julia against). What was worse, they were both opinionated and stubborn as goats. Still, a Madigan family legend had it that a cadaver's hand had slipped and grazed her chest in med school: "Julia could even raise the dead," the professor had said to the class. When she had arrived in Chicago looking for a flat, the landlady snapped, "I don't rent to Jews," to which Julia responded, "But I'm Irish!" and stormed off. She had high cheekbones, full lips, and black eyes. She was short. She was curvacious. And to George she was absolutely divine.[33]

· · ·

On the afternoon of May 7, 1945, Julia and George ran out of the lab and rushed to the quadrangle. With its armies in total collapse and its cities in rubble, the Third Reich had finally been defeated. The campus was abuzz. From the lofty gray stone tower of Rockefeller Memorial there floated a melody, the university magazine would later report, "more majestic than a thousand muffled drums." The earth literally "shook under the iron clangor of the mammoth Bourbon bells." In the chapel students crowded in the pews, seated, knees hugged to chin, before the altar, clinging to vantage points in the balconies. The sunlight poured into the vast chamber from the multicolored Gothic windows, as two thousand people stood to sing the "Star-Spangled Banner." The war in Europe was over.[34]

Just as people were getting used to the newfound peace in Europe, in August the United States dropped a uranium bomb on Hiroshima, and three days later, a plutonium one on Nagasaki. That week George's uncle and namesake died in the Midwest. "From what I have learned from the next world," Alice now wrote to him from New York about her brother:

> *He will be far happier there than here. I have already had some remarkable messages come thru relative to him, one from Dear-Dear who appeared to a psychic friend of mine who sat right beside me, described her, and repeated all she said to me. Aunt Julia has also been coming right into my body since I visited Cousin Bessie, also mediumistic, in Battle Creek. Aunt J. seized my right hand and arm there as we sat together, then took control of my left hand and arm, and when I asked her to demonstrate how she tramped thru the Heliker's house last summer on the first anniversary of her death (did I tell you?) she controlled my entire body suddenly and bent it over as she was bent, and whirled me right across the room. . . . At home she clapped my fists together and said "Goody, goody, goody, goody, goody" over and over again when I asked her if she was happy that the war was ended.[35]*

It seemed a rather strange way to express her thoughts about the most momentous event of the century. But then she added, "Hurrah, George, Hurrah! Wonderful to have it come so soon, and that is what all you chemists did for the world. I'm so proud of you, not yet 23, that you had something to do with it."

George was not amused. "Please do not ever again write to me or speak to me about 'vibrations,' 'batteries,' 'guides,' about dead people throwing you around the room, or anything else of that sort. I do not ever want to hear you refer to such things again, and I just won't reply to any letter in which you mention supernatural phenomena."[36]

A few months later the Coffee Shop on campus reopened.[37] Even though the war was over, George couldn't yet tell anyone about what he was doing. In August 1946 he handed in his thesis: "Fluorescence Studies of Uranium, Plutonium, Neptonium, and Americium." "I concluded my writing," he summarized to a friend,

> *with a 59 hour sleepless period, during which I also practically cut out eating. When my benzedrene gave out and I was too busy to get more, I had a surprising auditory hallucination for several hours. But now that's all finished.*[38]

The previous April the University of Chicago had accepted a letter contract from the government to operate an Argonne National Laboratory, as yet unnamed and not yet in official existence. On July 1 the name of the Met Lab was officially changed to the Argonne National Laboratory. And on August 1 President Harry S. Truman signed the Atomic Energy Act, ushering in the age of civilian nuclear power.

At convocation that summer at the end of the month, Vice President Lawrence Kimpton spoke on "Science and the Humanities" followed by the singing of the "Alma Mater." It was a happy occasion, but there was something weighty and somber about it, too. Between the verses ("Of all fair moth-ers, fair-est she / Most wise of all that wis-est be") George could take a moment to consider his future. He had taken a number of courses as an undergraduate in physiology and on the nervous system, and although the problems of biology interested him more, he felt a much greater aptitude for physics. Kicked out for his grades five short years earlier, he was set to return to Harvard as a chemistry instructor, and had secured an Argonne consultancy at $2.50 an hour. The arrangement would give him a lot of free time to think even if it wouldn't make him rich. He was ponderous as he packed up and got on the train for Cambridge. "Without the atomic bomb the prospects of civilization would be dubious enough," President Hutchins wrote that fall in his reports to the friends of the university. "Now that we have it, they are black," and civilization is "on the brink of catastrophe."[39]

· · ·

On June 28, 1947, Julia and George were married in a small ceremony in Munising. George's childhood auburn hair had turned
darker, and with his glasses and suit and tie, he looked more grown
up. He was dashingly handsome—in spite of the slacks pulled up
slightly too high. Julia's Irish family expressed concern that he wasn't
Catholic, and in the face of his atheism made him agree to raise his
future kids in the Catholic Church.[40] He had spent the last year as an
instructor of chemistry at Harvard, living like a student in the dorms
at Winthrop House. Now the newlyweds were on their way back to
Cambridge, to settle in a small home on Linnaean Street, just a ten
minute walk north from George's office at the Mallinckrodt Chemical Laboratory Building.

Almost everyone in the department was a big name: There were
Professors Woodward and Bartlett and Fieser in organic; Rochow and
Lingane in inorganic; and Kistiakowsky, Wilson, and Doty in physical. The atmosphere was serious, matter-of-fact. At least three young
fellow instructors of chemistry—Walter Gensler, Edward King, and
Leonard Nash—would go on to long and distinguished careers in
academia. But George was different. He'd left behind his strange
antics and was becoming more withdrawn. A "loner who kept to
himself," he was hardly known in the department. Academic life,
with its teaching and administrative work, wasn't for him. To his colleagues he seemed "wholly disengaged."[41]

Looking for a way out, he found an opening in a childhood pastime. The university chess team required two faculty members; his
old Harvard classmate Lloyd Shapley, now doing economics, invited
him to join as Top Board. It was a godsend. There was practice and
there were meets. And then there was the Argonne consultancy,
which would take him on weekends to Chicago—anything to keep
him away from the department.[42]

In January the Atomic Energy Commission had designated
Argonne National Laboratory its principal reactor-development cen-
ter. Work on a liquid-metal-cooled, fast neutron reactor, dubbed
"Zinn's Infernal Pile," together with the first nuclear submarine plant,
was already under way.[43] George was making only $3,500 a year at
Harvard, and the pocket change from Argonne wasn't much of a

boost. And so when Annamarie Louise Barbara Juliet Price came along on May 4, 1948, he began to feel the pressure of the provider. He knew that Julia didn't like him traveling so much—increasingly she let him know it. Besides, the real action at Argonne was in energy and reactors, not in measuring fallout in people's urine.

Other pastures suddenly seemed a lot greener. That summer of 1948 Claude Shannon published the first part of a general theory using mathematics to quantify information.[44] Immediately it caught George's attention. He liked things like that: It was elegant and precise. It was simple. It got rid of clutter, told it like it was, altogether cut to the matter. More important, it seemed to be leading in the direction the world was heading, and George wanted in.

In August he and Julia and Annamarie packed up at Linnaean Street and headed down to Morristown. With an offer doubling his salary, and far from the pressure and tedium of students and examination papers at Harvard, George was going to work for Bell Labs.[45]

. . .

Morristown, New Jersey, was a small town but its history was rich. It was there in 1777 that George Washington encamped with his Continental Army, to rest and spread the news to the world of their victories at Trenton and Princeton. There Samuel F. B. Morse, together with Alfred Vail, built the first telegraph at the Speedwell Ironworks in 1838, sending the first-ever telegraph message: "A patient waiter is no loser." And it was there that George and his family now settled, to join the new information revolution.[46]

Just a few miles down the road, at Murray Hill, Bell Labs were shaping the way people live, work, and play. In the 1920s the first public demonstration of the fax, the invention of the synchronous-sound motion picture system, and the very first long-distance transmission of television images (of Herbert Hoover from Washington to New York) had all been gloriously accomplished. Then came Karl Jansky and the amazing discovery that radio waves were being emitted from the center of the galaxy. In 1933 stereo signals were transmitted live from Philadelphia to the capital, and four years later

Bell researchers won the Nobel Prize in Physics for the discovery of electron diffraction, laying the foundation for solid-state electronics. Then, in 1939, came the world's first binary-digital computer, and, after the war, the transistor.

It had been John Bardeen, together with Walter Brattain, who observed that when electrical contacts were applied to a crystal of germanium, the output power was greater than the input. With the help of William Shockley, also at Bell Labs, the three men had conceived and invented the transistor, for which they would win the Nobel Prize. George was already hard at work mapping germanium surfaces, grinding off the surface, reetching, polishing, then remapping. His experiments were yielding "unambiguous information," he wrote, regarding the relative importance of surface treatment and bulk properties in determining transistor characteristics. This was electrical-engineering stuff, applied short-term research. He'd been contracted to work on long-term basic science in the Chemistry Department but was pursuing his own interests. Meanwhile Bardeen and Shockley wanted to know more about temperature effects on transistor properties, and had turned to George to do the measurements.[47]

The photos that survive from 1948–50 are a picture of small town living: A lawn, an asphalt path, a dog, a two-storied redbrick home with hatched roof and shuttered windows; George in white pressed shirt with tie, Julia in colored dress and lipstick, Annamarie with a balloon, everyone smiling. A second daughter, Kathleen Barbara Elizabeth, had arrived in late summer of 1949. The war was over, the fifties just under way. The Prices were living the American dream.

It was a lie. Beneath the facade, cracks had begun to emerge. George's quirkiness had started off endearing, but his unorthodoxy was growing maddening. Julia didn't like being called a "hippopotamus." She didn't like the idea of raising Annamarie and Kathleen in a Skinner box. George, for his part, was finding it hard to swallow her religion. "Better the girls become prostitutes than nuns," he would goad her, pushing Julia into a cocoon of bitter silence.[48] She was glum, he thought, and kept in her anger. When it came out in other places, its irrationality drove him up the wall. From the get-go it had been an unlikely union.

George never performed Bardeen and Shockley's experiments. When his old Manhattan Project boss, Sam Schwartz, invited him to come work with him again, this time in Minnesota, it didn't take much to persuade an increasingly depressed Julia to pick up and leave. Maybe a change could save their marriage.

• • •

Quietly the Prices settled in St. Paul in a quaint two-story cottage with a red roof and two small lawns up front and in the back. Just a short ride in the 1940 Chevy from Fortieth Avenue was the medical school, where George would now be working in the Radioisotope Lab at the Veterans Administration Hospital.[49]

Minnesota was a powerhouse in medicine. Dr. C. Walton Lillehei, who arrived at just about the same time, was pioneering open-heart surgery, and together with Earl Bakken, a medical equipment repairman, would soon invent the pacemaker and incorporate Medtronic. Work on X-ray diagnosis, radiation therapy, diseases of the liver, "deep-freeze" surgery, and the new specialty, oncology, was considered the top in the country. By the mid-1950s a *Minneapolis Star* editorial referred to the med school "as one of the greatest in the world—in the opinion of some, the best anywhere."[50]

For George it was just like during the war, back to uranium fluorescence. But there was now also the Porphyrins, Tumors, and X-Ray group, which met biweekly in Schwartz's office.

Porphyrins are a group of organic pigments with four linked nitrogen-containing rings that bind to metals, including the heme in hemoglobin. What Schwartz had found was that by localizing the fluorescence of intravenously administered porphyrin in tumor tissues, surgeons could get an idea of the extent of the metastatic spread of a cancer. He also found that porphyrins could either enhance or protect against the effects of ionizing radiation, depending on the type of porphyrin and its dose. If they could figure out a way to administer the right dose with the right pigment, controlled radiation combined with hematoporphyrin could be used to get rid of tumors. The implications were obvious, and Merck and Company were hot on the trail. Once again George's tinkering genius would be needed.[51]

It was like Display all over again. Using a Schott BG 12 primary filter, about 4 mm thick, plus a 405 mu interference filter, George went searching for porphyrins in slices of rat liver tissue. The usual secondary filter was the Wratten No. 15 gelatin, sometimes supplanted by a Corning didymium glass, Color Specification No. 1-60 (Glass No. 5120), to increase red-green contrast by removing yellow. His light source was an Osram HBO 200 mercury arc. The problem was that the fluorescence faded too quickly, so George constructed a slide-cooling system to reduce the rate of fading. He decided to try first a simple system in which rapid jets of air are cooled and dried by dry ice and then blown over and under the slide, with the microscope enclosed in a box to keep out moist room air. On occasion he'd take the girls to the lab to marvel at the sight of the mysterious freezing smokers.

Preliminary results seemed promising when a second difficulty arose: Most animal tissues showed comparatively weak fluorescence, except when stained with fluorescent dyes called fluorochromes. This made it difficult to recognize cell types and to observe the exact localization of the porphyrin. So George developed special equipment by which a phase contrast image of contrasting color and smoothly controllable intensity could be combined with the fluorescence image. A beam splitter was used to combine the irradiating violet or ultraviolet light and the visible light of contrasting color, and the intensity of the visible light could be controlled by a variable transformer.[52]

The problem had been solved, and Schwartz was ecstatic. George was "one of the most remarkable individuals" he had ever known, and the only person with whom he had worked who deserved to be called "a man of genius."[53]

. . .

A certain young man with porphyria
Whose existence grew drearier and drearier.
His agonized yells
Were due to phorphyrinized cells
And not psychosexual hysteria.[54]

His relationship with Julia was unraveling. Religion, temperament, and now sex, his limerick made clear, placed them in universes squarely apart. The move to Minnesota had failed to save their marriage. He was growing edgy; she increasingly, and understandably, hurt. He had met another woman, and over the summer, another from Kansas City, named Jan.[55] By the beginning of 1953 George left the cottage on Fortieth Street, saying good-bye to his two little girls. "This, I believe, automatically makes me owe you $20," he wrote to his old friend Al, whose skepticism had been right all along.[56]

George had come a long way from the leaky apartment on Ninety-fourth Street, where he and Edison and Alice had huddled together in the twenties and thirties to survive during the Depression. He had graduated with highest honors from a top university, worked on the Manhattan Project, married, started a family, and now separated. In a decade he had moved six times: from New York to Cambridge to Chicago and back, then to Morristown, then to Minnesota. Like a real-life Forrest Gump, he was present at every important juncture: the making of the bomb, the development of the transistor, the growth of modern medicine. Never at the center, he arrived to solve problems, winning admiration before disappearing like a ghost. Things that seemed to baffle everyone else came easily to him. But he was restless and unhappy. Maybe the people at the co-op had been right: Maybe he really was different from the rest.

When the Russians got their own bomb in the summer of 1949, the world seemed to change in an instant. There would not be years of a lag to enjoy. Russia had caught up and the Cold War had arrived. For the time being it was playing itself out in Asia, around the thirty-eighth parallel in Korea, and in the development of hydrogen bombs. When Truman left the Oval Office in 1953 just as George was settling into a dingy student complex north of the university, and Stalin died two months later, it was anyone's guess where the world was heading.[57]

Meanwhile little Kathleen had contracted polio but was recovering. The doctors recommended blowing air into her brain, but, reading up on the subject at the medical library and at her bedside, Julia refused. Soon Kathleen was recovering. Since supply had been short,

George was the only Price not to get his gamma globulin. Away from his family now, he had contracted the disease and was in a hospital bed, exhausted. Alice begged him to remember how much she loved him. Hoping to pick up her boy's spirits, she wrote of a lavish banquet at the Waldorf thrown by her Japanese boarder Mr. Washio in honor of the imperial prince of Japan. It gave him little comfort. He was thirty-one years old. His future was uncertain. Depressed and alone, George Price was helplessly roaming.[58]

John von Neumann (1903–1957)

Warder Clyde Allee (1885–1955)

5

Friendly Starfish, Selfish Games

I am violently anti-Communist," the man intoned in a low
accented voice, "and a good deal more militaristic than most."
It was January 1955, Capitol Hill, and not a senator in the confirmation hearing room stirred. John von Neumann was going to be sworn
in as a new member of the Atomic Energy Commission, and John
von Neumann was a man to listen to.[1]

The H-bomb was on everyone's mind. Back in 1952 the "Ivy
Mike" trial had destroyed the Enewetok atoll. It was official: The
bomb was terrible and viable. But it would take years to build
an arsenal, be massively expensive, and would have to be accomplished under a veil of complete secrecy. Still, in possession of a
large stockpile, America would unequivocally rule the world. That
is, if Russia didn't have its own program too. If it did, the arsenals
would cancel each other out, with the already costly expense and
effort incurred. Should she "defect," then, and build the arsenal,
or "cooperate" and hold off? Clearly each side would prefer that
no one stockpile, rather than both stockpiling for no net gain.
But each side might also decide to build its H-bombs either in the

hopes of gaining the upper hand or out of fear of being caught without them.

It was a prisoner's dilemma, and for von Neumann the solution was clear. The Soviets could not be trusted. To save itself and the world, America would need to wage a preventive war, to become, as Secretary of the Navy Francis P. Matthews had called it earlier in the decade, vicious "aggressors for peace." But when? Von Neumann was adamant. With the room hanging on his every word, he said: "If you say why not bomb them tomorrow, I say why not today? If you say today at 5 o'clock, I say why not one o'clock?"[2]

Why was a mathematician being asked by the U.S. Senate whether and when to use the most destructive weapon in history? Surely he was one of the few people who had the knowledge to make the crucial calculations that would make or break the project. But the real reason was different. The H-bomb dilemma hinged on the mystery of human nature. It had been a quest that traveled through economics and biology. And John von Neumann, people said, had finally cracked the nut.

. . .

The eighteenth century Scottish economist Adam Smith had a simple message to convey: Under certain conditions free economic competition will lead to the best allocation of society's resources. It sounded like a paradox, but it was unequivocally true: Unfettered contest will by an "invisible hand" maximize society's benefits. The more ruthless the competition, the greater the social good; individual selfishness leads to collective benefit and plenty.[3]

By the time Thorstein Veblen arrived as a professor at the University of Chicago when it opened its gates in 1892, this economic worldview was referred to as "classical." Welded now more strongly to the political theory of laissez-faire, Adam Smith's legacy beckoned a new name. Veblen called it "neoclassical economics" and didn't shy away from expressing his view: He absolutely hated it.[4]

It was said of Veblen—born in Cato, Wisconsin, to Norwegian immigrant parents—that taking one of his classes was like "undergoing a vivisection without anesthetic." A notoriously bad teacher, he was also a formidable critic. The basic assumption that individuals pursuing their own self-interest necessarily promote the good of society was to him both insipid and false Capitalism was leading to "conspicuous consumption" and "conspicuous leisure." Not only did this bring waste and inefficiency, it suppressed fundamental human instincts: acquisitiveness, workmanship, parenthood, and idle curiosity. Forms of social control could be used to reawaken them, but this could only be accomplished with the help of a broad science of human behavior. With its exclusive dependence on price theory, neoclassical economics was nothing but a narrow "hedonistic calculus." Based on "immutable premises," it had little to do with reality.[5]

At Yale political economy was associated with social science, at Johns Hopkins with political science, and at Columbia with politics. Chicago was the first university in North America to create an independent department of economics. Veblen had long been kicked out of the university for impropriety; girls liked him, it was said, and he didn't exactly object.[6] Gradually, a worldview almost directly opposed to his own became dominant.

"All talk of social control is nonsense," Frank Knight was often heard saying in his deep, magisterial voice. The oldest of eleven children raised in religious orthodoxy in McLean County, Illinois, Knight had grown up to become a Chicago professor of economics and a notorious slayer of sacred cows. Clergy and medics were quacks, the institutions of social order imposters, the prevailing moral norms—slaves to fashion. Knight was suspicious of the political system and even more of politicians. "The probability of the people in power being individuals who would dislike the possession and exercise of power is on a level with the probability that an extremely tender-hearted person would get the job of whipping master in a slave plantation." In *Risk, Uncertainty and Profit* he provided the first complete formulation of perfect competition—unfettered, unencumbered, uncontrolled. Perfect competition was important not because it was always most economically efficient, though it most certainly usually was. Perfect competition was important because it guaranteed individual freedom, and nothing—not the injustice of luck nor the trampling of acceptable standards of fairness—was more important than that.[7]

The way Knight saw things, societies have five economic problems: How to decide which goods and services to produce and how much of them; how to organize the available productive forces and materials among the various lines of industry and coordinate their use; how to distribute the goods and services; how to bring consumption in line with production; and how to ensure continued economic growth and improvement of the social structure. All five of the problems involve making choices, and there are two alternative mechanisms for directing how such choices should be made: at one extreme, central planning based on the command principle, at the other—a free-market system with voluntary exchange. More and more people in the Economics Department at Chicago had fewer and fewer doubts about which was the superior system. Central planning inevitably became linked with political totalitarianism; free-market went hand in hand with democracy.[8]

. . .

"The theory and teaching that there is a God is a lie."

The words hit Warder Clyde Allee on the head like an iron gavel hurled from a heavenless sky. He had not been prepared for this. There was silence in the lecture hall. He was confused, saddened. Most of all he was filled with a surge of pity. He felt sorry for the misguided man, his animal evolution professor, a controversial figure who would leave his post some years later on account of an ugly courtroom divorce. He had often heard of such people—infidels, atheists and that sort—but this was the first one he had met, and he planned to show him the error of his ways.[9]

It was the fall of 1908, Hull Court, University of Chicago. Allee had arrived that summer from Indiana, a strapping, broad-faced twenty-three-year-old, his burly frame and callused hands signs of years of labor on the family farm. Balding and sporting round spectacles, he looked like an intellectual football player, which, in fact, he had been as an undergraduate at Earlham College. Warder's father, John Wesley Allee, was the son of a Methodist minister from Parke County. When he fell in love with and married Mary Emily Newlin, whose forefathers had established the nearby Quaker settlement of Bloomingdale, he became a "convinced Friend," but still took the family to the Methodist church from time to time. At eleven Warder was officially converted at a revival meeting. Earlham was the pride of the old Quaker settlements south of Lake Michigan, which had played a role in the Underground Railroad, sneaking black slaves on "Tracks to Freedom" into Canada in the mid-1800s. When he graduated from high school it was only natural that Warder enrolled, joining the football team and becoming a "Hustlin' Quaker."[10]

Now he was at Chicago, a graduate student in the Zoology Department. The cloistered walkways, grass quads, and stone Gothic buildings were a far cry from the open fields and broad woodlands of Indiana where, as a boy, he had fallen in love with nature. Still, *Oekologie* had been a term invented by the German biologist Ernst Haeckel back in 1866 to designate the study of the relations of organisms to their environments, and Chicago was one of the few places in America where ecology could be studied. Warder had arrived a traditional believer. He was excited: Here he would study the nature God had instilled in all His creatures. Wide-eyed, he did not yet know that science would soon fix all that.

Isopods are ugly little creatures. Leggy and segmented, they look like a mysterious aquatic blend of scorpion and cricket. Allee was in love. Excited, he determined he'd crack the mystery of the tiny crustacean, abundant in shallow waters, the deep sea, and freshwater streams and ponds.

Creating artificial currents in the lab, he observed that stream isopods moved toward the current more than pond ones, except when their metabolic rate was low when they were breeding. Since oxygen and carbon dioxide affected metabolic rate, and differed from streams to ponds, it had to be the gases that explained the creatures' behavior.

Using depression agents like low oxygen, chloretone, potassium cyanide, low temperature and starvation, he could make stream isopods act like pond ones. Conversely, with high oxygen, caffeine, and elevated temperatures, lazy pond dwellers morphed into energized stream sprinters. Since all the isopods were from the same species, differences in behavior could not be due to heredity. Clearly it was all about interaction with the environment.

The discovery shook his religious foundations. Hadn't the Deity instilled behavior in His creations? If so, how could coffee be so powerful? The iron gavel, he now saw, really did fall from a heavenless sky. There was no "hand of God" to behold, only physics and chemistry. Science was winning out over the supernatural.[11]

After graduating with a doctorate, Allee was growing uneasy. Married now with a child, he was increasingly disturbed by the war. Why, for heaven's sake, this horrific bloodshed and carnage? At Chicago science might have laid his childhood belief in an all-powerful God to eternal rest, but at times like this roots provided comfort. If he couldn't pacify unbelief, he could sure as hell deify pacifism. That March he was appointed chairman of the Quaker War Service for civilian relief in Chicago, an outfit that had grown out of the Monthly Meeting of Friends.

It was early 1917, and the United States still remained on the sidelines. Already conscientious objectors were being humiliated and beaten in military training camps and prisons. Couldn't enlistment in the newly formed American Friends Service Committee, aiding relief and reconstruction work abroad, qualify as conscription, a form of noncombatant service during war? After all, liberal pacifism held both the individual and the state responsible for the welfare and rights of the citizen. Conscription to go kill and die in war was in direct violation of this most holy of commitments. Individualism was the bedrock of democracy because it meant the assertion of human freedom, not the vulgar triumph of egoism. The least government could do in times of war was to allow pacifists to provide their service in nonviolent currencies.

Allee had recently been appointed professor of biology at Lake Forest College. He had read Kropotkin and in his gut knew that he must be right. But he had yet to convince himself of the biological justification for peace and cooperation with an original scientific discovery of his own.

In the college chapel ne preached on the rights of conscientious objectors. When the administration forbade him to preach in the chapel again, he spoke up in the classroom. The college docked his salary. A few faculty and students were heard murmuring the word "traitor" under their breath. Traitor? The local *Springfield News-Report* wasn't so inhibited: "Sometimes war is unavoidable, and college professors are no more necessary to civilization than carpenters and cobblers." Allee's was a "most convenient theory." If a choice had to be made, "we should prefer to give up the professors."[12]

In April, Congress voted to enter the Great War. The requests of the American Friends' Service had been rejected. At Lake Forest, Allee waited for more peaceful times. If he wanted to find scientific proof to combat the folly of human warfare, he would need to go someplace else.

. . .

On the afternoon of December 28, 1917, the delegates shuffled into the Animal Morphology Building at the University of Minnesota blowing into their freezing hands. The newly established National Research Council could easily envision how physicists and mathematicians, chemists and geologists, might lend a hand to the war effort. But what about biologists? At the annual meeting of American Society of Zoologists, a special session on "The Value and Service of Zoological Science" had been hurriedly convened. The delegates settled quietly in their chairs.[13]

Darwinism had become about as German as liverwurst. Back in 1859, when he was under attack in England for *The Origin of Species*, Darwin wrote to a colleague: "The support which I receive from Germany is my chief ground for hoping that our views will ultimately prevail." And although he himself had carefully avoided any political implications for man, in Germany, Darwinism was interpreted as having repercussions for the future of civilization. It was Ludwig Woltmann, a German, who first gave the enterprise its name. Amounting to a revolt against Judeo-Christian and neo-Kantian ethics, "social Darwinism" advanced a set of biologized beliefs: The moral sense is a biological instinct, not a spiritual endowment; human races are unequal; biology is destiny; the welfare of the individual is subservient to the health of the group; the struggle for existence renders war and death necessary to progress; progress and biological purification are one and the same. It was this distortion of Darwin's theory, many American zoologists claimed, that was blowing wind into Germany's war sails.[14]

No one made this clearer than the Stanford entomologist Vernon Kellogg. Published earlier that year, *Headquarters Night* was an account of his war experiences as the chief liaison of the Commission for Relief of Belgium to the German high command in France. "The creed of the *Allmacht* of a natural selection based on violent and fatal competition is the gospel of the German intellectuals," Kellogg wrote; "all else is illusion and anathema." Former president Theodore Roosevelt agreed. In the preface to the book, he broadcast: "The man who reads Kellogg's sketch and yet fails to see why we are at war, and why we must accept no peace save that of overwhelming victory, is neither a good American nor a true lover of mankind." Formerly a pacifist, Kellogg had now changed his colors. Only a "war to end all wars" could save civilization.[15]

The delegates perked their ears in attention. If they could defeat German Darwinism on the scientific battlefield, they'd be lending their shoulders to the war effort. One by one, the big guns were paraded to make the cooperatist case: Did not Herbert Spencer argue that evolution led life from the simple to the complex, from the homogeneous to the heterogeneous? Had he not explained how the ensuing specialization of function necessitates cooperation, the better to bring about an integrated whole? And wasn't society just like an organism, comprised of individual parts each contributing to the community? Sure, he had later abandoned such ideals to paint a picture of a cutthroat struggle for survival, but he had been more insightful as a young man. And what about Kropotkin, that stellar exemplar of humanity: Had he not defeated Huxley's "gladiator" in Nature's glorious arena? Weren't the fittest, after all, not the fiercest or the strongest but those who acquired the habit of mutual aid and cooperation for the benefit of all?[16]

Answering such questions in the affirmative, they suddenly felt like physicists. They left freezing Minneapolis immeasurably more important than when they'd arrived just two days before.

. . .

"How much is sixty million, five hundred and fifty-three thousand, eight hundred and ninety-one divided by twenty-seven?"

"Two million, five hundred and one thousand, nine hundred and ninety-five point ninety-six."

"Good boy!"

Johnny von Neumann could divide eight-digit numbers in his head by the time he was six. Visitors to the Budapest family home of the successful Jewish banker Max Neumann were as stunned by his son's ability to memorize phone books as by the jokes he told in classical Greek. When he grew older he studied chemical engineering, physics, and mathematics at Europe's finest universities: Berlin, Zurich, Budapest, Göttingen. Soon the word was out: Von Neumann was a genius. By 1930, at twenty-six, he was sitting in the room next to Albert Einstein's at the Institute of Advanced Studies in Princeton. Einstein's mind, they said, was "slow and contemplative. He would think about something for years. Johnny's mind was just the opposite. It was lightning quick—stunningly fast. If you gave him a problem he either solved it right away or not at all."[17]

Unsolvable problems were rare, though. Living with his wife, daughter, and an Irish setter, Inverse, at a Princeton mansion on 26 Westcott Road, von Neumann was famous for hosting lavish weekly alcohol-fuming parties, and even more for scribbling mathematical formulas with pencil and paper in the middle of it all—"the noisier," his wife said, "the better." He wore prim, vested suits with a white handkerchief in the pocket, "an outfit just enough out of place to inspire pleasantries." Von Neumann was balding and porky; his diet consisted of yogurt and hard-boiled eggs for breakfast and anything he wanted for the rest of the day. He loved fast cars, hard liquor, classical music, and dirty jokes. He was a prankster. Once he offered to take

Einstein to the Princeton train station and then put him on a train in the opposite direction. He was known for scribbling equations on the blackboard in a frenzy, erasing them before students could get to the end. Klara, his wife, claimed that he wouldn't remember what he had for lunch, but could recall word for word books he had read twenty years before. He had produced groundbreaking papers in logic, set theory, group theory, ergodic theory, and operator theory. He had described the single-memory architecture of the modern computer, and performed the crucial calculation on the implosion design of the atom bomb. Along side these accomplishments, he loved toys and was observed unaffectedly scrapping with a five-year-old over a set of building blocks on a carpet. Though he was charming and witty in public, few felt that they really knew him well. People joked that John von Neumann was not human but a demigod who could imitate humans precisely.[18]

Above all he was fascinated by games, especially the kind, like poker, based on bluffing and deceit. "It takes a Hungarian to go into a revolving door behind you and come out first," he used to say. In fact John von Neumann loved games so much that he had decided to study them, seriously, as a mathematician. What he discovered amazed him: In games where two opponents were in absolute conflict, where the loss of one is the gain of the other and only one side can ever win; in such "zero-sum" games there is always an optimal strategy for both players to pursue. Tic-tac-toe is the simplest example, but here is an illustration from life: Imagine two sweet-toothed kids being given a cake and told to share it. When a grown up carefully divides the cake down the middle, one side always feels slighted, even by a crumb. It is best for one of the kids to divide the cake, knowing that the other can choose which piece he wants. Since both kids know that the other wants as big a part of the cake as possible, cutting the cake precisely down the middle is the optimal solution.

It was mathematically airtight. It applied to games that involved perfect information (both kids know that the cake must be divided), complete self-interest (both kids want as big a piece of the cake as possible), and rational decision making based on a calculation of the other side's agenda (the kid cutting the cake understands that the other kid wants as big a piece as possible, and vice versa, and both act accordingly). The issue was always how to maximize the minimum the other side strove to leave you with, or, in other words, to minimize maximal losses. The "minimax theorem" proved that there was always a best way to do this. With the relevant information, the right strategy could be known.[19]

· · ·

The economists at Chicago had picked a hard time to wage their battle. In Europe the Great War was now over, but Adolf Hitler had risen to become führer of Germany, and Mussolini ruled with an iron fist in Italy. The Great Depression had taken hold. Reeling citizens looked to their governments for salvation.

Enter John Maynard Keynes. "I have called this book the *General Theory of Employment, Interest and Money*," he wrote, Cambridge-style, in his magnum opus in 1936,

> *placing the emphasis on the prefix general. The object of such a title is to contrast the character of my arguments and conclusions with those of the classical theory of the subject, upon which I was brought up and which dominates the economic thought, both practical and theoretical, of the governing and academic classes of this generation, as it has for a hundred years past. I shall argue that the postulates of the classical theory are applicable to a special case only and not to the general case, the situation which it assumes being a limiting point of the possible positions of equilibrium. Moreover, the characteristics of the special case assumed by the classical theory happen not to be those of the economic society in which we actually live, with the result that its teaching is misleading and disastrous if we attempt to apply it to the facts of experience.[20]*

If government didn't intervene in the economy it would be betraying its citizens. Since employment was not determined by the price of labor but by the spending of money (what is called "aggregate demand"), the assumption that competition will deliver full employment in the long run was patently false. On the contrary, underemployment and underinvestment were the likely natural state of competition; unless active measures were taken, that is. Lack of competition, Keynes was arguing, was not the fundamental problem; the reduction of unemployment by cutting wages, no promised panacea. What was needed instead was active government intervention: Public works, deficit spending, redistributive taxation, the lowering of long-term interest rates. In short, if society was to be stable over time, it would need to abandon the Invisible Hand in favor of the welfare state.[21]

"Nonsense!" Knight wrote in the margins of his copy of Keynes's book, alongside more powerful expletives. Nonsense, Chicago colleague and expert on monopolies Aaron Director, a "mordant critic" of governmental and social institutions, concurred. Nonsense, agreed Henry Simons, down the hall, author of *A Positive Program for Laissez-Faire*. Most combative of all was Jacob Viner, "a strutting Talmudic Napoleon with facial expression alternating at express-train speed between joviality, challenge and utter contempt."[22] Challenging Keynes, Viner took welfare economics to the ring. Markets had to be left free, competition unencumbered. "Survival of the fittest" was the name of the game, intervention a curse of maudlin sensibilities. Man was selfish, but selfishness was not a curse. On the contrary, it was the true and only road to happiness.

It was said of Viner that his self-appointed task in life was destroying students' confidence. When in the fall of 1932, a young four-foot-eleven-inch Brooklyn boy, son of immigrant parents from Carpatho-Ruthenia, shuffled into the classroom to attend Economics 301, like everyone else he was trembling. Unlike his fellow students, though, Milton Friedman would one day teach the very same course, becoming the leader of the intellectual movement that would challenge Keynes and the welfare state. Others would argue about precisely when it had all started, but the Chicago School of Economics was on its way.[23]

. . .

Darwin had pondered over the mystery of sterile insect castes, and so now did a new Chicago recruit. After a decade training at Cornell, teaching in Pittsburgh, and knee crawling in the jungles of British Guiana, Alfred Emerson arrived at the Chicago Biology Department in 1929 with termites on his mind.

He was *the* premier expert; when he retired his personal collection consisted of 91 percent of the known species in the world.[24] His eyes were exceptionally dark, glistening like plums dipped in cold water. Easygoing and gregarious, he exuded the air of the confident son of a respected scholar of classical archaeology. Still, there was a problem that was killing him. Just like ants, termites were divided into castes, but how did such different castes arise? Were soldier and worker termites genetically determined before hatching, or did some environmental factor, like food, fashion individuals as they developed: for soldiers, ferocious fighting mandibles; for workers, special digestive systems for consuming cellulose. It was difficult to know. After all, at birth all termites are identical.

Good thing, then, that Sewall Wright was in the room next door. Recruited by Chicago in 1926 for his prodigious mathematical and statistical skills, he had yet to develop the evolutionary model that, alongside Fisher and Haldane would soon make him famous. But he did have a theory about how genes worked: Particular chemical cues activated particular genes in particular ways; they could mutate them or turn them on or off. The genes would then activate particular physiological gradients, depending on the chemical cue they had gotten, directing genetically identical organisms to morph into entirely distinct creatures. That's how development proceeded: The organism, Wright believed, was a "highly self-regulatory system of reaction"; it was a mass of disparate parts each doing its thing to bring about an integrated whole.[25]

Emerson liked analogies. Just like organisms, colonies were integrated biological systems. Like genes and chemical cues and physiological gradients, termite warriors and toilers and queens and kings each sang their own particular aria. Whether a newborn termite would develop into a soldier or worker depended on the relative complex of castes in the colony—the analog to Wright's chemical cues. It was one big coordinated opera. Far from a random aggregation of individuals, a termite colony was in fact a "superorganism," a system so well integrated that it assumed a life of its own.[26]

This had been the idea of a scholar Emerson had met on his knees in the jungles of British Guiana. In fact the Harvard entomologist William Morton Wheeler was so convinced that the colonies of social insects were like a single individual that he saw no difference whatsoever between sterile workers toiling for the nest and a heart pumping for the good of the body. With the help of Wright's fancy physiological genetics, Emerson could now apply Wheeler's superorganism to explain termite evolution. Cooperation was a by-product of specialization of function. Ultimately it existed because individuals were subordinate to the whole.[27]

"Homeostasis" was the key. Coined in 1926 by the physiologist Walter B. Cannon, the term referred to the properties of self-control, regulation, and maintenance that ensured the stability of internal environments, like body temperature in mammals. Once again turning to analogy, Emerson replaced "internal environment" with "population," fashioning homeostasis an ecological, not just a physiological, principle. "Just as the cell in the body functions for the benefit of the whole organism," he wrote, "so does the individual organism become subordinate to the population." Homeostasis was the solution to the conflict between part and whole; in the tug-of-war between the interest of the individual and the good of the group, the balance leading to stability was where the rope had to be fastened.[28]

. . .

It had been a cool autumn day in September 1921 when Allee stepped into his new office at the University of Chicago. In the spirit of the Economics Department, the chair, Frank Lillie, had declared the subjects of investigation in the Department of Biology to furnish the basic scientific foundations for social control. Nature was humanity's normative guide, the ecologist its healer. To help bring peace to the world he would plummet into her secret depths, rising with prescriptions. Allee's heart was racing. Lake Forest was behind him. At the University of Chicago he'd be going straight to work in the service of civilization. His scientific experiments, he told his new colleagues, would be designed to throw light on the subject of war.[29]

Everything revolved around animal aggregations.

Bats aggregate in caves, mussels at their bed sites, prairie chickens at their booming grounds, iguanas on rocks to stay warm. Countless examples existed in nature of individuals clumping up in groups. Why did they do it? Wouldn't such behavior put them all in grave danger, prey to a ruthless lucky predator, or otherwise doomed to an ill-fated turn of the environment? Dozens of studies showed not only that aggregating was dangerous but also that it led to stunted growth, low rates of reproduction, weak offspring, even unexplained death. If individual struggle for existence was Nature's avenue, why would a mussel seek out friends with whom to consort?[30]

The only possible answer was that groups offered individuals benefits after all. Allee began looking for them in his beloved isopods: Placing ten groups of the creatures alongside ten solitary specimens, all of them on filter paper, he was able to show that the groups retained water much more successfully. Here was proof that cooperative aggregation helps isopods deal with dry environments where retention of water is crucial.

Next he turned to starfish. When he put them in containers with no eelgrass, immediately they clumped together into groups; when he planted eelgrass, the groups disbanded. Was this not proof enough that assembly provides protection in the absence of natural cover? Sea urchins, too, demonstrated how beneficial group life could be: The more fertilized eggs present in one area, the faster did they all divide and develop, becoming free-swimming larvae at significantly accelerated speeds. Life in groups had enormous benefits for individuals.

Allee admitted that above certain numbers aggregations became dangerous, but as long as these numbers weren't reached, aggregations conferred benefits on every and all. "Stranger" mussels came together because it was good for them. Nature had appointed cooperation her chief executor.[31]

Neither tainted with German Darwinism nor plagued with biological determinism, aggregations were about unrelated animals cooperating to adjust to their surroundings; evolution and genetics were beside the point. But if the simplest of creatures were continually aiding each other, if cooperation was the rule of the sea cucumber and urchin, then surely there was a lesson to be learned by humans.

Animal communities and human communities, after all, were part of the very same developmental process: individuals reacting to their environments by coming together, slowly learning to tolerate one another, gaining attractions to one another, then acting in coordination, then finally cooperating. So universal was this succession that for Allee it became a part of the very definition of life. Living beings, he wrote, all reproduce their kind, and all are continually adjusting to their environment. But a creature had also to show "at least the forerunners of cooperation"; without this it wouldn't be counted as living.[32]

Finally, biology and Quakerism had united. Even though they knew better in their gut, early Friends could not counteract the claims of hard-boiled militarists; "they had no proof," Allee wrote, "of the correctness of their position on this fundamental point." Now things had changed. "There is abundant evidence from modern science that the centuries old Quaker attitude to war is correct." Those who believed that science lends support to the present war system in settling international disagreements were relying on a false, outmoded phase of the biological understanding of the nature of life. Far from being smothered by competition, cooperation was possible precisely because the individual was important. It was through his seeking out of others that all the goodness in the world was born.

Allee had melded politics, faith, and science into a comprehensive philosophy, good for planarians, insects, and humans. The social Darwinists were wrong because Huxley was wrong; it was Kropotkin who got it right. And, he added with a crack of home-grown Indiana pride, "It is to our glory that we Quakers attempted at all times to substitute cooperation for struggle."[33]

Then came World War II

. . .

Back in Princeton, von Neumann was as usual. He was not convinced by Quaker biology, nor, for that matter, by the view of human nature advanced by the advocates of the welfare state. Joining forces with Oskar Morgenstern, a tall, imposing Viennese émigré, he produced a twelve-hundred-page formula-filled tome, *Theory of Games and Economic Behavior*, to cover all eventualities. "Economists simply don't know what science means," Morgenstern, who claimed to be the grandson of Frederick III of Germany, wrote in his diary in 1942. "I am more and more of the opinion that Keynes is a scientific charlatan, and his followers not even that."[34] Morgenstern and von Neumann would change things. "We hope to establish satisfactorily," the authors wrote, "that the typical problems of economic behavior become strictly identical with the mathematical notions of suitable games of strategy."[35]

As surely as tic-tac-toe, economics demanded precise strategy. When von Neumann would travel from Princeton to visit the Cowles Commission for Research in Economics at the University of Chicago, all the professors lined up like children to present him with problems they couldn't solve.[36] There was a buzz when he walked in, briskly, as was his fashion. But this was greater, people whispered, than mere economics: Von Neumann was the first man to crack the code of human behavior.[37]

. . .

From Sewall Wright, Emerson had learned that natural selection acted not just on individuals but also on populations; mathematically speaking, "group selection" was possible. Wright's math was complicated and its implications counterintuitive: Sometimes what was bad for the individual could actually be good for the group. Once again Emerson turned to analogy. In the Kaibab forest of Arizona,

pumas were known to prey on deer. At the individual level the competition between every puma and every deer was a ruthless struggle
for survival: Either a puma killed or starved, either a deer escaped or
was eaten. But at a higher level pumas were actually regulating the
deer population; when man hunted too many pumas, deer numbers
quickly got out of control, and this was bad for everyone. If natural
selection could "see" populations, choosing more successful ones for
a given environment over others, evolution could flip intuition on its
head: Competition could be considered beneficial, even cooperative.
It all depended on the perspective.[38]

And, of course, on homeostasis. Only homeostasis had allowed
organisms, aggregated in groups, to take control of their lives in
face of hostile environments. It was homeostasis that led termites
to triumph and the Kaibab forest to flourish and grow. The more
homeostasis, the more evolution moved away from conflict and competition. In "The Biological Basis of Social Cooperation," Emerson
offered his prescription: If the probability of survival of the population depended on the degree to which individuals had adjusted
themselves to one another and to the environment, then a little bit
of sacrifice on the part of individuals was not all that much to ask for.
The motto was "United we stand, divided we fall"; if individual freedom threatened group stability, in the interest of "peace, well-being
and progress" it would have to be carefully controlled.[39]

George Gaylord Simpson was having none of it. Recently the
goateed, pointy-eared paleontologist had been measuring his walks
across Harvard Yard in political teaspoons: There was the landing
on Normandy, the march on Berlin, the disintegration of the Nazis.
Finally the Third Reich had been felled. But what was this pabulum
coming out of Chicago? Hitler was dead in his bunker, and one Hitler
had been enough. "The evolutionary analogy suggests," he wrote,

> *that the epiorganism will and should evolve in the direction of greater
> integration (i.e. less individual freedom and responsibility), and that
> its units (i.e. you and I) should become more specialized (with less
> scope for activity and change), more interdependent (less self-reliant),
> and more a part of the whole state (less individual). . . . Then the*

biologist finds himself face to face with the fact that this is a totalitarian idea.[40]

The "aggregation ethics" of these Chicago men was all the same: the "superorganism," group selection, social control. It all just sounded too eerily familiar. For him groups were nothing more than a collectivity of individuals; they had no life of their own. It was the *individual* who was the fundamental unit of selection, not the group. The *individual* had motored all that was good and noble in evolution—intelligence, knowledge, and of course responsibility. Emerson believed that nature should be mankind's moral beacon, and Simpson concurred: Ethics needed to be naturalized. Still, the Chicago man thought stability could be bought only at the price of constraints on individual freedom; for Simpson freedom was the cornerstone of democracy.[41]

Difficult questions lingered. What if democracy wasn't the best system, after all? What if totalitarian regimes maintained "homeostasis" more effectively? Fascism had led to destruction and been defeated. But could man compete in ways that were socially cooperative? Kropotkin and Huxley had sparred over the true nature of mankind. In the absence of a clear verdict, was it safe to abandon society to the ideal of individual freedom?

Emerson leaned back in his chair, unperturbed. Now that the war was over, it was time to return to the wiser ways of nature, where a hopeful balance had been born. "The issue is clear," he declared just a few months after Nagasaki. "It is cooperation or vaporization. It is a struggle for existence by means of the cooperation of all mankind, or extinction through unnatural destructive competition between individuals, classes, races and nations already incorporated into a larger interdependent whole." His conviction resounded through the halls of the Department of Biology at Chicago like the din of a sea of termites marching up a woodland hill.[42]

Over in the Met Lab, bending over uranium, the young George Price remained completely oblivious.

.   .   .

Milton Friedman, Viner's former trembling student, arrived back at the University of Chicago in 1946, just as George was leaving.

During the war he had worked on tax policy at the Treasury Department, and emerged with a one-track mind. "Everything reminds Milton Friedman of the money supply," economist Robert Solow complained. "Everything reminds me of sex, but I try to keep it out of my papers." Still, Friedman was formidable. Everyone loves to argue with Milton, his Chicago colleague George Schultz used to say, especially when he isn't there.[43] He had a winning smile, a balding head, and oversize glasses. The nineteenth-century liberal regarded an extension of freedom as the best way to promote welfare and equality but nowadays a liberal was someone who saw welfare and equality as either prerequisites of freedom or its alternatives. Friedman was on a crusade: He was going to reclaim the mantle of classical liberalism.

It wouldn't be easy. Memories of the Great Depression and of war still lingered, and men looked to government for answers. Keynes was a hero. When Friedman published *Capitalism and Freedom* more than a decade and a half after World War II, neither the *New York Times, Chicago Tribune, Herald Tribune, Time* nor *Newsweek* even reviewed it.[44] Still, as John Stuart Mill's *On Liberty* had been for the nineteenth century, so would Friedman's book become for the twentieth: a manifesto on freedom. The problem, just as for Allee, was one of integration. Literally millions of people were involved in providing one another with their daily bread, let alone their yearly automobiles. How could such interdependence be reconciled with individual freedom?

In Russia the solution had been to concentrate economic power and political power in the very same hands. To Friedman this spelled catastrophe. All forms of collectivism necessarily lead to tyranny; economically and politically speaking a centralized economy was "the road to serfdom."[45] On the other hand, viewed as a means to the end of political freedom, the economy could be made to disperse power and do away with coercion. The key was for economic power and political power to be made separate so that they could offset each other, and the best way to do that was to free markets and encourage private enterprise exchange. Adam Smith had been right all along: The Invisible Hand born of individual self-seeking was the key to collective prosperity.

Friedman would have loved to be an anarchist, like Kropotkin. Unfortunately he was all too well aware that individual freedoms often collide. "My freedom to move my fist," the Supreme Court Justice William O. Douglas had put it, "must be limited by the proximity of your chin."[46] Government was necessary to provide a means to modify and mediate the rules of the game, and to punish anyone who would break them. Still, the hand of intervention had to be circumscribed as much as possible, just as Kropotkin had wanted. Their reasons for agreement were diametrically opposed: The Russian prince thought creatures were naturally cooperative and therefore needed neither coercion nor direction. The son of Russian immigrants, on the other hand, thought men were naturally competitive and that competition always leads to the best results. It was the Invisible Hand of the market, after all, that would safeguard welfare and equality. More fundamentally, it would protect mankind's most precious treasure: freedom, that "rare and delicate plant."[47]

. . .

Back in the Zoology Department, Allee was breathing in the winds of change. He had always argued that cooperation was all about the individual: the individual adjusting to its environment, the individual adjusting to other organisms, the individual becoming a part of the group. It was physiology that drove things, the challenge of reacting to the environment. Huxley had taken the Darwinian struggle for existence to be nothing but a bloody war. But Darwin was softer: "A plant on the edge of a desert," he had written, "is said to struggle for life against the drought." Like Kropotkin, Allee trained his gaze there: to where the chronicle of nature bespoke organisms coming together to fight the elements to survive, not of a ruthless *bellum omnium contra omnes*. Cooperation was the happy result of organisms reacting to things like the chemistry of gases and the physics of camouflage.

Down the hall, though, Emerson had gotten to him. Far from simply being about individual physiology, aggregation was a problem in population biology. Perhaps it was the group, after all, which counted more than the individual. Perhaps a starfish aggregated not simply to adjust to a lack of cover but because natural selection favored groups that clumped together over those that didn't. Signs of paralysis had appeared in Allee's legs back in 1930, and operations were performed to remove the growth. In 1938 a third operation had left his legs paralyzed. Now in a wheelchair and more dependent than ever, he wondered about the "superorganism" and the meaning of integration of the individual with the whole.[48]

Emerson had argued that the individual was to the population as the cell was to the body. But if integration was to be studied properly, Allee thought, analogies would not suffice. If populations had their own unique traits, if the "superorganism" really was alive, he wanted to know how it worked. Chickens would provide him with answers.

As with many social animals, chicken life revolved around a hierarchy: Low-ranking hens lost a great proportion of their fights to higher-ranking ones, all the way up to the dominant hen at the very top.[49] Allee knew that dominant hens laid more eggs than subordinates. This made it look as if selection were simply working on individuals: the stronger and more aggressive gaining a larger representation in future generations. If that were true, integration would be meaningless. Just as Simpson had argued, groups would be collections of individuals and nothing more.

Impressed by Wright's group-selection equations, he searched for a different explanation and together with a student soon made an encouraging discovery: Socially unstable flocks ate less, were scrawnier, laid fewer eggs, and had smaller combs than flocks where hierarchy was well established. Not only that: Once dominance-subordination relationships were stable, overall aggression dropped dramatically. Hierarchy was a group property, after all, a wise and benevolent integrating mechanism. The road to Mount Harmony traveled through the Lowlands of Competition.[50]

It was a satisfying solution but it left Allee cold. Competition, after all, wasn't supposed to be part of the natural equation. And yet here, and in the "higher" vertebrates of all creatures, it seemed a cardinal ingredient for survival. What would the Friends think of that?

Just two months before Hiroshima and Nagasaki, in June 1945, Allee had made the problem clear in an article in the *New Republic*. Not only was competition a part of the game, it necessarily would lead to instability and bloodshed. "Sooner or later," he wrote,

> *on the international stage as among our groups of mice, or fish, or hens, or other animals, a subordinate always seriously challenges the alpha individual or nation. Although the challenger may be beaten back, often many times, eventually alpha rank is taken over by a new despot and the cycle starts again. In so far as any new international organization is based primarily on a hierarchy of power, as are the peck orders of the chicken pens, the peace that follows its apparent acceptance will be relatively short and troubled. Permanent peace is not to be won following the precedent established by the dominance of vertebrate animals.*[51]

If Nature was mankind's moral compass, what was to be done? Was civilization condemned to eternal cycles of bloodshed?

He cringed at the thought. To be sure, sea urchins and chickens had taught him that there were two forces acting in nature: "the self-centered egoistic drives which lead to personal advancement and self-preservation, and the group-centered, more-or-less altruistic drives that lead to the preservation of the group." But how, ultimately, was the circle to be squared? Weighing the alternatives, he finally came down on the side of goodness. "After much consideration, it is my mature conclusion . . . that the cooperative forces are biologically the more important and vital." Hierarchy and dominance, after all, were a mark of the social vertebrates alone; if one considered the bigger picture, they were just one tiny branch on the great tree of cooperative life. Competition had arrived late in the evolutionary game. Man was a vertebrate, all right, but he needed to learn from the isopods.[52]

To rise to the challenge he would have to break down barriers, not erect them. Gender, family, races, nations: all only ever led to division. Like isopods, independent humans didn't have to be related to come to one another's aid. They needn't be of the same race or nationality. All they needed to do was to act toward one another *as if* they were closest of kin. It was the cooperative instinct of the individual, not the exclusive politics of the family, where goodness and altruism had come from. But now it was time to graduate to the next level. Once individuals came together, the "superorganism" could be born, gaining a life of its own. Then, like all good things, it would be selected by evolution.[53]

Sitting slouched in his wheelchair, the "Hustlin' Quaker" took a moment to reflect. It had been a long journey: from Indiana to Chicago, from believer to scientific pacifist, from ecologist of the individual to evolutionist of the group. His friend Emerson saw cooperation as a mechanism of homeostasis, a means to accomplish an end. But by now Allee had learned that life was much less utilitarian. Far from a means to an end, peaceful cooperation had been Nature's initial design, her honeyed Garden of Eden. With such thoughts in his heart, he wheeled his chair down to the South End of Chicago, where he'd be helping to build interracial summer camps for black and white children. At long last he knew what he believed in: education as deliverance, biology as morality, the peaceful group, mercifully, both the end and cherished goal. Man had come a long way but mustn't abandon the wisdom of his evolutionary past. It was time to return to the ways of the starfish.

· · ·

Behind guarded gates in terra cotta and putty white midrises just a block from the Santa Monica beach, they sat in swivel chairs in cubbies twiddling pencils and making jokes.

There were no rules: Surfing, semantics, outer space, Finnish phonology, neurosis, the Arab class system, a hermeneutic study of the writings of Lenin, an analysis of the popular toy-store puzzle "Instant Insanity" . . . they could study whatever tickled their fancy. The official air force contract called it "research on intercontinental warfare," but mainly they were being paid to "think the unthinkable": How many casualties would there be if a nuclear bomb were dropped on Cleveland? How would Washington know? How should intercontinental ballistic missiles be developed? What were the odds of a Soviet attack, the pros and cons of an American one?[54]

The RAND Corporation was a civilian nonprofit think tank chartered in March 1948. For an elite cadre of physicists, mathematicians, economists, and political scientists, it was the Cold War equivalent of Los Alamos. *Pravda* called it the "American academy of death and destruction," and there were some wary Americans who concurred. Amid crew cuts, pipes, and practical jokes, a strange zeitgeist pervaded: the worship of the rational and the quantified, a geopolitical obsession, and a "weirdly compelling mix of Olympian detachment, paranoia, and megalomania." Game theory was central. The head of the mathematics division, John Williams, wanted the very best men at his disposal. John von Neumann was at the top of his list.[55]

His services had already been called for in battle. In World War II his students had devised bombing strategies for the air force designed to minimize the chance of pilots being shot down over enemy territory.[56] Von Neumann himself advised Gen. Leslie Groves, military chief of the Manhattan Project, on where best to drop the atom bombs in Japan. (A note in his handwriting dated May 10, 1945, reads: "Kyoto, Hiroshima, Yokohama, Kokura.") Whether it was a poker player staring down an opponent, a couple arguing over going to a film or the opera, firms bidding at auctions, or two nations building stockpiles of atomic bombs, von Neumann provided solutions. The bounce of the dice, the flip of the card, the raised eyebrow of a totalitarian ruler—all divulged an elemental truth: Human beings are self-seeking, rational agents out to maximize their gains in a fierce, competitive world. Game theory would teach them how best to wage their wars.[57]

And so now, between "high-proof, high I.Q." parties at Williams's home in Pacific Palisades, the brilliants of the division went to work. John Nash, Paul Samuelson, John Milnor, Lloyd Shapley—all were there beside von Neumann. In September 1948 the young Kenneth Arrow was given the task of demonstrating that it was okay to apply game theory to nations even though it was formulated in terms of individuals. What his "Impossibility Theorem" showed was not encouraging for integration: It is logically impossible to add up the choices of individuals into an unambiguous social choice under any conceivable constitution. Except, that is, dictatorship. Just as people were beginning to swallow Arrow's frog, Melvin Dresher and Merrill Flood devised a game that did not bode well either. The Princeton mathematician Albert Tucker, also at RAND, named it the "prisoner's dilemma." A version of it goes like this:

> *Two members of a criminal gang are arrested and imprisoned. Each prisoner is in solitary confinement with no means of speaking to or exchanging messages with the other. The police admit they don't have enough evidence to convict the pair on the principal charge. They plan to sentence both to a year in prison on a lesser charge. Simultaneously, the police offer each prisoner a Faustian bargain. If he testifies against his partner, he will go free while the partner will get three years in prison on the main charge. Oh, yes, there is a catch. . . . If both prisoners testify against each other, both will be sentenced to two years in jail. The prisoners are given a little time to think this over, but in no case may either learn what the other has decided until he has irrevocably made his decision. Each is informed that the other prisoner is being offered the very same deal. Each prisoner is concerned only with his own welfare—with minimizing his own prison sentence.*

	B refuses deal	B turns state's evidence
A refuses deal	1 year, 1 year	3 years, 0 years
A turns state's evidence	0 years, 3 years	2 years, 2 years

The prisoners can reason as follows: "Suppose I testify and the other prisoner doesn't. Then I get off scot-free (rather than spending a year in jail). Suppose I testify and the other prisoner does too. Then I get two years (rather than three). Either way I'm better off turning state's evidence. Testifying takes a year off my sentence, no matter what the other guy does."[58]

The problem was that if both prisoners were rational and self-seeking, both would reason exactly in the same way. What that meant was that they would both "defect" and get two years in jail, whereas had they "cooperated" and kept their mouths shut, they'd only have to serve one year—a better solution for everybody. It was a maddening contradiction of Adam Smith's Invisible Hand: The pursuit of self-interest does not necessarily promote the collective good. Nash, a handsome but strange genius who would soon fall into schizophrenia, had just proved that there was an optimal solution to games played by many people in which interests were overlapping, not just diametrically opposed. It was an important extension of the "minimax theorem," but was contradicted by the prisoner's dilemma. Dresher and Flood figured that either Nash or von Neumann would solve the paradox. Neither ever did. The conflict between individual and collective rationality was real.[59]

· · ·

War or Peace, the Individual or the Collective: Where had and would true "goodness" come from? As always, man and animal, civilization and the wild, were helplessly entangled. New vocabularies had been developed by game theorists, economists, and ecologists: "integration," "regulation," "optimization," "homeostasis," "group selection," "efficiency." Each offered confident prescriptions. And yet the hard questions still remained: What was the natural state? Was it noble? Should it be followed? How and, ultimately, why? The battle of the nineteenth-century gladiators had not yet been decided. Huxley and Kropotkin's legacy was alive.

In February 1946 Allee had wheeled himself by mistake into an open elevator shaft, landed on his head, and cracked his skull. Soft spoken and gentle before the accident, he became domineering and tempestuous. As he recovered and returned to the lab, the world outside grew ominous: Capitalism and communism were locked in battle;

the threat of thermonuclear destruction loomed; prospects of world government and peace seemed vanishing. Naturalizing ethics, too, felt more dubious than previously suspected. For a peaceful integrator the "superorganism" now looked more and more like a monster: Wasn't democracy, after all, about individual autonomy and freedom?

Quaker biology was a farce, "integration" a bogey. However much Allee would have wanted him to do so, man could not simply become a planarian. "It is fine for you to say that the study of animal population problems is the key to establishing the peace of the world," a reply to one of his grant proposals to the National Research Council now read. "If you could prove that, there ought to be loads of money to help you do the work. But as it stands now, there seem to be too many links in the chain of reasoning connecting research in animal population and the peace of the world."[60]

When he reached retirement age from the University of Chicago, Allee moved to Florida, and on March 18, 1955, succumbed to a kidney infection. At the funeral someone said that by showing that cooperation could arise between unrelated organisms, he had brought the greatest word from science since Darwin.[61] Shuffling their feet, loving mourners and even Friends tried to feel encouraged. But as they looked at the world around them an unmistakable glint of doubt had sneaked into their eyes.

Just a year after Allee's death and his own Senate confirmation hearing, von Neumann was invited to the White House to receive the Medal of Freedom. As always he was dapper with a white handkerchief in the breast pocket of his dark suit and a shiny war medal on his lapel. But he was not well. Shaking President Eisenhower's hand from his wheelchair, he mentioned how he wished he could be around long enough to deserve the honor. "Oh yes, you will be with us for a long time," the president replied, adding, kindheartedly, "we need you."[62]

The golden age at RAND had passed. Real problems, people were now saying, were simply too messy to be solved in a matrix.[63] Science had not been a panacea after all. It had failed to deliver human nature.

Von Neumann was dying of bone cancer. As his body deteriorated, he began to lose his mind. In a hospital bed he mumbled nonsense in Hungarian. At night terror-filled screams echoed from his room throughout the ward: Dementia had set in. To prevent secrets from being accidentally divulged, air force personnel with special security clearance were stationed outside his door.

His brother Michael was at his bedside, reading Goethe's *Faust* to him in the original German. Michael paused to turn the page. His eyes closed, von Neumann whispered the next few lines from memory. He died the next day, on February 8, 1957, convinced, as Orwell put it, of the "bottomless selfishness" of mankind.[64]

Researcher Devotes His Life to Science

By JOHN EMMERICH
Minneapolis Tribune Staff Writer

Dr. George R. Price looks more like a crew-cut college youth than a chemist and writer whose articles have caused a stir in the world of science.

Price, 34, is a research associate in the University of Minnesota d e p a r t m e n t of medicine.

He received his B.S. degree from the University of Chicago in 1943 and his Ph.D. degree in 1946.

WHILE STILL working on his doctor's degree he was a researcher on the Manhattan project, the program which led to the wartime atomic-bomb development. Later he was a full-time project consultant.

Before joining the University of Minnesota in 1950, he was a chemistry instructor at Harvard u n i v e r s i t y and a transistor researcher at Bell Telephone laboratories in New Jersey.

In August 1955 Price attracted nationwide attention when he debunked the theory of extrasensory perception in an article published in Science, a monthly magazine of the American Association for the Advancement of Science.

‎ **HE DREW** counter-attacks

from both sides of the Atlantic when he challenged believers in mental telepathy to show "just one experiment that does not have to be accepted simply on a basis of faith in human honesty."

The issue inspired letters from scientists and psychologists from some of the nation's leading universities.

Price's interest in science started when he was a youngster. He was the son of the head of a stage-lighting company which patented some 25 electrical inventions.

A New Yorker, Price in his teens attended a public high school which specialized in science training. Later he attended Harvard one year before transferring to the University of Chicago.

THE BLOND, bespectacled chemist speaks in the precise manner of a scientist. He lives alone in a two-bedroom apartment at 516 University avenue SE.

In his present job, Price has concentrated on an instrument for studying fluorescent spectra. For the past two years he has been rebuilding a complex instrument which

he said is the only one of its kind in existence.

Price began thinking about his Design machine some 10 years ago.

"I was new at the game and I was impressed by how much of a researcher's time was taken up in drafting work," he explained.

"A FAIRLY intelligent ma-

chinist can make a simple piece of equipment from a rough sketch faster than the time a draftsman requires to finish his drawing."

Price's answer is his Design-machine system. He anticipates something on the order will be achieved in the next few years. Within five years, he said, his system can be in operation.

MINNEAPOLIS TRIBUNE PHOTO BY DON BLACK

DR. GEORGE R. PRICE
Researcher shows how to speed up invention

"Dr. George R. Price—Researcher shows how to speed up invention," from the Minneapolis Sunday Tribune, *January 20, 1957*

6

Hustling

The American Society for Psychical Research had been founded in Boston in 1885, just three years after its mother society in London. William James, Harvard philosopher and psychologist, brother of the novelist Henry, and one of the city's most illustrious sons, was a proud patron; the scientific study of so-called psychic or paranormal behavior was the society's mandate; and its validation and broadcast its spur. Astonishing feats of levitation, clairvoyance, and telepathy had captured the American imagination. Bedazzled journalists reported from chiaroscuro inner sanctums on "materializations," or the appearance from thin air of lost brooches, misplaced wills, hidden family heirlooms. There were "veridical" apparitions, "crystal visions," and "hallucinogenic trances." Was all this for real? people wanted to know; and could science somehow explain it?[1] Two men, Joseph Rhine and Samuel Soal, would be the ones who would provide the answers.

Even though the society's own days were short-lived, the supernatural continued to gnaw at the nation's mind. In 1911 Stanford University became the first major academic institution in America to pick up the challenge, followed by Duke in 1930. It was there, in Durham, North Carolina, that a former preministerial hopeful who had seen the light of science, abandoned theology, and in turn been disappointed by materialism, turned to the enigma of "psi." Joseph B. Rhine had flip-flopped from faith in miracles to faith in physics to faith in something science could not account for. But amid these acrobatics one thing now seemed clear: Parapsychology was real.[2]

Across the Atlantic in England, a first-class mathematician from Queen Mary College became interested in communication with the departed. Samuel Soal's brother had died in the war, and like many grieving loved ones he turned to the mediums. Impressed by a particular instance of telepathy, he wrote a long entry on "spiritualism" for the *Encyclopedia of the Occult*. But Soal had exacting scientific standards, and moved methodically to test them. More than 128,000 card-guessing trials with 160 participants later, skepticism had emerged the victor. ESP, he reluctantly but also mockingly now pronounced, was "miraculously" an American phenomenon. Rhine was by this time the doyen of parapsychology, celebrated author of the best selling *New Frontiers of the Mind*. Crushingly, he soon became the butt of Soal's relentless ridicule.[3]

But then, in 1939, Soal took a second look at his old data, and what he found left his mouth dry and jaw dangling. Refusing to believe what his eyes had witnessed, he set up the most meticulous ESP experiments ever performed. The results, he now claimed, proved beyond a shadow of imaginable doubt that precognition and telepathy were bona fide. Two individuals, the celebrated London portrait photographer Basil Shackleton and a Mrs. Gloria Stewart, had beaten the odds against chance by enormous margins. Even when sender and receiver were miles apart, Shackelton and Stewart could predict future card picks. Statistics couldn't lie nor, Soal claimed, could twenty-one prominent observers. Whatever the explanation, whatever the device, the regular laws of physics had been fabulously violated.[4]

On both sides of the Atlantic believers finally got what they had asked for: a foolproof corroboration of the miraculous. Rhine and Soal were neither quacks nor impostors nor swindlers nor cheats; they were respected members of the scientific community. By 1955 their two-punch combination had entirely silenced the opposition, or so, at least, they believed. To anyone following the proceedings, the chilling implications were plain: Modern science would need to come up with an explanation; if it couldn't, its entire edifice would collapse.

. . .

Meanwhile in New York City Alice Price was communicating with the dead. "My dear dear hardworking wife" she wrote to herself from her husband who had died twenty-five years earlier. "I am so sorry that you must go through this awful struggle for your daily bread but soon your worst will be over." Another letter, addressed to "Dear Friend," pledged intervention on an impending Display deal: "Alice is under such strain, and I am doing all I can to influence everyone connected to the deal. . . . Tell her that I am working overtime to bring things around as they should be. Sincerely, W.E. Price." A third communication promised salvation: "I am sure you will be rewarded soon for your patience and Christian spirit." it said in scribbled Scripture, and ended: "I am with you always, your Billy."[5]

Back in Minnesota the polio had left its mark on George: a limp that unsettled his gait and a right shoulder that forced him to bring his cup to his mouth southpaw to avoid completely soiling his face. He was back in his student quarters, alone, in the Minnesota winter. Most days he stayed at home. At the lab there was plenty of work on porphyrins, and a new project constructing a mechanical heart-lung. But he had lost any real interest. His heart was somewhere else. His work was so technical that only a handful of people would ever read it. The more he hibernated, the more he craved an audience, the more he wanted to write about things that people cared about. And so, limping and brooding and altogether searching; trying to get his body to the bathroom and his cup to his lips, George came to the rescue of modern science.[6]

Or maybe it was to get Alice to stop writing those letters to herself.

Whether he acted out of filial concern or gallant scientism, one thing was clear to George: Rhine and Scal were frauds. There were no two ways about it. Espionage agencies knew it. Earthquake watchers knew it. Houdini knew it. Even the great dead Scottish philosopher David Hume knew it, all the way beneath his Calton Hill tombstone. For if parapsychology were real, secret messages could be teleported by agents from the Kremlin. Catastrophes could be averted. Magic could be performed without trickery. If Rhine and Soal were right it meant that knavery was less probable than miracles, a possibility that Hume had found highly unlikely. Growing up with Alice, George had believed in ESP. He had even written to Rhine as a young undergraduate from Harvard to suggest clever ways to help prove it. But gradually, with science, incredulity had replaced faith, and for years now he had been internally fuming. "Is it more probable," Tom Paine had asked in his *The Age of Reason*, "that nature should go out of her course, or that a man should tell a lie?" To George the answer was obvious.[7]

And so, in the pages of *Science*, for all the world to see, he suggested six ways in which Soal could have cheated. Rejecting the peddled notion that parapsychology and science were compatible, he demanded "not 1000 experiments with 10 million trials and by 100 separate investigators giving total odds against chance of 10^{1000} to 1." What George Price wanted was "just one good experiment"; one convincing experiment that didn't have to be accepted "simply on a basis of faith in human honesty." The essence of science was *mechanism*. The essence of magic was *animism*. Until Rhine and Soal could show a mechanism to explain their findings, George would not be impressed. And, he hoped, all thinking people, too, would withhold belief in such pabulum.[8]

Who this George Price was no one quite knew, but he sure had excited a furor. In an exposé in *Esquire*, Aldous Huxley, the grandson of Darwin's "bulldog," called it "almost unique as a piece of bad manners." Lambasting the author's "fetish for facts" and his shamanlike belief in his "favorite metaphysical hypothesis," Huxley churlishly apologized that the human mind wasn't as tidy as the physicist's "molecules." Was the essence of science really mechanism and nothing more? "No date, no qualifications of any kind—just a flat statement of the Eternal Truth by direct wire from Mount Sinai to the University of Minnesota." If Price was after repeatability and would not acknowledge ESP without it, then why acknowledge Bach or Shakespeare or

Wordsworth? After all, such men had beaten all odds against chance, and even *their* brilliance couldn't be summoned at a coin drop.[9]

The muckraking writer Upton Sinclair, too, was unenlightened by George's diatribe. Arriving in Chicago at the turn of the century, he had exclaimed: "Hello! I'm Upton Sinclair, and I'm here to write the *Uncle Tom's Cabin* of the Labor Movement!" His classic study of the corruption of the meatpacking industry, *The Jungle*, had stunned America and won him a Pulitzer Prize. But Sinclair himself was most stunned by his wife's clairvoyant abilities, powers that became apparent when she sensed Jack London's impending suicide from afar. In *Mental Radio* from 1929, he and his wife described three hundred carefully controlled experiments in which she had guessed what doodle he had placed in an envelope in another room. The book was such a hit that it played a role in Duke University's creation of Rhine's department, and even received a preface in its German edition from Albert Einstein. Now nearly an octogenarian, Sinclair wrote to Price excitedly challenging him to explain *that*![10]

Thousands of readers who had seen the write-up in the *New York Times*, wrote to express their thanks, advice, or outrage. A reverend from Vallejo, California, reminded George politely that "there are many things in heaven and earth that scientists do not know." A woman from France animatedly shared how her dead husband teleported which kind of spaghetti sauce to make for dinner each night. Another, from New Haven, Connecticut, puzzled over how it was that she had performed Rhine's experiment on pigs and gotten the same results as he had in humans. Mr. Chalmers of Chalmers Oil Burning Company in Chicago suggested how cards could be rigged at their edges ("Thanks a lot!" George replied). And Fern Clarke from Los Angeles wondered why George "could not see and talk to God," and then offered her complete psychological evaluation ("Thank you," he replied kindly, "but your guesses about me are not particularly accurate").[11]

The public reaction was so great that *Science* decided to dedicate its next issue to rebuttal in the winter of 1956. Here Rhine and Soal and even some Minnesota colleagues came at George like clairvoyants after a scrap of the future. The editorial had called for "skepticism . . . on both sides of the argument," but Soal found George "grossly unfair," and his Minnesota colleagues deemed his attack "pointless" and "irresponsible." Could any one really believe that respectable scientists were mere mountebanks and swindlers? Price had offered no shred of evidence. His unlikely "act," Rhine suggested, must be a deliberate undertaking to sell parapsychology to the public in the guise of a slanderous critique. After all, George had done parapsychology an unheard of service: "Yes, either the present mechanistic theory of man *is* wrong—that is, fundamentally incomplete—or, of course, the parapsychologists *are* all utterly mistaken. One of these opponents is wrong; take it, now, from the pages of *Science!*" [12]

Only Harvard's emeritus professor of physics, the Nobel Laureate Percy Williams Bridgman, expressed any doubt about the claimants. "The paradox inherent in the application of a probability calculation to any concrete situation," he wrote in a dry academic demeanor, "is well brought out by Bertrand Russell, who remarked that we encounter a miracle every time we read the license number of a passing automobile." If a calculation had been made for that happening, the chances against odds would be overwhelming. "Probability" was a confused concept. Until it was untangled Bridgman would pass. [13]

George was unperturbed. He had the uttermost respect for Bridgman, but his probability argument didn't provide an escape from having to choose between ESP or fraud. Human psychology was a strange and curious beast: However he detested the thought of unpredictability, man would rather believe in a suspension of natural law than countenance the possibility of deception. A strange mixture of credulity and incredulity was our lot, but it needn't take over our reason. "Where is the definitive experiment?" George stubbornly demanded. Nothing else would satisfy him. [14]

Preparations were being made. Born Orlando Carmelo Scarnecchia, John Scarne had come a long way since a local shark taught him three-card monte on the streets of Fairview, New Jersey. He was now America's most famous magician and authority on gambling, and many times over a millionaire. His signature trick, "Scarne's Aces," was a dazzler, and his "Triple Coincidence," too. Sure, he would be glad to take part in George's challenge. In fact he would even pay for flying *the* expert antitrickster, the Argentine Ricardo Musso, all the way from Buenos Aires for the event.[15]

It was just the kind of attention for which an awkward outsider yearned. First the bigwigs at the Manhattan Project, then Bardeen and Shockley at Bell, now Bridgman and also his hero, Claude Shannon, to whom George had written for his thoughts about "Science and the Supernatural" (Shannon replied that if ESP were real, it would "undermine everything").[16] Aldous Huxley, Upton Sinclair, Albert Einstein. . . . His name was right up there with the big ones. The "definitive experiment" would make him famous.

It never happened. There was the business of translation problems of Ricardo Musso's excited letters from Argentina. There was an old porphyrin paper to get off. There was fatigue. And then there were Julia and the kids. "All I can say," George wrote to his buddy Al in their usual oddball humor, "is that I got up to a 5 mile run (run?), and last spring I did 140 in a clean and jerk. Oh yes, also I was divorced January 20."[17] The "definitive experiment" had been killed before it was born, a victim of alimony, inertia, and Babel. In truth these were all just excuses. George never quite finished what he started, and he knew it. Besides, his heart was already somewhere else.

.   .   .

In November 1952 America had obliterated a Pacific island with an H-bomb. A Central Asian desert was rocked by a Soviet trial just nine months later. By that time the number of CIA agents was ten times greater than it had been only three years earlier, and the budget for secret activities had grown from $4.7 to $82 million. "You have a row of dominos set up," President Eisenhower waxed metaphorically in the spring of 1954. "You knock over the first one, and . . . the last one will go over very quickly." The Cold War race for the allegiance of the unaffiliated nations was well into its blistering noon. George Price had already come to the rescue of modern science. Now, from his cramped little apartment on Sixth Street in South East Minneapolis, he was hatching a plan to save the world.[18]

It started with economics. He had been wondering whether poverty leads to communism, as many were claiming. Since he could find no negative correlations between per capita income in different countries and the degree of sympathy toward communism, it didn't look as if it did, after all. Communism, he thought, was rather usually brought about by professional agitators, and agitation flourished under conditions of unhappiness. Correlations between poverty and unhappiness were known to be weak, but George thought that if they could somehow be measured, unhappiness and communism would track. What America needed to do was to tell this to the world, especially to those "domino" nations dangling precariously in the balance.[19]

With no training in the field, but with confidence that hardly disclosed it, he had sent an article on the matter to *Science*. Miraculously his demolition of a Marxist theory by an economist named George Altman was accepted for publication. Altman had argued that since markets ebb and flow according to fixed laws of capacity, governments needed to intervene and take control; socialism was the only answer to the laws of economics. George disagreed. "To try to reduce economic cycles to a simple question of 'too much capital' or 'too little capital,' is like trying to explain all of chemistry in terms of the four elements of the alchemists." He proceeded to review Altman's calculations one by one with a steel-trap logic. Economic booms and busts were not the result of investment above some imagined "capacity of the economy," a function of a Malthusian "ecological law of nature." In free economies they were rather always the sum result of decisions of entrepreneurs based on expectations of gain or loss. And expectations themselves were determined by all sorts of things, ranging from sophisticated mathematical models to—George knew all too well—"communication from the spirit world." Economic behavior was complicated; no single-cause mathematical model stood a chance to be of any value. It was high time, he thought, to approach such problems with new tools. And since economics was still at a developmental stage from which the natural sciences had largely emerged, George suggested that it would be worthwhile to see what contributions the natural sciences could make to economics. Maybe there were deeper natural laws pertaining to behavior. "Perhaps the time approaches," he wrote rather mysteriously, "for a new Boyle to produce a *Skeptical Economist*."[20]

Was he talking about himself?

Next he turned to the second part of the plan: research and development. Possession of a superior economic system would not suffice to win the Cold War, nor even would a deeper understanding of human nature. What was needed further was superiority in technology, the means by which to produce faster and better and more. Now, in "How to Speed Up Invention" in *Fortune* magazine, George presented the answer.

He called it the "Design Machine." Certainly, industry needed to plot "optimum strategies," to set "impossible goals," and to reward engineers more handsomely. But the Design Machine would be the true panacea. A machine to take over the mathematical and mechanical operations of the drafting department and model shop, it would revolutionize American industry. How it worked was simple:

> *An engineer will first describe the shape of a mechanical part, intro-*
> *ducing this information quickly by pressing keys and moving levers.*
> *The machine then translates this into its own internal mathematical*
> *language, and within a few seconds presents to the operator a stereo-*
> *scopic picture of the part viewed from any direction specified. Or,*
> *within minutes, it will machine the part from metal.*[21]

It was a system for dealing with models—models constructed out of mathematical equations stored in the computer memory—and nothing like it yet existed. If marshaled on a rational scale, it could become a repository for all the design and engineering information able to be programmed. And though intended for mechanical design, analogs for electronics and for chemistry could easily be imagined. Finally, here was an idea to make his old teachers at Stuyvesant proud.

Shown the proposal, a leading computer expert was skeptical. In reply George quickly prepared a seventy-five-page, single-spaced supplementary memorandum showing how an IBM 704 computer could be incorporated into a Design Machine, how a complex part could be described to the machine, and how the machine could display the part—in 3D. The skeptic conceded, and other IBM experts did, too. Not only could it be done, but it could be done in three to five years for less than 5 million dollars. The Russians, with their Bison bomber, took only four years to go through the eight-year development cycle that Americans needed for the B-52. "The tempo of U.S. technological progress," George wrote, "is not an academic matter." Time was of the essence. No less than the "fate of the non-communist world" was at stake.[22]

George was growing nervous. The "Reds" had just launched Sputnik II, and marched machines and men in an awesome celebration in Moscow to mark forty years since the Bolshevik revolution. Russia was gaining on America. In a strongly worded essay in *Life* magazine titled "Arguing the Case for Being Panicky," he now detailed the precise steps by which the United States would become a member of the USSR by 1975—if it didn't wake up and smell the kvass. Americans were like the people in the Hans Christian Andersen tale "who stood and watched their emperor parade naked though the streets, and then turned to one another to praise the beauty of his clothing." America was like Babylon, Baghdad, Constantinople, and Rome: the rich, proud, luxuriant nation, smugly confident and dangerously oblivious to the "tough barbarian adversary, poorly provisioned and shabbily dressed but high spirited and strong in its drive to conquer."[23]

The article appeared interspersed by a full-page ad from Bell touting the "Seven Ages of the Telephone." There was a photo of a smiling mom holding the receiver to her blond baby's ear: "Hello, Daddy!" the caption read. Another, of a "Dynamic Teen" resting on a sofa with plaid skirt and varsity letter, obviated the footer: "Girls talk to girls. And boy talks to girl. And there are two happy hearts when she says, 'I'd love to go!'" And what about "Just Married," with a brunette in an office chair with phone and adoring hubby reclining above her?: "Two starry-eyed young people starting a new life together. The telephone, which is so much a part of courtship, is also a big help in all the marriage plans."

What more did Americans want? George asked, indignant: "A Cadillac? A color television? Lower income taxes?—or to live in freedom?" Would it be luxury or liberty? The World Series or the Nobel Prize? And who would play the part of a Franklin or a Hamilton? "Optimism talks" felt good but were confusing. If America didn't double its defense budget now, it wouldn't be long before it became a Soviet province.[24]

. . .

He had parachuted from nowhere to the center of a debate about the foundations of science. He had jumped into the fray over world economics. He had invented a "Design Machine." And now he had used a premiere stage to warn of impending national disaster. With his crew cut, steel-framed round clear-rims, pursed lips, and bow tie, the unknown thirty-four-year-old from Minnesota cut an original figure. Whatever you said of him, he was hustling.

Still, there was a weird duality to these disparate interventions: What seemed like genuine concern for the welfare of America and the world also had the panatela reek of egotistic smugness. Was he a cocky chemist? A sober economist? A restless engineer? A prophet? Somehow George Price was simultaneously all of these—and none.

Whether this was altruism, patriotism, or diarrheic self-promotion, people were beginning to take notice. "We are proud of you," Minnesota senator Hubert Humphrey wrote to him following a full-spread write-up of the Design Machine in the *Minneapolis Sunday Tribune*. It was a welcome piece of encouragement: George was supporting two apartments, two cars, two telephones, and two attorneys. It was freezing. There were no women to meet, and he was sex starved. His shoulder still bothered him. Already he considered himself an "ex-chemist," yet this was still his job. No, he wrote to Al, it wasn't due to his usual "masochism" that he had yet to leave Minnesota; it was due to his reluctance to work as a chemist, and others' reluctance to employ him "as a physicist, economist, writer, or anything else for which I have little training."[25]

Finally he quit, in the winter of 1957, leaving porphyrins, Schwartz, a steady salary, and his daughters, who had since moved with their mother to Marquette, all behind. He was moving to Kingston, New York. A respected researcher at one of America's premier hospitals, he would now become a rather low-rung subcontract technical "reviewer," working for Stevens Engineering Company on instruction manuals for IBM computers.

George brushed aside accusations of self-destruction. Whether others believed in his reality meter or not, he was on his way to turning his fortunes around. He hadn't yet secured a "big" job, but prospects seemed encouraging. From IBM's headquarters at 590 Madison Avenue in New York City, Emanuel Piore had expressed interest in his *Fortune* Design Machine. A Jewish immigrant from Lithuania, Piore had risen to become the navy's top-ranking scientist, winning its Distinguished Civilian Service Award and serving on President Eisenhower's Science Advisory Committee. As IBM's director of research, he was leading the corporation into the era of digital computers. Would George mind coming down to the offices, he wrote to him, to discuss some of his ideas?[26]

The day before the meeting with Piore, on July 15, 1957, an old girlfriend from the time he was breaking up with Julia appeared rather suddenly in town. Her name was Anne, she was from the Midwest, and, like Julia, was a Roman Catholic. He had thought of marrying her before he fell ill with polio, but she had broken it off for another man. Now, she made it clear, she was once again for the taking. Still jealous, George said he'd think about it.

The next day he settled into a large camel-colored sofa in a plush office on the twenty-third floor of the IBM Building. It was still premature for the corporation to start developing the Design Machine, Piore told him, smiling. But if he was interested, George and his imaginative ideas would be welcome at Research and Development. This was quite an offer to a subcontracting technical reviewer, and from the director of research, no less. But unknown to Piore, George had already contacted a fancy lawyer from midtown to inquire about a patent. If Piore was not interested in developing his machine, George wasn't going to bite. After all, if he joined IBM he wouldn't be able to work on a private patent, and millions of dollars were at stake. They parted with a friendly handshake. George had to run to make his vacation flight to Puerto Rico—his first trip ever outside America.[27]

The following week George went down to the train station to pick up his girls, who were living now with their mother in Washtenaw County, Michigan. Julia was a frustrated woman who "never saw the beauty around her." She considered her marriage to George "unlucky"; after all, she'd been a med-school prospect, and worked on the Manhattan Project—now she was a third-grade teacher. These days, when their mother's dull schoolteacher friends came over for coffee and cookies, she demanded good manners and hushed voices of her daughters, and, under no circumstances, any talk of Daddy. When Annamarie and Kathleen refused to go to church she gave them her untempered piece of mind. There were rants about Daddy leaving because "they were so awful." Worst of all, there were Aunt Edith's hideous boiled dinners, followed by dreaded stewed prunes. Life was not exciting.[28]

New York was a far cry from Ypsilanti. Quirky and fun and just the opposite of in-lockstep, George was showing the girls the big world. Carefree and boundless, they went for hot dogs and ice creams, climbed the Statue of Liberty, visited Niagara Falls, and listened to Johnnie and Joe's one-hit wonder, "Over the Mountain, Across the Sea." And then there was Yul Brynner starring in *The King and I* on Broadway. When George waved good-bye at the end of the week as the train pulled away from Grand Central, a crying nine-year-old Annamarie and eight-year-old Kathleen couldn't have known that it would be one of the last times they'd spend a week with their father.

He was considering proposing to Anne that week, when he received a notice of termination from Stevens. IBM was cutting ties to subcontractors, and he would need to leave his office by Friday. Meanwhile, the Patent, Trade Marks and Copyrights division at the law offices of William R. Lieberman at 551 Fifth Avenue wrote to explain that a patent needed to be applied for before November, when the *Fortune* piece would turn public domain. Since he didn't have the money for this or even a prospect that could promise collateral, it began to sink in with George that his patent was slipping away. Frantic, he wrote to Fiore, asking to be considered again for the IBM job. But Piore was on a month's vacation, and his replacement showed no interest in a fired subcontractor technical editor of whom he had never heard. George had made up his mind by this time—he wanted to marry Anne. But he was out of work and in debt, and she was back in the Midwest and drifting. If only he had proposed to her that day before he met Piore: Surely he would have been focused on finding a stable job then, and grabbed the offer handed to him so generously by the director. It was on that fateful day, he would later claim—July 15, 1957—that his downward spiral began.[29]

John Maynard Smith (1920–2004)

William D. Hamilton (1936–2000)

7

Solutions

*J*BS was in a London hospital bed. Rectal carcinoma, the doc-tors said. He was back from India to get the very best treatment available, though the "auto-obituary" he taped, sitting draped in a sari, wasn't a very good sign. Haldane was about to leave this earth, and he knew it. Meanwhile he asked his student John Maynard Smith to go out to the bookshop to get him something to read.[1]

It wasn't easy for Maynard Smith to see his mentor in such a state. He was born in 1920 in Wimple Street, London, and childhood had been a lonely affair. He was eight when his stern and distant ex-military physician father died; an absent mother provided little comfort beyond the winter home in Berkshire and summer home in Exmoor, the means for which her wealthy Edinburgh stockbroker family could afford. Endless hours watching birds in the countryside confirmed both his love for nature and his isolation.

At Eton, by way of lore, he soon learned that there was one graduate who had attracted the particular hatred of some of his teachers by betraying his class and religion. There was a noxious blend of privilege and prejudice at the school, but to its credit

J. B. S. Haldane's writings were in the library, and seeking them out, John was captivated. Haldane's blend of atheism and reason, he thought, "never left you wallowing in a sense of misty profundity." Almost naturally, scientific and political commitments blended: John requested *Capital* for a school prize he had won, delved into mathematics, and made peace with the absence of God in his life. He would be a "puzzle-solver," he hoped, and a socialist. Shirking his maternal grandfather's wish that he join the family's stockbroker firm, he read engineering at Trinity College, Cambridge, and joined the Communist Party in 1939.[2]

After spending the war years making stress calculations for Miles Aircraft near Reading, John was ready for a change. Poor eyesight meant he'd never be able to fly the planes he designed, so he could never really love them either. As for politics, it was either that or science; one's heart could be in two places, perhaps, but the brain was more demanding. Since theoretical physics seemed too difficult and chemistry a chore, it was to biology and his old love for nature that Maynard Smith now turned. When he discovered in October 1947 that his hero Haldane was professor of genetics at UCL, he applied forthwith. "Dear Comrade," his letter began,

> My interest is mainly in evolution and genetics. My main existing qualification is that I am a competent mathematician, and in so far as I have shown any ability as an engineer, it has been in expressing physical problems in terms of mathematics. I read Huxley's "Evolution: the New Synthesis," and there seemed to be plenty of scope for a mathematical approach to the subject of natural selection, the origin of species, and so on.[3]

Whether he was in a good mood that day or genuinely impressed, JBS fired back a letter of acceptance. It proved a smart decision: As an aircraft engineer John had learned to trust models and the necessary simplifications they demanded. After all, if RAF fighter planes could stay in the sky even though dry calculations on land assumed incompressible air (refuted by the pneumatic tires and inflated life jackets), simplifying assumptions could be valuable.[4] Of course, JBS had known this ever since he set foot in biology.

Maynard Smith made his way to the bookshop. Please don't die, he thought. Not yet.

. . .

He returned from Dillon's with a big fat book, *Animal Dispersion in Relation to Social Behaviour*, by an author with a long name. The son of the headmaster of Leeds Grammar School, who retired to become rector of the picturesque country parish of Kirkland in the Vale of York, Vero Cope Wynne-Edwards had grown up chasing rabbits in the Pennine hills. At Rugby his buddies called him "Wynne" and together, he'd remember, they "collected plants and Lepidoptera, found birds' nests, hunted for fossils in the local cement pits, 'fished' in ponds for aquatic life, made drawings of 'scratch dials' on medieval church walls, and 'excavated' for pottery in a Roman camp on Watling Street."[5] By the time he arrived as an undergraduate at New College, Oxford, in 1924, his heart was set on zoology. His teachers were giants: E. S. Goodrich for comparative anatomy, Gavin de Beer for experimental embryology, E. B. Ford for genetics, Julian Huxley for general zoology, Charles Elton for ecology. When Huxley left for King's College, London, in 1925, it was Elton who loomed large in his education. Elton was now systematically trapping voles and wood mice in Bagley Wood near Oxford, trying to figure out their dynamics. Does the size of wild populations of animals fluctuate in a periodic cycle, he wanted to know, and if so, what was the cause?[5]

Elton saw that size did fluctuate, but he never solved the mystery: Was it due to weather, the food supply, migration, predator-prey oscillations? No one seemed to know. It was just around then that he gave Wynne-Edwards a book to read, *The Population Problem*. Written in 1922 by a former student of Huxley's who would go on to become the director of the London School of Economics and be knighted for public service, it made a revolutionary claim. Contrary to Malthus, Alexander Carr-Saunders argued, humans could do without disaster. Neither plagues nor war nor famine nor any form of "natural corrective" was necessary to provide the perfect fit between what the world can feed and the number of hungry mouths. When there was plenty, man procreated generously; when there was dearth, he procreated less. Neither a slave to the elements nor a victim of the earth, he was perfectly attuned to his environment. This much evolution had taught us: The primitive tribes that survived into modernity were exercising population control. And density was always at its optimum.[7]

Forty years later, when Wynne-Edwards was writing the book Maynard Smith now brought to the dying Haldane, he suddenly understood. He was a professor at Aberdeen, an Englishman with a Welsh name who had lived half his life in Scotland. Like Kropotkin, he had made expeditions to northern lands, where the harsh elements had driven animals to cooperate.[8] If competition existed in nature, the Arctic taught him, it was directed at the environment, and animals had developed a myriad of social mechanisms to cope. Already in 1937, on the Baffin Islands' coastline, he observed that only between one-third and two-fifths of the fulmars in the breeding colony mated while the rest were pushed into marginal territories and often died.[9] What a clever way to prevent overexploitation! Even more ingenious was the chorus of singing accompanying each breeding season: Short of using a calculator, it was the best way for the flock to assess its size and, surveying its resources, reproduce accordingly. Now, in 1962, as he sat down to explain such phenomena after years of thought, he fathomed his debt to Carr-Saunders. Birds, just like primitive man, were regulating their numbers.[10]

It was an idea, some thought, that flew smack in the face of Darwinism. Man might exercise birth control, but birds? Surely their brains were no match for the inexorable natural imperative to procreate, the ultimate arbiter of the survival of the strong. In the 1930s Julian Huxley had sent another of his students to the tropics to study Darwin's old finches. What David Lack saw there was that competition for food was rampant, but that slight differences between geographically isolated groups might reduce it.[11] Each group of finches specializing in a particular food in a particular habitat made for a wonderful mechanism to get out of the others' way; gradually, genetically, the populations became distinct. Nature, in other words, would go to great lengths to avert conflict, even to the end of divining new species. But as powerful as the force of competition, more powerful—since more fundamental—was the instinct to procreate. Individuals were out to maximize their fitness, to sire just as much as they could. The idea of altruistic birds passing up a number for the greater good made absolutely no sense.[12]

Unless, of course, natural selection was operating on the group, which was exactly what Wynne-Edwards was arguing. Drawing on Wright's model of group selection, Wheeler's superorganism, Emerson's homeostasis, and Allee's fowl hierarchies, he made a nonmathematical case for the collective. Individuality was important but subordinate: When the physiology of the singleton came up against the "viability and survival of the stock or race as a whole," group selection was bound to be the victor and individual reproductive restraint the result. It was just like with fishing: If every fisherman set his net to catch just as many fish as he could, the village folk would quickly find themselves "entering a spiral of diminishing returns." If, however, an agreement was reached over the maximum catch for each, depletion of the villagers' maritime food source could be happily avoided.[13] As with fishermen and their catch, so with fulmars (and red grouse and many other birds) and their environment: To prevent exhausting limited resources, numbers could be regulated by social convention for the benefit of all.[14]

Wynne-Edwards was certain that he was walking in the path of a giant. Darwin had translated Malthus back into nature as he had translated Carr-Saunders. Darwin had used the analogy of artificial selection to explain natural selection as he had used fishing to explain population regulation. But Darwin had also written:

> *Whatever the cause may be of each slight difference in the offspring from their parents—and a cause for each must exist—it is the steady accumulation, through natural selection, of such differences,* when beneficial to the individual, *that gives rise to all the more modifications of structure, by which the innumerable beings on the face of this earth are enabled to struggle with each other, and the best adapted to survive.*[15]

Now Wynne-Edwards felt, after more than a century, that he'd understood what even the great master had failed to fathom: Adaptations work for the good of populations, not of individuals.

On his deathbed JBS just chuckled. "Smith, do you know what this book says?" he asked his devoted student, with his usual mischievous air:

> *Well, there are these blackcock, you see, and the males are all strutting around, and every so often, a female comes along, and one of them mates with her. And they've got this stick, and every time they mate with a female, they cut a little notch in it. And when they've cut twelve notches, if another female comes along, they say, "Now, ladies, enough is enough!"*[16]

When it became clear that the best medicine in the world wouldn't save him, Haldane returned to India. Clearly he hadn't been impressed by the biological argument for the greater good, a fact that did not stop him from dying a devoted Marxist on December 1, 1964.

. . .

Maynard Smith was less of a mule than his teacher. The Lysenko affair, the purges, and the invasion of Hungary in the fall of 1956 had been enough for him. Without giving up the sentiment, he gave up the Party, increasingly turning to model evolution and nature. It wasn't easy: "Why do theory when Haldane is sitting in the room next door?"[17] In the beginning he stuck to fly genetics. But gradually, with Haldane's move to India in 1957, John's confidence had grown. He was ready to take on evolution on his own.

John hadn't read Wynne-Edwards's book before he brought it to Haldane, but now decided to pay attention. Reviews of the tome had been mixed. Lack, of course, hated it. So did Wynne-Edwards's beloved teacher Elton who judged it "messianic" and "rather wooly."[18] Many, however, found the notion of a "balance of nature" plausible, and, more importantly, deeply relevant to man. Wynne-Edwards himself egged them along. That summer he had written an article for *Scientific American* that began: "In population growth the human

species is conspicuously out of line with the rest of the animal kingdom."[19] Man was virtually alone in showing a long-term upward
trend in numbers. It was a bright red warning sign: However highly
he thought of himself, compared to fulmars and red grouse his social
skills were retarded. Modern individual freedom, alas, had trampled
tribal homeostatic wisdom.

Just like Darwin, he was trying to bring animals and man closer
together; he had shown convincingly, an anonymous reviewer in the
Times Literary Supplement wrote, that "social life in man . . . is no
unique affair, but the culmination of a very widespread biological
phenomenon."[20] The trouble was that they were drifting apart. This
much, at least, was clear to the anxious reviewer in *The Nation*:

> *In Wynne-Edwards' proofs we can see reflected the breakdown of rela-*
> *tions between parents and children, the male's and female's dimin-*
> *ished attachment, the constant migration of peoples, the female's*
> *objection to being just a breeder, the male's resentment of being just*
> *a provider, smaller families, divorces, desertions, minorities escaping*
> *from "ghettos," elites struggling to keep out the invaders, the increase*
> *of homosexuality and neuroticism, alcohol and drugs, and above*
> *all, the evidence that young people, the group most sensitive to social*
> *stress, desire violence, especially if it is unprofitable and senseless: in*
> *all this we see that human society is reacting just as Wynne-Edwards*
> *says a crowded society should. It is giving a warning which nobody*
> *heeds; even when they see the Sunday cars jamming the highways, as*
> *in a dance of gnats, or a swarming of locusts.*[21]

Maynard Smith took a cool look at the data. Despite the vogue
of population-explosion hysteria there was no need to get excited.
Theoretically, though, if group selection worked, short of decreasing
homosexuality and clearing up traffic jams, it could be an important
mechanism in evolution.

He himself had been contemplating the phenomenon of aging.
Why, for heaven's sake, would evolution select for the degrading of
the body: Wouldn't it be better to be able to reproduce indefinitely?
The nineteenth-century giant August Weismann had considered the

conundrum and thought the answer lay with the collective: Evolution pushed individuals into old age to make room for the next generation—in the absence of unlimited resources this was surely the best solution for all.[22]

But Maynard Smith wasn't convinced by the logic of the generous-hearted doddering for the good of the replenishing tribe. In the wild, animals hardly ever made it to old age; reproduction declines with survival, not necessarily aging, since the more time lived the greater the chances to die. If natural death rarely occurs in nature, selection couldn't have produced the trait of aging. Perhaps this was a genetic story, then, rather than one of "group selection": The very same genes that helped an organism's fitness when young might with time contribute to aging, having sneaked past the watchful gaze of selection when they were being passed on to the next generation.[23]

This kind of "gene's-eye" view of evolution he had inherited from Haldane, which was perhaps why the editors of the *Journal of Theoretical Biology* had sent him a paper by Bill Hamilton that had left two other reviewers completely baffled and perplexed.

• • •

Thanks to the unusual social life of the *Hymenoptera*, "The Genetical Evolution of Social Behaviour" introduced an original idea. For the better part of the century classical population geneticists had defined "fitness" as the measure of an organism's reproductive success: The more offspring an organism sired, the greater its fitness. A corollary of this definition was that the persistence of any behavior or gene responsible for it which reduced an organism's fitness would be difficult to explain: in time, as was her mandate, natural selection should see to its demise. But Hamilton now showed that if fitness were redefined to include the progeny of relatives rather than just one's own, in many cases the Darwinian difficulty of explaining the evolution of altruistic behavior would simply disappear. The key was adopting a "gene's eye" point of view, and the *Hymenoptera* showed precisely why.

In the ancient order—comprising sawflies, wasps, bees, and ants—female workers are more closely related to their sisters than to

their offspring due to a genetic quirk in their sex-determining heredity called "haplodiploidy." All females are born of eggs fertilized by
sperm and are *diploid* (meaning that they possess a full compliment
of chromosomes from both their mother and father), whereas all
males are born of unfertilized eggs and are *haploid* (meaning that
they have no father but rather only a single set of chromosomes from
their mother). Since female sister workers share all their father's genes
and half their mother's, their coefficient of relationship is $r = 0.75$,
whereas since they pass only half their genes on to the next generation
the relationship to their offspring is $r = 0.5$. Brothers and sisters, on
the other hand, only share one common ancestor, their mother, and
are therefore only half as related as normal siblings, with $r = 0.25$.

The implications were strangely illuminating. Lazy male drones
were not moral slackers but genetic opportunists: They spent their
time seeking to mate rather than work since their relatedness to
offspring would be double their relatedness to any sisters they might
oblige. Female workers, too, were shrewd genetic calculators. Neither
shirkers of parenthood nor mindless toilers, farming the queen as a
"sister-producing machine"[24] served their interests more generously
than did siring their own brood. Everything depended on perspective: From the point of view of the gene trying to make its way into
the next generation, it made absolutely no difference in whose body
it was being carried. In one stroke, a century after Darwin, "inclusive
fitness" unveiled the mystery of the ants.[25]

But Hamilton didn't stop there; the *Hymenoptera*, after all, were
just an illustration. Trudging through pages of algebra he had come
up with an elegant equation. It sounded almost economic: Every
altruistic act—such as a rabbit thumping its legs to warn its friend of
the presence of a predator, or a howler monkey doing the same with
its call—would entail both a fitness cost to the altruist (the chance
that he might attract the predator himself), and a fitness benefit to
the receiver (the chance to live another day and produce more offspring). What Hamilton showed was that, for every social situation,
if the benefit (B), devalued by the relatedness between the two (r),
was greater than the cost (C), genes responsible for altruistic behavior
could evolve. The greater the relatedness the greater chance for altru-

ism. With the help of the mathematics of genes, rB > C formalized what Darwin and Huxley and Fisher and Haldane had all intuited: altruism was a family affair.

. . .

Boom!

Thrown to the ground, he took a moment to notice the blood soaking his shirt below the chest and oozing from his mangled hand. There was sawdust, and smoke, and the unmistakable bittersweet taste of gunpowder. He could hardly see through the haze and was growing woozy.[26]

William Donald Hamilton was born in the summer of 1936 on a small island in Egypt's Nile to Bettina, a physician, and her husband, Archibald, a military engineer, both of them New Zealanders. After they discovered that army life was neither their inclination nor priority, they'd settled in a cottage on Badgers Mount, at the rural edge of the North Downs, County Kent, England. Bettina quit medicine to raise a family—there was his sister Mary and four more children on their way, and Archibald set up a bridge design business, gaining his growing family a steady income from the prefab steel Callender-Hamilton used by army tanks to gallop across ravines in World War II.

Archibald had taken away the rusty tins filled with explosive powder Bill unearthed in a rabbit burrow a few days earlier and stashed them in a shed. But it was too little to stave off his son's curiosity. Now, just as he was clamping a powder-filled brass cartridge with a vise, it blew up. With his last ounces of strength, he dragged himself into the house and collapsed on the kitchen floor. He was twelve years old. There was shrapnel in his lungs, and the tips of three fingers were missing. He would live. Risky business was a Hamilton family trait.

"Oaklea," named after Archibald's old New Zealand dwelling, was a cramped home whose inhabitants were tinkering and practical, and "as far as possible, a self-supporting republic." You wouldn't find a plumber invited, nor a carpenter, a welder, or an electrician. Nor, for that matter, could you find a proper room; Bill usually slept on an army cot in the corner of the dining room, or, when the frosty winters gave way to the cool evening breezes of spring, in one of the many tool sheds surrounding the cottage. In a dry creek or beneath a large piece of birch bark or in a cozy burrow would do; sometimes he'd forget to come home at night from his lost-to-the-world natural explorations. Walking barefoot and oblivious down woodland paths "chasing clouded yellow butterflies through fields of red clover," Bill was happiest among plants and insects. The Hamiltons were fiercely independent, highly spiritual, unsentimental people. Bettina had a keen interest in natural history, and clearly she'd passed it on to her son.[27]

But young Bill had inherited his father's aptitude for math, too, even if he thought he was no good at it.[28] When he arrived at Cambridge he decided to read Genetics, the better to combine the two, but not before he'd requested the *Origin of Species* as a school prize.

He was following in Fisher's footsteps. Soon, accompanied by awe and trepidation, he discovered Fisher himself, white haired in carpet slippers in his room at Saint John's. The meeting between the two never developed into a relationship, however; Fisher's mind, it seemed, was by this time somewhere else. Hamilton's encounter with *The Genetical Theory of Natural Selection*, on the other hand, had been positively transformative. "This is a book," he later wrote, "I weighed as of equal importance to the entire rest of my undergraduate Cambridge BA course."[29] It was also a book not taught at Cambridge, where one of his professors representatively acknowledged Fisher's importance as a statistician but wondered out loud whether he had any qualifications in biology. In zoology, it seemed, the "good of the species" had conquered everything, another professor's book starting: "Insects do not live for themselves alone. Their lives are devoted to the survival of the species whose representatives they are . . . ," and so on. Fisher's emphasis on natural selection always working at the level of the individual was ignored. Even natural selection itself was scarcely respected anymore at Cambridge.[30]

And so Bill Hamilton did what he had always done as a boy in the woodlands: strike his own path. To his sister Mary he admitted that Fisher's mathematical reasoning was "rather beyond me," but he plugged away nonetheless, spending months in solitude on each chapter. The eugenic bits on civilizations, especially, had snagged him. "I now know why they decline, and how to stop them," he wrote to Mary. Cultural progress, he had suckled from the bosom of his hero, depended on natural selection.[31]

A convert to evolution by natural selection working to maximize individual fitness, he'd need to tackle the kind of behavior Darwin himself said would annihilate his theory completely if even one instance of it could be convincingly documented.[32] He had read Fisher on butterfly distastefulness and Haldane on drowning cousins, and yet, he'd later remember, "one reads and forgets."[33] After all, however comprehensive the two giants' treatments of evolution, they had devoted only a few paragraphs to the problem, and surprisingly not attempted a comprehensive mathematical model. Besides, treating it as a phenomenon related to Military Cross winning, as both had tended to do, seemed to confuse things. Still, clearly, Hamilton thought, the genetics of behavior was a subject relevant to animal and plant as well as man. If the geneticists at Cambridge didn't see the point, or the social anthropologists what connection it had with genetics, he would go elsewhere to pursue his puzzles. Just twenty-two, a gentle and shy loner with a shock of thick hair, workman's hands, massive jaw, and the general appearance of a Neanderthal, Bill Hamilton had found his problem: the evolution of altruism.

Already he knew: Feelings for family are not the same as for strangers. As the badgers for whom the hill of his childhood was named made famous, protecting kin was a risky but prevalent affair. Why should this be so? Pondering the question, he put in an application to become a teacher, just as Fisher had done before him. But his genetics degree, the good people at the School of Education at Moray House teachers' training program in Edinburgh informed him, would only qualify him for junior high. Cuddling the snub, he thought about becoming a carpenter.

Finally he walked into the office of Lionel Penrose, Galton Chair of Genetics at University College London. Haldane was by now working at the Indian Statistical Institute and living in an ivory tower in Orissa. No, Penrose intoned coldly to Bill, he didn't see the connection between a moral trait like altruism and genetics. Recently he'd changed the name of the *Annals of Eugenics* to the *Annals of Human Genetics*, and would have no talk of genes for human behavior. At best the genetic evolution of altruism was a waste of time; more to the point, it was pernicious. Morphology was one thing, but hadn't the Third Reich been enough?[34]

Whether Penrose really believed this, Hamilton was not certain. The times had seen a strong environmentalist backlash, but to him it all seemed maudlinly political. He had a sneaking suspicion that, as in the old *Punch* cartoon, Penrose was trapped and wishing it all to go away: "Have you heard that Mr Darwin says that we are all descendant from the ape?" one shocked Victorian lady asks the other. "Oh, my dear—that surely cannot be true! . . . But, if it should be true, let us pray that at least it will not become generally *known!*"[35]

To contemplate the biology of character was to think "the socially unthinkable"; after *Mein Kampf* and the Final Solution it was positively verboten. At stake, rather than religion in Darwin's day, was the egalitarian premise of democracy, not to mention the delicate politics of race. Penrose acknowledged a Cambridge genetics degree, and offered Hamilton work on classical fly genetics; for the evolution of altruism he'd need to go elsewhere. Luckily, unlike Penrose, Professor Norman Carrier of the Department of Human Demography at LSE seemed "quite unaware of even a possibility that I might be a sinister new sucker budding from the roots of the recently felled tree of Fascism, a shoot that was once again so daring and absurd as to juxtapose words such as 'gene' and 'behaviour' into single sentences." The friendly Carrier didn't quite grasp the project but was encouraging and helped Bill secure a studentship.[36]

He was deeply lonely. Enrolled as a doctoral student in two separate departments and institutions, he spent hardly any time at either. With no office space and virtually no one in the corridors or libraries knowing his name, he did most of his work alone in his bedsit in Chiswick, or, when the sun came out and—struggling to keep the pages from blowing away in the wind—on a bench at Kew. At UCL Penrose projected a "kind of gloomy avoidance," so the friendly but rather absent Professor Cedric Smith was assigned as his formal supervisor. At LSE Carrier was replaced by the mathematically apt but biologically innocent John Hajnal. Bill was on his own, walking the narrow bridge of obscurity. Unable to distinguish the conviction that he had seen something that others had not seen from the suspicion that he was nothing but a crank, Hamilton teetered on the brink of despair.

Some nights he'd stop on a bench at Waterloo Station, back from Holborn Public and the Senate House—the late libraries—not wanting to return to his depressing quarters. There, at least, the waiting passengers in the main hall gave him some comfort. Scribbling out his mathematical model with pencil in notebooks, Bill felt less alone.

Partly to save money, partly for exercise, and partly to conjure somehow, "below the bricks and mortar around me, the marshy fields that had once been there, the streams dividing them, and the heaths behind—to seek, in short, elemental forces underlying the city," he walked London through and through:

> *Even a lonely child on the street did not tug my heart as hard as a bracken fern when I saw it, for example, in the valley of the stream once called the Fleet. . . . I recall also the soul-piercing yellow from the flat, star-like flowers of a ragwort. What plant was this, daring to flower under such hostile cliffs, where was it from? Was it a growth form of the more crinkled and robust plant I knew so well on the Kentish fields, there crawled upon and stripped by the stripy caterpillars of the cinnabar moth? But if the leaves in London might be laxer and greener because of a lack of light or to an excess of lead in these wells, why would the flower be larger? . . . I simply imagined that, like me, the plants were longing to be away from these dark concrete canyons, back in their own countryside. Fixed in their cracks, as I thought, what else was there for them but to signal desperately with their bright petals to the rare and equally lost London bees crossing (again like me) the chasm of the Farringdon Road? Was it through such petals and then such an insect's aid that they hoped to create the seeds which an autumn storm could carry away?*[37]

Bill Hamilton felt an intimacy with the outdoors that blurred the boundaries between himself and nature. Those who knew him couldn't help feeling that this gentle giant was more comfortable in the presence of spiders than humans.

And then, from beneath the despair and reams of equation-filled notebooks, $rB > C$ appeared. Others thought the idea behind it a "solvent of a vital societal glue," but imagining himself a gene in the body of a toiling ant, Hamilton welcomed it as a "love child." Finally, after Darwin and Fisher and Haldane, not to mention Penrose, he had cracked, in his painful solitude, the mystery of kindness to others.[38]

.   .   .

The paper sent to him by the *Journal of Theoretical Biology* was brimming with mathematical equations and natural history. The name Hamilton sounded familiar from UCL, but he remembered it rather vaguely.[39] Not following the math too closely—the notation was unconventional and the typesetting confusing—Maynard Smith recommended splitting it in two. It was too long and needed revisions. Still, unlike the other two reviewers who hadn't understood it, he was impressed by Hamilton's insight. "Of course, why didn't I think of that!" he exclaimed to himself, just as T. H. Huxley had done when encountering Darwin.[40] Since he was presently writing a review of Wynne-Edwards's book, he wanted to distinguish it from group selection. After all, a family is a kind of group, and "inclusive fitness" might be interpreted as just a form of group selection. A better term was needed.

Meeting with Lack and a few of his colleagues at Oxford to discuss the annoying excitement of many biologists over *Animal Dispersion*, Maynard Smith hit the jackpot.[41] In "Kin Selection and Group Selection," published in *Nature*, he proposed a combined verbal and mathematical model showing that the conditions necessary for group selection to work were so stringent that one would be hard pressed to find them in nature.[42] A major obstacle was invasion: If amid a population of birds exercising population control suddenly appeared a bird who didn't, little could stop it from procreating. Assuming its antisocial nature—since genetic—is passed on to offspring, before long the entire population would be comprised of self-seekers.

No, group selection was not the way social behavior evolved, the "greater good" nothing but a hopeful, unworkable illusion. Kin selection, on the other hand, was a viable evolutionary mechanism.

. . .

Besides ants and monkeys and thumping rabbits there were moths. The theory predicted that camouflaged (cryptic) and brightly colored (aposematic) species should act differently immediately after siring kin. If caught by a bird, a cryptic mother would unwittingly be helping the bird learn the moth's disguise, thereby making the bird more successful in catching the now exposed younglings. An aposematic mother in a bird's beak, on the other hand, would only be doing a service to her kin: Since her colors are true indicators of a noxious toxin, the bird will have learned to stay away from similar-looking progeny. It made sense, on the "gene's-eye-view" logic of inclusive fitness, that postreproductive cryptics should die much faster than postreproductive aposematics. Hamilton was delighted when he learned from the lepidopterist A. D. Blest that postreproductive cryptics have significantly shorter life spans, but only when they are surrounded by kin. The greater good was a fiction. And happily, more natural examples of his theoretical predictions kept rolling in.

Hamilton wasn't a crank, then, after all, hallelujah!—though he'd convinced nobody except himself. It was 1963, and his Leverhulme studentship was ending. He had yet to publish a word. Eager to show something for his years of toil, he sent a three-page account that was rejected by *Nature* but accepted by *American Naturalist*.[43] The much longer, detailed article he sent to the *Journal of Theoretical Biology* would cover all the bases when he got around to addressing the anonymous reviewer's comments—the most important of which, he now read, was to split it in two. Meanwhile, following in the path of Darwin's codiscoverer Alfred Wallace, he was off to Brazil to study social wasps in the jungle.

Walking nature's trails, Hamilton was once again where he wanted to be. It would take nine months to resubmit the corrected manuscript, "The Genetical Evolution of Social Behavior, Parts I and II." Oblivious to the world, he did not know that in that space of time John Maynard Smith, citing only the shorter *American Naturalist* article, wrote and published "Kin Selection and Group Selection," giving "inclusive fitness" its catchy name. Nor, of course, did he know that Maynard Smith had been the anonymous reviewer responsible for its delay. When he did find out, it would ignite in his humble soul a feeling it rarely countenanced. But this was still far off in the future. "Very probably," he later wrote of the present,

> the sun of the day that saw my paper going into the post from JTB . . . would have seen me weaving my old American jeep between the corrugations, stones, and potholes of the Belém-Brasília road. . . . At midday it would have blazed near vertically on the top of my head as I stopped at the roadside and collected wasps from some nest; later at sunset, if still able to pierce the haze, it would have seen me and my Brazilian companion . . . slinging our hammocks between low cerrado trees not far from the stony or sandy piste where occasional lorries still groaned on into the night. For sure, both that day and that night I was blissfully untroubled about the finer points of measuring relatedness.[44]

.   .   .

No one hated group selection more than George Williams. A lanky American fish specialist with the beard-but-no-mustache look of what some described as an Easter Island statue, he was determined to fight Wynne-Edwards. As a student he'd heard a talk by Emerson: Death, the Chicago man argued, was an adaptation for the good of the group. Williams wouldn't believe it. Organisms dying just to "get out of the way"? That couldn't be right. "If this was evolutionary biology," he said to himself, "I wanted to do something else—like car insurance."[45]

Across the Atlantic in Oxford, meanwhile, Lack was doing his damnedest to put the notion of the "greater good" to rest. There was no reason to assume with Wynne-Edwards that birds restrain their reproduction to strengthen the group. If a mother bird produced a smaller clutch of eggs, it wasn't because she was in cahoots with all the other mothers to lessen the number of progeny. On the contrary, if resources were slim that year, it made perfect sense to sire fewer younglings in order to be able to care properly for those who were born: In a season of dearth, a mother might feed three goslings to health, whereas if there were seven all might die of hunger. No, natural selection didn't fashion adaptations for the good of the group; always, inexorably, it was "looking out" for individuals.[46]

It was 1966. Nine years earlier Williams had worked out a model, better than Wright's, that revealed the conditions under which group selection might work.[47] But like Maynard Smith he'd become convinced that while possible theoretically, it was improbable in nature. Lack was right: If behavior could be explained by selection working at different levels, there was no reason to choose the higher level over the simpler one. Walking Occam's razor, the evolutionist should be guided by parsimony.

However powerful a tool, though, parsimony was a matter of interpretation; it was difficult to settle unequivocally whether selection was working at the level of the individual or the group. What Williams really needed to kill group selection was a prediction that could generate a natural test. And he found what he was looking for buried in a paragraph in the pages of *The Genetical Theory of Natural Selection*.[48]

Fisher had argued that sex ratios in any species should always evolve to an equal number of males and females. The reason was simple: A male born into a species with a female-biased sex ratio is at an advantage, since he has more mating opportunities than a female. If genes determine sex, then in a population with more females, the male-making gene will be favored by natural selection. Eventually the frequency of that gene will reach a point where the sex ratio of the species is now male-biased, whereupon its alternative, a gene promoting the production of females, will be favored. The dancing dialectic would ensure a symmetry of sexes: The wisdom of selection fashions males and females one-to-one.

Even though he didn't know it, Fisher was playing a game, just like the ones John von Neumann loved. Whether a mother should produce a son or a daughter, after all, depended on what other mothers were doing—the requisite condition for the application of game theory. But beyond the convergence of economics and biology, what Williams now saw was that Fisher had done a blessed service: Figuring out the logic of the symmetry of the sexes, he'd delivered the deathblow to the notion of the "greater good."

Here is why: Imagine a mother parrot mating and then flying off alone to a deserted island in the middle of the sea. Imagine that she lays 10 eggs, half of which hatch into males, the other half into females, and imagine that her daughters do the same. What will happen? The population will grow rather quickly: From 10 in the first generation, it will balloon into 50 in the next (25 females and 25 males), 250 in the generation after that (125 of each sex), and so on.[9]

But what if the mother parrot lays 10 eggs that hatch into 9 females and just 1 male, and her daughters do the same? How then would the population grow? In the first generation, as before, there would be 10 parrots (9 females and one male). But in the next generation there would be 90 (81 females and 9 males) and in the one after that 810 (729 females and 81 males). Compared with a 1:1 sex ratio, a female-biased ratio would run much faster. Before long the island would be teeming with parrots.

But there was a twist. Even though siring more females would benefit the growing group, it would actually reduce the individual fitness of the mothers who did so. Imagine two mother parrots arriving on the island: Linda sticking to the 1:1 sex ratio, Barbara to the 9:1. Together they will produce 20 offspring in the first generation: 5 + 9 = 14 females added to 5 + 1 = 6 males. When these offspring mate among one another, while each female continues to lay 10 new eggs, the average male sires 23 children (140 divided by 6), since he fathers the offspring of more than 2 females. The results for Linda and Barbara are surprising: After three generations, Linda, who stuck to the slower 1:1 sex ratio but had more males, will have 165 grand-offspring whereas Barbara, who went for the 9:1, will have only 113.

When it came to the sex ratio there was a conflict between community and individual: If the group counts, "altruistic" Barbara is the winner, but if the individual is selection's client, "selfish" Linda prevails. Wynne-Edwards had shown that in times of dearth it is not always to the advantage of the population to breed as much as possible. But assuming sex ratio was an adaptation, when the times of dearth were over one would expect a switch to a female-biased sex ratio. Scouring the literature, Williams saw that this never happened—not in flies, not in pigs or rabbits, not even in humans.

Over time and across all vagaries the 1:1 sex ratio remained stable—the very prediction that selection working to maximize individual fitness required.

It was a mortal blow to the doctrine of the "greater good." Practically overnight, *Adaptation and Natural Selection* became a classic and group selection a theory to stay away from like the plague. Finally "Wynne" had been felled, though, like Sisyphus, he would keep on fighting his battle. Williams ended with a tip of the hat to Fisher: Beyond a doubt, fulmars and red grouse notwithstanding, natural selection produces adaptations optimizing individual fitness. "I am convinced," he wrote triumphantly, "that it is the light and the way."[50]

Back from Brazil, Hamilton had news of his own: In countless species of ants, thrips, wasps, beetles, and mites there are *extraordinary* sex ratios, major departures from Fisher's pristine one-to-one. Hamilton, unlike Fisher (and even Williams), would know: He had found them peeling back bark from capirona and kapok trees, stripping weeds of larvae beneath shallow jungle streams, and munching on figs with internal surprises. *Mellitobia acasta* was an example: A tiny parasitic wasp, its female lays her eggs inside the living pupae of bumblebees. When the eggs hatch they eat their way out of the pupa, but not before each female engages in sex with the sole male, their only brother. After all, it makes sense for their mother to use the confined body of the pupa to lay as many female eggs as possible and only one male to inseminate them: Once they've been fertilized, they can fly away to lay their own eggs in another poor bumblebee pupa, while their exhausted brother and lover, the wingless male, wastes away in the abandoned cocoon.[51]

By now Hamilton had a research position at Imperial College Field Station near Ascot, an institution eminent in the study of insect ecology. Around its rotten-wood grounds in Silwood Park, in the asphalt cracks of his cramped Berkshire home, and on walks in the royal forests around Windsor he discovered the Western European cousins of his Amazonian friends. Parasitism, he was learning, could get complicated—and fast: Sometimes two females laid their eggs in the same caterpillar or pupa, sometimes progeny dispersed and sometimes they didn't; sometimes ruthless males, like "maniacal homicides armed with knives," fought over amorous sisters to the death. Each situation changed the rules of fitness entirely. How would a miniature *acasta* know what precise proportion of male-to-female eggs to lay?

To figure out himself what the answer should be, Hamilton journeyed into London many days to use the UCL computer center at Gordon Square. It wasn't a pretty affair: Programming in the dinosaur FORTRAN language, which still preserved its nineteenth-century Hollerith punched-card six-space markings—"those once-useful claws that are now diminished to useless toe nails,"—Hamilton would queue along the old kitchen staff's route to the basement, roll of five-punched tape ready in hand, watch as the technician set the toggles to guide the "magic monster" Mercury computer, walk across to collect the spaghetti-like tape spewing chatteringly out the far side, and rush to an adjacent room to view the results via a teletype. It was a far cry from sitting in front of a laptop.[52]

But despite the hassle and beneath the natural complications, the teletypes disclosed a pristine river of logic. Whether you were an ant in a weevil egg, a wasp in a wild fig, or a mite inside your own mother, there always existed an optimal solution to the problem of the precise proportion of female to male eggs to lay. Hamilton called it the "unbeatable strategy," and discovered to his great astonishment that, with no Mercury and toggles and teletype and FORTRAN, the little critters always found it on their own.

It was mind-blowing. Mites "calculating" fractions as precise as 3/14ths? Hamilton was godless, but this was as close to faith as anything. Now, he knew, he'd discovered a powerful tool that would help to fathom nature's wonders. Back when he was a student at Cambridge he had idly read von Neumann's great book, never imagining that it might have anything to do with biology. But there was no mistaking that here was an evolutionary analog to the prisoner's dilemma: Fitness was a "payoff," the opposing sex-determining genes the "players." In parasitic wasps and sex ratios he was dealing in the theory of games.[53]

No one seemed to notice at the time that extraordinary sex ratios provided the very test that Williams had sought to apply to group selection.[54] Hamilton himself contributed to the confusion: Extraordinary sex ratios couldn't be an adaptation for the good of the population, he wrote, since even a tiny bit of outbreeding would destroy any "altruistic" genes.[55] No, once again this was a family affair; small spaces like pupae, after all, were like isolated cottages. In line with the times, parasitic wasps were interpreted as just another nail in the coffin of the "greater good": It was gene frequencies that were being maximized, after all, and they could only do it by manipulating the behavior of related individuals.

Laying down his pencil, Hamilton might have smiled. First $rB > C$ and now sex ratios: The mathematics of the "gene's-eye-view" was turning biology into a "hard" predictive science. Fisher had been right: Selection *was* a mechanism for generating an exceedingly high degree of improbability; even altruistic behavior was fashioned by its unsentimental, ruthless cull. He ended his equation-filled article, in the grand natural-history tradition of humble cooperation, by thanking a Dr. Bevan "for information about bark beetles" and a Dr. Lewis "for information about thrips."

.   .   .

Group or Individual, Optimization or Chance: Where had true goodness come from? Wright and Haldane and Fisher had begun the work, uniting Darwin and Mendel, selection and genes, in a bold evolutionary synthesis. They had argued among themselves over the relative role of mutation, drift and selection, but their project had been one of foundation. Now, on the back of Wynne-Edwards and the "greater good," younger men like John Maynard Smith, George Williams, and Bill Hamilton were penetrating further into Nature's puzzles. Theirs had been a thought experiment as audacious as Einstein's, only from the point of view of the gene rather than a man traveling on a beam of light. Finally, they thought, they were solving the mysteries of behavior.

In the Amazonian jungle in 1964, Hamilton received a letter from a girlfriend in London. She was breaking up with him. That week in the forest he had met two Brazilian kids, Romilda and Godofredo. Alone and morose, he was having serious doubts whether he'd ever marry or have children of his own. Turning to their parents, he offered to educate and take care of them as a foster parent back in England.

Now, four years later, his glum misgivings had proved premature. He was married to a wonderful woman, Christine, and thinking of starting a family. But a promise was a promise. The year after publishing "Extraordinary Sex Ratios" in *Science* he and Christine were back again in the forest and would be taking Romilda and Godofredo to Berkshire. Evolution hadn't selected him to behave this way toward nonkin, he was now more than ever certain. But lying in his hammock listening to the beat of the wings of an iridescent dragonfly and a seriema pealing its dawn call like cracked bells from the hills, he knew it was in his lonely heart, still.[55]

New York
Nov.6,1963

Dearest aGeorge,

Hear Ye, Hear Ye, Hear Ye.,

The birds are xxxxstarving to death fast in h this
city of pleaty. It is terrible, horrible to hear and
almost impossible to believe.

There are signs up down town near wall st. and the
park forbidding peopleto ffeed the birds at all-anything.
One man who feed them the correct seed was given a summons
and had to go to court. He was given a suspended sentence
altho he had not disobeyed any law at all.

The ASPCA man told e not to be afraid out if
anyone made me ay trouble to call him at once, i chat with
the police at time who drive by but there are no threats
of any nature.

Here's how you can help and Alice in Wonderkand
begs you to help, and so do I. Help is needed at the
ASPCA because so few people s em to know what to do to show t
their sympathy. Mr. Metcalf there told me so. He wants as
much help as possible from sympathetic folks like you and me
as he is to appear opposite the bird haters in court in a vey
few days. It may be this week so please give 20 min. to a
brief letter, George, just expressing yourself as you did
to me when here. Be sure to state your station in science.
I can't think of the proper name. It will help him so much
if you will so please don't refuse.

It is almost time for the last mail pick-up so
must close without further news. Please mail the letter to

Mr. Metcalf
ASPCA
441 East 92 St.
New York 28, N.Y.

Love and the

A letter from Alice Price to her son George, November 6, 1963

8

No Easy Way

*I*n the spring of 1957 Senator Hubert Humphrey had introduced a bill offering income tax credit for tuition paid to institutions of higher learning. George wrote a letter of thanks. It was all about education, he thought, and the bill was a step in the right direction. Still, really, intervention was needed much earlier. Why not exempt the first six thousand dollars received for teaching in grade school? After all, the greatest weapon against the Soviets was undoubtedly the first-grade teacher. "I belong to the 'as the twig is bent, the tree grows' school," George explained.[1]

He had struck up a correspondence with his old senator from Minnesota, and there were further suggestions to help win the Cold War: Why not send every Russian two pairs of shoes if their government would free Hungary? Or ten million dollars worth of polio vaccine if the Hungarian police would stop torturing prisoners? Or what about radio messages broadcast in Russian extolling the advantages of butter over guns? And could the senator please send a copy of his "U.S. Foreign Policy and Disarmament" speech to Congress in April? And the hearings of the Johnson Preparedness Subcommittee? And had he seen the piece by Donald Harrington yesterday in the *Times*? "Of course! Why didn't I think of that myself?": an international court of law to help restrain the Chinese in Formosa. Could the senator perhaps "plant the idea tactfully within Mr. Dulles's mind?" George would very much appreciate it.[2]

The senator replied politely that he would welcome his advice and counsel. George's views were "eminently sane," the idea about creating a strong UN police force just along the lines he had imagined. Price's overanxious manner might have struck the senator as somewhat odd, but his replies made George exceedingly proud. Nothing felt better than a man of stature recognizing his potential.

Increasingly, writing popular science articles became the perfect vehicle for gaining notoriety. It was the journalist's prerogative, after all, to get in touch with famous people. And so, between "The Physics of Bowling," "U.S. Begins Search for Beings in Other Worlds," and "What We're Learning from Animals" (all for *Popular Science Monthly*), he drove to his old hero Claude Shannon's home to write a profile on him for IBM's magazine, *THINK*—the two "paddling around" his lake in "glorious" silence. Between "Achievements of American Science," "The Real Threat from Red China" (government brainwashing), and "How to Hatch an Egghead" (motivate him), he contacted the Nobel laureate Hermann J. Muller, to ask his views about evolution on other planets. On each occasion the excuse developed into a conversation, with George trying his very best to impress. Muller, who was considered by many to be the greatest living geneticist, remembered "Science and the Supernatural" and complimented him on it; George replied with quotes from Poe and Keats, and his thoughts about people reading too much science fiction. Muller thanked him kindly, and the correspondence ended at that. It was a pattern: the brushup, the exchange, the titillation.[3]

If IBM had been stupid enough not to hire him, George was not about to give up and roll over. On the strength of the piece in *Life* he had been offered a contract by Harper & Brothers, and was hard at work on a book now that he hoped would help to save the world.

It was right about then that Skinner came along.

• • •

Adored as a messiah and abhorred as a menace, Harvard professor Burrhus Frederic Skinner was the most influential psychologist in America. He had an elongated face and an unsetting smile. The leader of the "behaviorists," who likened man to machine, he was reviled by Freudians and humanists. Some called him "totalitarian," others "Orwellian"; others thought his ideas man's only hope.[4]

George was intrigued. Early Christian thinkers thought it was the "soul" that set man apart from animal: God-given, immaterial, impalpable, otherworldly. But what if man were just an animal, firmly planted in the natural world? And what if men and animals were not all that different from machines?

Such thoughts had been afoot since the Diderots and La Metries of the eighteenth century, and even earlier, with Hobbes. Now, after Darwin and the rise of psychology and genetics, Skinner was arguing that freedom and free will were no more than comfortable illusions. For him autonomy was a "feel-good" invention, morality a sinister sham. The belief in an "inner man" was like the belief in God, a superstition, nothing but a symptom of humanity's failure to understand a complex world. Whether he liked it or not, man was already controlled by external influences, some haphazard, others evil, others merely in step with "convention." Where things went bad was when man made a fetish of individual freedom, seeking to give life to the internal "soul" at the expense of orderly society. In reality, Skinner preached, it was environments and not people, actions and not feelings, that needed to be changed. The "behavioral technology" called "operant conditioning" was civilization's hope for deliverance. Where moral arguments had failed, it could create a world where man would refrain from polluting, from overpopulating, from rioting, from hating, even from waging war. George was a "as the twig bends" man. Visiting Skinner's Harvard lab on the pretext of writing a story, he soon befriended "Fred."[5]

The lab was a scene out of wild science fiction. Pigeons playing Ping-Pong, rats balancing balls on their noses; there was even the pigeon-guided missile, the "Pelican." But these were animals. What about humans: Could they be similarly programmed? And was it punishment or reward that would be more effective in controlling behavior?

Skinner was convinced that it was reward, and had designed the "Teaching Machine" to prove it. A question was posed on a screen: If the child answered correctly he'd be immediately rewarded, not with a grain of corn but with a printed statement of approval—just as satisfying and effective. Walking George through the lab, Skinner claimed that kids could be taught arithmetic just as rats were taught circus acts and pigeons Ping-Pong.[6]

George was captivated. "Very quietly," he wrote for *THINK* later that month, "almost unnoticed amid the fanfare over thermonuclear weapons, earth satellites and moon probes, an important new invention has made its appearance." To Skinner he offered: "No doubt the most immediate application for such machines would be to teach reading and writing to the illiterate masses of Asia and Africa." For the last hour he had been sketching out a way to make "a very cheap machine using an acoustic phonograph with spring motor, plus a paper disc, with controls to automatically position the phonograph pick-up at the beginning of the question, stop the phonograph at the end of questions, do the same for the answers, repeat questions and answers when desired, and skip questions previously answered correctly." It could be mass-produced at just five dollars a pop.[7]

However useful Skinner's machines, they got George wondering about larger questions. Was man in control of his destiny? Was freedom really just a fantasy, or worse—a trick? Skinner argued that man believes in free will only so that he can take credit for his "good" behavior. But George had been free to leave his family, and was taking no credit for good behavior. Clearly there were traits buried deep in his nature, things that had not necessarily been learned. The problem was how to manage them, especially self-centeredness and egoism, the most entrenched natural trait of all.

He translated personal deficiencies into public affairs, the easier, perhaps, to flee them. Writing drafts of his book for Harper, George remembered Thucydides' famous description of Athens falling to Sparta on account of selfishness. De Tocqueville too, he recalled, had lamented how through vain self-absorption and greed men "lose sight of the close connection that exists between the private fortune of each of them and the prosperity of all."[8] He might be a bad family man, George thought, but he cared about his nation. True self-interest lay in strengthening the community, in devotion to the country, in paying a personal price for the good of all. Russia had done it. China had done it. Could America do it too? Could limits be put in a democracy on the individual pursuit of happiness? Skinner was hopeful, and George tended to agree. Americans just needed to be taught that personal sacrifice meant communal reward.

. . .

Meanwhile, Harper had reneged on its contract, finding some of his suggestions "brilliant" and "unexpected," but the overall book unsalable. With a change of title, and the help of a talented literary agent, George secured a new publisher and advance. It was April 1958, and Doubleday would want *No Easy Way* no later than July.[9]

He was living now in the West Village in New York City. The small loft at 88 Bedford Street was just down the way from the historic town home of Edna St. Vincent Millay, whose poetry Thomas Hardy called America's second attraction, after skyscrapers. Jack Kerouac, Allen Ginsberg, and William S. Burroughs were regulars in the coffeehouses. Dylan Thomas had collapsed and died a few years earlier drinking at the White Horse Tavern, just a few blocks away, and in a few years a young Bob Dylan would show up, playing "neo-ethnic" songs in the coffeehouses in the style of Woody Guthrie. There were writers and poets and artists and "beats," and, dapper in his bow tie and crew cut, George was an anomaly. But he liked the idea that he was an "ex-chemist." In a strange way, bohemia appealed to him. There was a kind of prurient satisfaction in being a straight arrow in a world of chaos. Proudly he had "writer" stamped in his passport.[10]

Still, he could not escape his past. Julia was after him for alimony, and his money was disappearing fast. He had taken another technical-editing job with IBM, this time in Poughkeepsie, and was only getting home on weekends. The Doubleday deadline was killing him. Besides, the world was changing faster than he could write about it: He had started off calling for armament, but now Russia's position on *disarmament* seemed much more honorable. Should America compete? Should it withdraw? Could the Russians be trusted? He was confused. He hardly had time to eat. The world around him was spinning.[11]

Loosening his bow tie he popped them in, one by one: iproniazid, Dexedrine, ephedrine, Seconal; uppers, barbiturates, psychostimulants. Everyone else around him seemed to be taking them, so what the hell? The effects, he soon learned, were less than exhilarating. "I've spent most of the day so far lying down," he wrote to his psychiatrist, Dr. Nathan Kline. The drugs had taken away his panic but had left him tranquil and sedate. He wasn't getting anything done. He was loafing. He felt neither pain nor joy. Worryingly, he was taking "some measures" to try to release adrenaline from his adrenals; he just couldn't quite remember whether it gets through the blood-brain barrier. Did Kline?[12]

Then, as the winter of 1960 rolled in, a tumor was discovered in his throat. At first his internist thought, Hodgkins. But an old Manhattan Project friend who was chief of X-ray at Memorial Sloan-Kettering sent him to a cancer expert who diagnosed a nonmalignant thyroid tumor. Recovering from the operation in March, George was handed a new type of medicine: He had a thyroid imbalance, and would need to take tablets from now on if he wanted to stay alive and healthy.[13]

. . .

There was an instruction manual for a GE electricity and magnetism kit to finish, and another one on gyros and accelerometers for Sperry-Marine. There was a new love interest, Joan, a teacher at Bennington College for Women in Vermont. There was Alice, still with her Japanese roomers, but growing old and infirm. There was his estranged lighting-expert-inventor brother, Edison, who had taken over Display, renaming it Edison Price Lighting Company Incorporated and operating out of 409 East Sixtieth Street. There was Julia, seeking contested alimony in domestic relations court. There were his daughters, whom he hadn't seen for quite some time. There were desperate letters to friends asking for loans. There was his health. There were the drugs. And then, on April 10, 1960, there was a letter from Fred Schneider of the Advanced Systems Development Division at IBM. Regarding his old article in *Fortune*, IBM was beginning to invest in computer-aided design (CAD). Would George be interested in joining the project?[14]

That fall a surgeon friend from his days in Minnesota wrote a kind note to say that his wife had a bad thyroid, too. Don Ferguson was now at the University of Chicago, and really, George, the thyroid was no big deal. He'd be fine. Still, didn't he think that he was wasting himself on those damn manuals? After all, he had the kind of mind that needed to be applied to real scientific problems.[15]

In fact George had been working on two scientific papers, with the hope of once again gaining a university position. The first was about the fallacies of random neural networks as they pertained to the organization of the brain, the second a theory of the function of the hymen.[16] Of course he had absolutely no training in either of these matters. But he had gone through the literature thoroughly and was planning to send the papers to *Science* all the same. He needed one big breakthrough—one truth, he thought, just one. IBM had toyed with him before, and by now he had lost interest in his Design Machine. Still, until the papers were accepted and made him a name, he'd need to take a job to settle his finances.

There could have been worse job opportunities. "The IBM company," the Annual Report of 1961 declared,

> is engaged in the creation of machines and methods to help find solutions to the increasingly complex problems occurring in business, government, science, defense, education, space exploration and nearly every other area of human endeavor.[17]

Gross revenue from domestic operations alone amounted to $1,694,295,547, an increase of more than $250 million from the previous year. His "unfortunate *No Easy Way*," George wrote to his editor, Dick Winslow, at Doubleday, would need to be put aside for now. Almost four years after that fateful interview with Piore, George was joining IBM.[18]

A few months after he had become an "official IBMer," *Science* wrote back rejecting his papers. The hymen theory was way too speculative, and a damning report had slammed "Fallacies" to the ground. "This crotchety, verbose diatribe has no place in a scientific journal," an anonymous reviewer had written. George was "merely a biased reader of other people's papers."[19]

The last thing he needed was another blow. His shoulder had been dislocated on a recent climb (his friends called him "kinetic") and again a few months later, swimming in rough surf. It was the old injury, compounded by the polio, and it would need to be taken care of. Recovering at home from the operation in the fall of 1962, he was gloomy. The *Science* report was a humiliating rebuke, a slap in the face from the professional to the impudent amateur. He hadn't seen his girls in more than five years. Joan was out of the picture. A Tatiana whom he had met in the public library was in and then out again. He was almost forty. He had yet to make his mark. He was starting to wonder about the merits of free will.

As always the only fixture in his life, Alice, came to his side. "You will succeed in a big way before you know it," she wrote to him, trying to be encouraging. She had her own troubles. There was a Mr. Aramachi, thank God, renting the southeast room, and a Mr. Ishida in the southwest. But then there were the "bird haters" from the municipality who plastered signs across the park forbidding feeding them. "The birds are starving to death fast in this city of plenty," she wrote to George, despondent. A diminutive octogenarian, she was already well known to the authorities. Municipal fines and court subpoenas had failed to stop her. Intransigent, at war with all the cruelty and lack of mercy in the world, Alice was sneaking pigeons to her home to mend and feed them before releasing them in Central Park.[20]

. . .

Back at IBM George had worked a bit on CAD but quickly lost interest entirely. It was hard for him to get excited about a brainchild he felt had now been stolen from him in broad daylight. The New

Product Line, on the other hand, was doing a market survey concerning programmed instruction, and George figured he could help with that. After all, it had to do with Skinner.

They had been friends, but George's sympathy had soured; in a "market requirements memorandum" he came down hard on his former pal. Falsely analogizing from pigeons and dogs to humans, Skinner had presented a simplified and therefore skewed theory of learning.[21] To him learning was a simple stimulus-response (S-R) pattern, and any intervening steps should be analyzed into the basic S-R components. But what if learning in humans really looked more like this: S-A-B-C-D-E-F-R, and what if A-B-C-D-E-F could not be collapsed into either stimulus or response? George was certain that this was the case: Perception (A), attention (B), understanding (C), belief or acceptance (D), memorization (E), recall (F), and performance (G) were distinguishable components of learning, and reinforcement worked differently on each of them. Skinner's notion that reinforcement led to learning was simple-minded and misleading. If teaching machines for programmed employee instruction were to work, one would need a better understanding of how reinforcement affects each of the components of learning. One would need to know what was innate in man, and what could be acquired. There were many layers lurking beneath the mystery of behavior. Free will was more complicated than Skinner thought.[22]

In fact George already had ideas on the matter. So much so, he wrote to Winslow at Doubleday, that he was thinking of writing a book. With No Easy Way not yet dead and buried, he had risen again like a phoenix from the sand and turned, as was his pattern, to a new project. "The Reformation of Psychology" seemed too colorless a title, but he would come up with something better, he was sure. The main thing was to explain how all the current theories in psychology were unsupported by masses of current data. Such theories still abounded because a replacement had not yet been formulated. Reviewing animal and human data relating to brain function and anatomy, introspection, learning, motivation, memory, love, and the nature-nurture controversy, George's book would provide the missing context. Here he could include the rejected papers on the fallacies of random neural networks and the function of the hymen, as well as his thoughts about everything from where memory resides in the brain to how mathematicians can be useful to psychology. He had left the Village now, he told Winslow, and was living in (and hating) Poughkeepsie. But the IBM job gave him a good salary and an enormous amount of freedom, and, given the thumbs-up, he could set to work on the book right away.[23]

The more covetous he grew of his liberty at IBM, the more his coworkers became suspicious. Who was this George Price: A journalist? A scientist? An inventor? A quack? And why was he often working from home? Some suspected that he might be trying to steal IBM secrets for further articles in *Life* or in *Fortune*, or maybe even an "intelligence machine" of his own. Others wondered why he was working on a book about psychology when he had been contracted to work on the development of a new computer. Believing in his abilities, his boss, Fred Brooks, was doing his best to cover for him. It wasn't easy. Brooks's own secretary forced George to buy stationery supplies with his own money. On one of the rare occasions that he had come in to the office, someone mentioned that there would soon be a public announcement of the new System/360 computer. "What's the 360?" George asked. "I never remember these machine numbers, you know." The people in the office shot incredulous glances at one another: George was working for the man who was in charge of the entire project.[24]

He didn't mind. He was after one big breakthrough and would do what it took to make it. "Many of the most imaginative and important inventions," he had written in an old article for *THINK*, "have been made by outsiders not employed in the field concerned, who have followed their own schedules and worked according to their own plan." This became his motto.[25]

. . .

There were two fields where he felt he might make important strides. The first was a new type of mathematical optimization system, the second a neurophysiological coding scheme to explain color vision. Neither of these had any direct connection to work at IBM, but offbeat as always, George went around trying to convince the people in Research and Development that they might bring the company glory.

The optimization system stemmed from work he had been doing on the development of procedures for solving linear problems with variables limited to 0 and 1. The objective had been to develop rapid methods for finding nearly optimal solutions of multiple-activity, multiple-factor situations—a computer programming problem. But it soon became obvious to George that the really interesting implication had nothing to do with the "register problem" in computer programming but rather with economic cycles. What he had found was that the 0,1 restriction resulted in models that were much more helpful than the usual linear models for understanding the behavior of price and profit-based economic systems. Immediately he wrote to the eminent MIT professor Paul Samuelson. Without having intended it, George had found himself with a skeletal model of a profit-based economic system in which complex phenomena became surprisingly transparent.[26]

"In order to show you that I am not as ignorant in economics as you may think," he explained to Samuelson in a long letter, he should know that during his one year at Harvard in 1940 he had been in the same class as Bob Solow, and that later in 1947 Andy Papandreou had used his Winthrop House digs for tutorials. "Finally, most impressive of all, when I was about thirteen the Chairman of the Department of Economics and Social Sciences of the Massachusetts Institute of Technology told me I was talented. . . . I met with him at a meeting of some organization to which we both belonged (no, not the Communist Party), and talked with him about his paper for a while, at the close of which he remarked: 'You ought to go into economics. You're good at it.' (Instead I went into chemistry. I wasn't good at it.)"[27]

It was classic George: trying to ingratiate himself by means of an original idea mixed in with some oddball humor. Samuelson was impressed. George's profitability algorithm sounded "interesting and novel." Still, there was a vast literature on business cycle dynamics that he would do well to look at. Picking up on the curious mixture of George's originality and somewhat autistic penchant for reinventing the wheel, he wrote: 'I am sure that much of what you are doing would interest economists, particularly if it could be related to earlier work.'[28]

Meanwhile the second "breakthrough" was increasingly grabbing his attention. Again it started off related to IBM work: a memo on "Approaches to Object Recognition for Complex Graphic Images." Soon, however, it had shifted in another direction, with George trying to crack the basic mysteries of animal vision. How precisely did the neural wiring in the eye work to produce a picture of the world? Delving into the literature with abandon, he had read all he could get his hands on relating to retinal microanatomy and electrophysiology. He was working on two papers he planned to send to *Science*—"Structure and Function in the Invaginated Synapses of Retinal Receptor Cells" and "Cone Pigments and Spectrophotometry Artifacts"—when the insight finally hit him. "I think I may at last have the key missing piece in the puzzle," he wrote to his boss, Fred Brooks, the computer man: It was the glial cells, nonneurons in the nervous system, and George had discovered their role in vision.

He had "neglected everybody, including my kids." Annamarie and Kathleen were teenagers; he hadn't seen them for almost ten years. The price had been steep, there was no doubt about it. But could he have done it? Could it all have been worthwhile? Could the difficult road from chemistry to economics to writing to computers finally have led him to glory? It was "a discovery of major importance," he thought, as original as it was bold. Finally, he had found his piece of truth. "Very optimistic," he confided to a friend. "Think it will make a big difference in my situation in the world."[29]

. . .

It was not to be. Like the big ESP experiment with John Scarne and the Argentine, Ricardo Musso; the amazing Design Machine; the vital *No Easy Way*; and like his economics-rattling optimization models—the eye "discovery" just fizzled away. The Nobel Laureate in Physiology or Medicine Sir John Eccles, to whom George had sent his paper, replied politely that it was "remarkable" but that he himself could not follow. "Your physical treatment of this important problem," he wrote, "is beyond my mediocre attainments."[30]

Meanwhile George's thyroid was acting up again, and he was contemplating an operation. There was a fine surgeon at Memorial, but he decided to go with Don Ferguson in Chicago instead.

Ferguson had overseen a thyroidectomy on his own wife, who had recovered completely. He was known to be conservative. He was an old friend.

Only when George was in his hospital gown at Billings did Ferguson tell him that he always did a radical neck dissection when he found thyroid cancer. There was no need to worry, though. The surgery would leave no more that "a slight deformity.' Too cowardly to change his decision now, George gave his okay and went under his friend's knife. It was February 1966.[31]

Back in New York following the operation, he wrote to Ferguson that he was "recovering well." He couldn't quite yet do mathematics ("that will be the sign of true recovery") but he was on his way. He was very grateful "for all your kindness to me—for the flowers, the radio, the phone call to my mother, the books, the music lessons, the hospitality at your home, and the beautiful stitching."[32]

Into the summer and through to the fall, the recovery had begun to reverse itself. He was doing physical therapy at the NYU Institute of Physical Medicine and Rehabilitation, but he could hardly move his right arm. What's worse, much of the right side of his neck and face and shoulder were completely without sensation. He was contemplating a nerve graft but was told by an expert that no one ever obtained significant motor recovery through a graft of ten centimeters, even if an autograft were used. He had stopped going out, stopped dating women, hadn't been to the theater for months. Then, in January 1967, Ferguson wrote a startled letter to say how shocked he had been by yesterday's package in the mail: With only a note to say who it was from, there in the box were George's ice-skates.[33]

"I wish he would go to someone for a hernia repair and have his penis denervated," George wrote to a friend. Ferguson was a "butcher." How could he have done this to him? It would make George much happier if he could sue him "under the law of Moses" instead of merely for money. Julia was after him again for alimony. Kathleen needed him to foot the bill for her final year at a fancy private school in New York City, St. Hilda's and St. Hugh's. He could hardly feel half of his face. Doubleday lawyers were writing to ask that he return the $2,500 advance for *No Easy Way*. Things were getting darker and darker. Depressed and neglecting to take his thyroid medicine, he was admitted for a short while by his brother, Edison, to the Payne Whitney Psychiatric Clinic.[34]

Everyone who knew the story at IBM thought George was over-reacting, but he was quitting his job all the same. There has been "a radical change in my purposes, attitudes, motivations," he explained to Mr. Brocker of IBM Department 630 in his letter of resignation. To a colleague from Chicago he expressed his interest in writing some papers "on the evolutionary origin of a number of the peculiarities of human anatomy, physiology, and behavior." To his daughters he wrote that he was off to Europe for two to three months "to do some article magazines [*sic*]." And to a friend he told the truth: "My family will think that I am going over for a couple of months to write some magazine articles, but then I will keep postponing my return." He had saved enough to last him more than a year on account of winning a rather generous health insurance payment. One might not have guessed it from his behavior, but George was now interested in the evolution of the family. Maybe that would be where he'd finally make his breakthrough.[35]

. . .

The decade had passed like a manic scavenger hunt, but what, in God's name, was George Price after? From sleepy downtown Minneapolis he had moved to the center of beat culture in New York City's Village, and then to gray Poughkeepsie. Along his route he had made hopeful visits to disparate worlds: porphyrins and cancer, telepathy, computer-aided design, economic cycles, the Soviet-Sino challenge, brain neural networks, the hymen, optimization systems, vision. He had been a chemist, economist, writer, mathematician, psychologist, physiologist, and self-appointed prophet. He had brushed against great men seeking to impress them: three Nobel laureates in physics (Bardeen, Shockley, and Bridgman), one in genetics (Muller), one in physiology (Eccles) and a sixth on his way to the prize in economics (Samuelson). He had engaged America's most influential psychologist (Skinner), the father of information theory (Shannon), the respected senior senator from Minnesota (Humphrey), and two of the day's

greatest writers (Huxley and Sinclair). He had thought about rationality and about learning. He had thought about human nature and
about the workings of the brain. He had studied the latest theories in
psychology and economics. He had worked hard.

All this, and what did he have to show for it? His family was
broken. His body was broken. He had switched countless jobs. He
was unemployed. He hardly spoke to his brother. He barely knew
his daughters. His mother was writing herself letters from his dead
father and feeding pigeons illegally. He was forty-five years old. He
had utterly failed to make his mark.

And yet, strangely, beneath the despair, the ingredients had somehow assembled: Optimality thinking; the penchant to break problems into their components; the consideration of individual versus
community; the problem of free will; a search for a Boyle-like law
of human nature; an interest in evolution. Most of all there was a
tenacity: a singularly dogged, bizarre, egoistic resolve to find a piece
of truth whatever it took.

On the windy Monday morning of November 13, 1967, he
stepped onto the deck of the *Queen Elizabeth*, ready to sail away from
America. What lay in the future was as faceless as the transatlantic
night. Unknowingly he was following in the footsteps of those who
had come before him: Kropotkin and Huxley, Allee and Fisher, Haldane and Wynne-Edwards, von Neumann, Hamilton, and Maynard
Smith. George Robert Price was on his way to England to crack the
problem of altruism.

Part Two

George Price's flat on Little Titchfield Street, above a butcher shop (now a delicatessen)

9

London

It was the end of 1967, and London was on fire. There were the trendy nightclubs—Ad Lib off Leicester Square and the Bag o' Nails around the corner from Carnaby Street. There were the offshore pirate stations, Radio Caroline and Radio City, and the Friday-night music-TV show *Ready Steady Go*, hosting the newest and grooviest in crackling black-and-white broadcasts (slogan: "The weekend starts here!"). There were the Stones and The Who, and Rory Gallagher's new band, Taste. The Beatles themselves, contra outfitters and clothing stores, had just opened their very own Apple boutique off Baker Street, where a psychedelic mural by the Dutch trio The Fool greeted shoppers hungry for the very latest in trend.[1]

Richard Hamilton and Peter Blake were giving pop art a homegrown slant; David Hockney was refining ironic understatement for the age; Mary Quant and Ossie Clark inventing new worldwide fashions. Men-about-town sported hair to their collars, sideburns, and three-quarter-length leather coats. Following Twiggy and Jean Shrimpton, women redefined skinny and the miniskirt, bell-bottomed pantsuits, and little coats with oversize buttons and high-heeled go-go boots. Pulse, vibration, bustle, groove—the city was a jungle overflowing with human spirit. No one could say where the name had come from, but after years of postwar austerity it was Swinging London: the outrageous, flamboyant world capital of cool.[2]

Not that George was paying any attention. Stepping off the boat into a world apart from Poughkeepsie, his life was in disarray. Everything outside seemed a blur.

"I continue to be anti-social," he wrote to his old girlfriend Tatiana back in the United States. He had moved into a cramped sublet on Whitfield Street, near the University of London. Amid the clutter of his landlord's bric-a-brac and, due to high storage costs, his own boxes from the United States, he ate 99 percent of his meals alone. Already on the boat he'd consigned himself to enjoying the last central heating he would have for a very long time; now American Express and Foyle's bookstore were the only places that offered a modicum of warmth. Traffic was confusing ("I don't like the driver-versus-pedestrian situation"), food was from a different planet, flats were expensive, service slow, and everything annoyingly, oddly inefficient. On November 21 George applied for a post office box at the Overseas Visitors Club on Earl's Court Road and was told he would get one, if all went well, by late January. Of cultural London he had "seen or heard practically nothing."[3]

Most of all the "flame of hatred" for Ferguson continued to burn brightly. He could hardly use his right arm, not to mention his neck and shoulder, and with little to do and no one to see, hating his old surgeon friend had become a major pastime, even a philosophy. Ferguson's butchering was an extension of the Protestant ethic into the Puritan ethic—an unconscious though deliberate attempt to improve him morally by showing that what really matters in life is not appearance or trivial sports but work, family, and the cultivation of the arts. George was a divorcé who had abandoned his girls, enjoyed exercise and dapper clothing and dabbled in drugs; severing the nerves to his trapezius was the perfect way for a frugal family man from Minnesota like Ferguson to correct his ways for good. In fact George was so obsessed about this that he was planning to publish a theory along the lines of "Unconscious Factors in Radical Surgery."[4]

His bitterness was growing. "I have not reacted to his butchering," he wrote to his editor at Doubleday by way of deferring once again his return of the advance for *No Easy Way,*

> *in the way that surgeons feel patients "ought" to react. (The "good patient" sits in his wheel chair paralyzed from the neck down, saying fervently, with tears rolling down his cheeks, "I thank first my brave surgeon, and then God, that I am still alive to enjoy the beauties of the sunset and hear the singing of the little birds").*[5]

In fact, since he had derived little enjoyment from the state of being alive since Ferguson "fixed" him, George was pretty sure that he "would not live for more than a rather small number of additional years, and perhaps for quite a bit less, depending upon how things go during the next year." He had a $25,000 life insurance settlement, which was more than Annamarie and Kathleen, now nineteen and eighteen, required. Since experience had taught him that selling magazine articles depended on scientific notoriety—the *Fortune* and *Life* pieces having panned out thanks to his 1955 article on ESP in *Science*—George was planning to do some scientific work in his everlasting hopes of making a miraculous breakthrough; then he'd surely be able to return to freelance writing in order to pay his bills. If he was successful, Doubleday should hear something about him in the newspapers a few months from now. Otherwise they'd be getting a call from the Manhattan Life Insurance Company at 111 West Fifty-seventh Street.[6]

As he enveloped himself deeper in a cocoon of self-pity, the world outside was burning. On March 17, 1968, ten thousand people crowded into Trafalgar Square to protest the war in Vietnam. Led by Vanessa Redgrave, the hordes decided to march peacefully down to Grosvenor Square to lay their petitions at the gates of the American Embassy. Anarchists, hard-boiled Trots, and flower kids all marched together hoping to bow America's crown. As they turned left on Alderney Street singing "Ho Ho Ho Chi Minh," the sea of demonstrators came to a halt before a wall of police. Soon rabble-rousers took control. Amid cries and waving red flags, rumors began to filter back that those close to the embassy had been machine-gunned. In an instant the peace demonstration had turned into a riot: Mounted police on white steeds charged into the crowd with long clubs in hand, like sabers. Crouching demonstrators kicked in the head, women screeching, coppers pulled off their horses, tear gas, mayhem—London had rarely seen anything of the kind. By the evening the toll of injured policemen was 117, the count of injured demonstrators unknown.[7]

"Did you get yourself a nice sturdy English girl to serve you a hot meal once in a while?" Tatiana wrote to George, trying to be encouraging. He hadn't. He was depressed, alone, hating Ferguson, unemployed. Trying to live on $5 a day, he cursed his brother, Edison, who was married now to Laura, a daughter of original General Motors stockholders, for not having repaid a loan. Ever since childhood they had vied for their mother's affection; now, when she was getting old, years of pent-up anger threatened to explode what little family unity remained. It was a dour conclusion. Wary of filial neglect, George sent Alice a book on "pigeons and people."[8]

.   .   .

And yet, beneath the despair a passion had already been planted. He'd been going to the libraries: the Senate House at UCL just across the way was his mainstay, but he roamed to thirteen others—including Camden Town, Highgate, the Zoological Society, the British and Natural History museums. When he was kicked out at closing time he'd run over to the late-opening Holborn Public. He had cast his net widely: anthropology, linguistics, medicine, neurophysiology, psychology, behavior—anything that might provide a clue.[9] The hours were long and lonely, but he pushed ahead all the same. Then one day, in early March, he came across something that caught his eye.

"I have just been reading your very interesting paper on 'The genetical evolution of social behaviour,' " his letter to Bill Hamilton began, "and would very much like to have reprints if you still have any to spare." It was the 1964 kin-selection paper, the one Maynard Smith had asked to split in two, and although George had yet to master all its mathematics, he was planning to use it as a basis for a paper of his own. In particular, did Hamilton know of any evidence for genes that enable those who carry them to detect the presence of exact copies of those same genes in the bodies of others? Could genetic similarity, in other words, somehow be sensed? It seemed a far-fetched idea for most species, but perhaps in humans—a species highly developed both in genetic and cultural inheritance—"some interesting effects could, in theory, occur."[10]

It was a road he had traveled before: Alice's letters from the dead morphing into a rant against ESP in *Science*; qualms about free will giving rise to papers about Skinner's "Teaching Machines" for IBM; fear of egoism translating into a global philosophy of imminent national disaster. Now, once again, George Price was turning his most personal, existential quandaries into a scientific project. His father had died when he was four; he had abandoned his daughters in America; and the Depression-era trio, Alice-Edison-George, was finally unraveling after long years of jealousy and neglect. His thought process was confused, still searching for an anchor to set in stone. Like a storm, a question deep inside him now began to roar: What was the meaning of family, and how had it been born in the first place?

He'd been consuming the new cadre of popular books on human evolution—Desmond Morris's *The Naked Ape* and Robert Ardrey's *Genesis Flood*, the kind that explained behaviors like aggression and sexuality by natural selection adapting prehistoric humans to the vagaries of hunter-gatherer existence. Morris, a Wiltshire-born zoologist, surrealist artist, and author, had already put in print about half of George's own "new" provocative theories, but ideas just kept on coming, and soon anticipation had dropped to only 5 percent. To Kathleen he wrote that he was thinking about the evolutionary origins of fatherhood and family.[11]

Then, a week after the Grosvenor Square riot, he found a letter of reply from Bill Hamilton in his mailbox.

No, Hamilton was sorry, he'd no reprints of his 1964 paper left. He was sending the latest—his 1967 article on sex ratios, instead;

after all, it too, in its way, dealt with the issue of genetic recognition. Of course he had no idea who this George Price from the Overseas Visitors Club was; still, he himself had not thought hard about the interaction of genetic and cultural inheritance in man, but he'd be interested to know what George made of it.

> *So far I haven't arrived at any clear idea even as to what sort of "game" the genes are expected to be playing when operating together (on different chromosomes or linked on a particular pair). Something like socialism (or is it racialism—can't tell), admittedly, seems indicated, but I have only vague ideas as to the mechanisms by which biological and cultural evolution interact. With man culture did once, in the form of primitive religions, reinforce socialism, but now what we take to be highest in culture has swung strongly against nepotism and the like. Can this be just a higher hypocrisy induced by the need which civilization creates for genetic diversity—and perhaps even racial diversity—in human groups? I wonder whether this is the field in which you think you see some light.*[12]

Hamilton's sensitive soul had recognized the shock to George's sensibility, penetrating the short lines of a reprint request from a complete stranger directly to his gut. For, reading Hamilton's 1964 paper on $rB > C$, George could not help but wonder: Was nepotistic altruism really the very best Nature could muster?[13] If so, prospects for humanity seemed bleak. Worse, Hamilton, in his reply, was painting an even emptier picture of morality: The very feeling of repugnance from a limited kin-directed goodness may itself be a trick of the genes to bring about a favored behavior; broadening altruism beyond the family would have the salutary effect of increasing overall genetic diversity, a biological end that would place "pure" moral sentiment in a rather darker, functional light.

Was there no way to win? Was goodness either true but limited or broad but nothing but a sham? Or was Hamilton suggesting something even more sinister: that no matter what the level or scope, humans were ultimately slaves to their true genetic masters? "Your letter is exceedingly interesting," George began in reply; he was a "as the twig is bent, the tree grows" man, but perhaps genes really were what controlled twig growth in the first place. It was extraordinary how much Fisher had pioneered and yet how much he had left after him. Still, had Hamilton recognized the flip side of kin selection: nepotism toward family meaning ill intention toward strangers? And would he be interested in a joint collaboration along the lines of "Natural Selection for Malevolence Toward Non-Relatives"?[14]

It was the end of March 1968, and Hamilton was already on his way to study multiqueen wasps in the jungles of Brazil; there was no address to which to post a reply.[15] With the orphaned letter in hand and heart racing at its dire contents, George took a second look at Hamilton's letter: "what sort of 'game' the genes are expected to be playing," it read. *Game?* What was this about?

"I am sure that prisoner's dilemma situations," Hamilton had written, 'are common and important in biological evolution." Precisely what this meant, George was not yet sure, but the stranger whose paper he had found in the Senate House Library had provided a clue that he might follow. Perhaps a morsel of goodness could be found, after all, to help allay his greatest fears.

He kept toiling away in the libraries: half a dozen papers started and given up. "My big one," he wrote to his daughter Kathleen back in America, "will be on the evolutionary origin of the human family," explaining:

> In many bird species, but only comparatively few mammals, the biological father contributes directly to the care of his offspring. In most mammalian species, the father just mates with the mother and she does all the child rearing herself. In a smaller number of mammalian species, there is some joint care by all the adult males in a group of all the young, but not individualized preferential care by fathers toward their own offspring. For example, dominant male baboons are intensely protective toward infant baboons, but do not differentiate between their own offspring and other infants. But in the human species, the dominant pattern in most or perhaps all cultures has

involved preferential care by adult males toward their own children. Problem: why did our species evolve in this way . . . ?[16]

He signed it, without a hint of irony, "With love, Daddy."

. . .

George continued to sway here and there, still lacking focus and searching for some unknown intellectual breakthrough. Shortly afterward, in the Senate House Library once more, he came across a paper in *Nature* that grabbed his attention. Already in the *Descent of Man* Darwin had recognized the problem: Deer antlers are "expensive" designs and yet highly ineffective weapons for inflicting injury on an opponent.[17] How then had they evolved? A certain G. Stonehouse now claimed to have the answer: Elaborate in form, often gigantic, antlers had evolved to dissipate heat by means of the flow of blood through the vascular covering of velvet during summer; they were the expensive but obligate result of the demands of thermoregulation. Males were usually the ones who grew them because males are larger and therefore more in need of dissipating heat. Stonehouse was confident that he had solved Darwin's mystery: What looked like a burdensome decoration was actually a physiological necessity.[18]

George was not convinced. Antlers had been on his mind since the summer of 1967, and still he hadn't cracked the mystery. He mulled it over in his head. He thought about Hamilton's clue. He tossed and turned. And then, in a flash, it hit him. Games! Of course! He had already been there! Yellowing away in some drawer, *No Easy Way* was after all about the Cold War dynamic of American and Russian disarmament; the logic of détente based on the threat of deterrence. Suddenly it all connected: If deer really needed to cool off, skin flaps and large ears were surely a less expensive route to follow than seasonal renewal of antlers. Besides, roe deer were in velvet in spring and without velvet in summer, and lowland tropical species would have to dissipate all year round. No, deer antlers could not be a solution to the problem of thermoregulation. Rather, they were ingenious accessories to nature's invention of limited combat. It was a classic von Neumann game.[19]

Here was its logic: If a group of male deer varied in both fighting ability (E signifying greater fighting ability than e) and ability to deescalate combat (D signifying greater ability to deescalate than d), one could go about calculating just how each deer (De, DE, de, and dE) might fare against another. Attaching probabilities to injury, survival, and victory, George discovered a fascinating result: Extended over four or five rutting seasons per generation and a hundred generations or so, limited combat strategies proved successful and should evolve. One obvious way to do this would be to grow ornate antlers: Since locking heads with such appendages is clearly less deadly than ramming one or two powerful sharp horns to the body, antlers would act as the biological analogy to a boxer pulling punches.

But wouldn't a deer with malicious intent and a sharp set of forward-pointing horns always stand to benefit? This is where the game came in, and the notion of an evolutionary stable strategy:

> *A sufficient condition for a genetic strategy to be stable against evolutionary perturbation is that no better strategy exists that is possible for the species without taking a major step in intelligence or physical endowment. Hence a fighting strategy can be tested for stability by introducing perturbations in the form of animals with deviant behaviour, and determining whether selection will automatically act against such animals.*[20]

When George did this, he discovered that antlers were the winners, and besides that they entailed some simple rules: First, an animal should avoid battle with a stronger animal. Second, it should be aggressive against a weaker animal. And third, when fighting an equal opponent, it should try an occasional "probe"—an escalation of combat meant to judge the adversary's reaction. Most important of all, however, was the principle of "getting even."[21]

It was an unbeatable strategy. An animal deficient in retaliatory behavior would in iterated encounters lose to it, but so would a third animal with a reduced tendency to deescalate when the score is even, and a fourth with a reduced tendency to probe. Most important of all, it turned on the fundamental game-theoretic rule: One's best strategy always depended on what the other player was doing. It would be to the advantage of an animal possessing a territory, for example, having more to lose, to fight longer if a challenger is likely to quit earlier; conversely, it would be to the territory seeker's advantage to quit earlier (and occasionally perform a probe) if the territory possessor was likely to fight to the death. Of course it was an oversimplification: Nature might hold the possibility of a lightning-quick fatal blow, or more than two categories of aggressive behavior might be in practice. Still, in a species that did not form coalitions, the basic strategy couldn't be bested. It was the very same strategy, George explained, that characterizes human "two-person game" conflict "at all levels from kindergarten children to nations."[22]

Pushed to its logical end, the limited-combat model ultimately resulted in the sublimation of all-out battle into the harmless domain of symbolic threat at a distance. For even in species that could discern two distinct levels of physical combat, such as locking antlers versus attacking the body, not fighting at all would always be safer than fighting gently. Of course, everything depended on the ability to discern the character of your opponent: An evolutionary arms race had been put in place between signals for strength (and "wildness" and "unpredictability") and the ability to judge their honesty. Still, the greater the variation in antlers in a population of males, the greater the chance that fighting will be avoided: A glance from afar (perhaps aided by some roaring and bellowing) would suffice to exclude most of the combat.

The flip side of kin selection, he had begun to write to Hamilton before learning that he was off in the jungle, was malevolence toward nonrelatives—a less-than-encouraging thought. But combat, too, was a Janus-like construction, and *its* flip side was the more hopeful promise of altruism. Kin selection could account for parental care and, perhaps, when the mechanisms responsible for discriminating degree of relationship were faulty, for "good deeds" to strangers. But George found it implausible that it could account for all cases in nature.

The literature he was reading was now beginning to finally settle in his head. George C. Williams, he now discovered, had made a suggestion on the matter in his 1966 book *Adaptation and Natural Selection*: Animals that live in stable social groups and that are intelligent enough to form personal friendships and animosities beyond the limits of family could evolve a system of cooperative behavior. But reciprocity, George now saw, actually demanded much less: In a species where cooperative behavior is important, the logic of games would suffice to ensure cooperation. There was really no need for the ability to form friendships and hates: the trick, rather, was for non-cooperative behavior to be retaliated against.

African hunting dogs were an example: Occasionally, it had been observed, a pack member is "mobbed," tumbled and rolled to the ground by the multitude. To George this seemed like the perfect punishment (and background threat) to ensure the remarkable cooperative behavior the dogs usually exhibit as a group. The logic was simple: Since an individual would increase in fitness both by helping others and thereby avoiding attack, and by attacking deviants enough to cause them to help him, both the tendency to cooperate and the tendency to attack those who did not cooperate would be selected for in evolution. Policing and punishment were necessary requirements for cooperation.[23]

It was not only an original application of game theory to animal behavior; it was a startling reflection. George had yet to figure out the evolutionary origins of the human family, fatherhood, and love, but these game-theoretic evolutionary asides were a beginning. Tidying up the rather long paper, "Antlers, Intraspecific Combat, and Altruism," he sent it off to *Nature* in August 1968.

That very week he got an address from Imperial College for Hamilton's whereabouts in Brazil and decided to send his belated reply. He'd been positively surprised to learn that the prisoner's dilemma had an application to genetics, Thank you. Now, though, he was working on "a more transparent (though less rigorous) derivation" of the 1964 kin-selection math. When the time came would Hamilton mind checking it? His approval would obviously be valuable.[24]

Shortly afterward George moved into a flat on the corner of Little Titchfield Street and Great Titchfield Street, just a five-minute walk north of Oxford Circus.

. . .

The neighborhood had the feel of a Sherlock Holmes locale. There was the butcher shop just below, owned by a German Jew, and Frank's Coffee House serving espresso across the way. Down the street on the corner of Mortimer and Great Portland was the local pub, the George, established in 1799 and still sporting its alabaster ceiling lamps, creaking wooden floors, and regulars slouched over ales and half-true yarns. Great Titchfield housed the unwealthy but comfortable—physicians, solicitors, journalists and the occasional eccentric bachelorette writer. It was a narrow and sleepy road flanked by red-and-brown four-story white-windowsill-painted Victorians, strangely providing the illusion of colluding to crowd out the sky at their crowns. Just a hundred yards to the west, Regent Street ran majestically north up to All Souls Church in Langham Place, before curling around the massive white stone BBC Broadcasting House into broad Portland Place with its Jaguars and Rolls-Royces and posh Royal Institutes of Physics and British Architects. It was a far cry from New York City's Greenwich Village.

Tucked away behind Regent Street, the flat on Little Titchfield was a spacious three-bedroom on the second floor of a landmark four-story redbrick with an attic and black thatched roof. George would be paying the last two of a seven-year lease that had been made during a down market.[25] It was a find.

Still, as peaceful as the environs were, there was a distinct but unexplained feeling he simply could not shake off. Whether walking home from Oxford Circus on the main, winding through a narrow back alley, or sipping a coffee at Frank's, the ever-present stone spire of All Souls was watching him, forbidding and portentous, like a hawk its unsuspecting prey.

. . .

Meanwhile, sweating beneath the August sun at the Royal Society/ Royal Geographical Society Expedition Base Camp in Mato Grosso, Hamilton sat down to pen his reply. George was a faceless correspondent, but Bill always tried to get back to those who wrote to him.

Besides, he was one of the few who seemed excited by the notion of the genetic origins of altruism. Sure, he'd be very interested in the intended paper, but if George owned only a single copy he shouldn't dare send it to Brazil. The post was very unreliable, but he could try a relatively safe address at their next stop in Belén. Otherwise, he and Christine and Romilda and Godofredo would be back in England toward December or January. Posting the letter, he returned to his wasps and soon forgot about George.[26]

Back in England, with the chilly winds of fall blowing, George hunkered down in the libraries. The idea of kin selection just wouldn't let go of his mind. Even in the terms of his own mathematics, could Hamilton really be right: "altruistic" genes able to spread only via family? If that was true, would the opposite of altruism—malevolence and war—be the fate of the unrelated? The bleakness of it depressed him.

He began to wonder: Was relatedness really the sine qua non? What if shared genes for altruism sufficed? Clearly relatedness could be *one way* to share such genes, but it needn't be the *only way*. In that case Hamilton's rule would be just an instance of a wider phenomenon. This had been the essence of the question about genes recognizing themselves in other bodies that he had written to him in early March. If altruists could somehow find one another, they needn't necessarily be related to help propagate their kind.[27]

He thought it through carefully. The problem was one of tracking the change in a character over time: What was the most transparent way to do this? Say there is a group of ten people with different heights, and a second group is formed, with the same number of people but a different sample of heights. To do this you are allowed to take only the heights that existed in the first group, but in a different proportion. Say the average height of the first group is 5.5 feet. How best to predict the average height of the second group? The answer was intuitively simple: The average height of the new group would be determined by the relationship between the height of each individual and the number of "copies" made of that individual in the second group, divided by the average number of copies. Scientifically speaking, that relationship was called a "covariance." So if half of the members of the first group are 5 feet tall and half are 6 feet tall, but the new group contains many more 6 footers than 5 footers, the new population will now have an average height much closer to 6 feet. A covariance equation

$$\Delta \bar{z} = \frac{Cov(w,z)}{\bar{w}}$$

captures this relationship by explaining how the number of copies made (w) of the different heights (z) determine the average height of the new group ($\bar{z}$).

This was a general-selection equation: It would hold true for everything from a child choosing radio channels to the earth preserving fossils to the culling of chemical crystals in far-off galaxies. But it could also be applied to biological traits like baldness and strength and crooked teeth, and, most profitably, to the evolution of social behaviors like altruism. All one needed to do was have z stand for the trait and w for its fitness ("copies" of traits simply mean their fitness), and, like a rabbit pulled out of a conjurer's hat, the covariance equation told you how it would evolve from one generation to the next. George hadn't done all the work yet, but already saw that his was a more abstract approach than using coefficients of kinship and could be made to work just the same: The spread of altruism could be tracked via statistical covariance of the character with fitness rather than calculations of the pathways of relatedness. Hamilton's $rB > C$ notwithstanding, altruism depended on association, not family.[28]

The reason was immediately clear to George: Natural selection is indifferent to why individuals end up together in groups; whether it's due to common descent, or similarity in traits, or any other pretext doesn't matter. On the other hand, since covariance could be made to treat relatedness as a statistical association rather than a measure of common ancestry, relatedness could actually be negative. What this meant mathematically was that while under conditions of a particular association altruism could evolve, under the conditions of another association *spite* could evolve: Everything depended on the environment. Spitefulness wasn't just the selfish harming of others to help oneself; it was doing harm to oneself in order to harm one's enemy even more. Explaining its evolution was therefore a similar problem to explaining the evolution of altruism: Both behaviors reduced fitness but existed nonetheless. It was a possibility Hamilton had entirely overlooked.

George pondered the larger meanings. Was there really nothing special about altruism? In evolutionary terms only a thin blue line seemed to separate it from spite. What determined whether a living being should act kindly or with malice had nothing to do with an "essence" or "inner core"—both, after all, resided within us. Instead, if the surrounding creatures were similar altruism could evolve; if they were different, spite was the solution. Pure unadulterated goodness was a fiction.

It was a shocking thought. And yet strangely, on further reflection, it was also laden with hope. For the adaptive success of altruism depended on the social environment, on society. Short of some unsullied divine morality goodness could flourish if it was recognized as important. Institutionalize cooperation and you kill competition; valorize self-interest and you penalize altruism. Virtue was already within us but needed to be helped along. Perhaps Skinner held a piece of the truth after all: Create the right conditions and goodness would see the light.

There was still no word from *Nature* about the antlers paper, but George was optimistic. Most important now was this new selection math. Finally he'd been sticking to one problem and not jumping around as he'd always done. "I think this work I'm doing," he wrote to Alice, emphasizing the adjective, "is really going to lead to something *important*."[29]

Then, on September 24, he relayed the news:

Dear Mother,
Something wonderful and totally unexpected happened to me an hour or so ago. I have been working on a paper on mathematical genetics and evolution, and I obtained a mathematical result that looked very interesting, but it was so simple that I felt sure someone must have discovered it before. So this morning I went to talk to a Professor Smith, an expert on mathematical genetics in the Depart-

ment of Human Genetics in University College of the University of London. He looked at my result and said it was interesting, very pretty, and he had never seen anything like it before.[30]

"He liked it so much that he took me to meet the department chair," he carried on in an excited letter to Annamarie. "90 minutes later," he continued to a friend, "I walked out with a room assigned to me, with keys, plus request for curriculum vitae so that they could make it official about giving me an honorary appointment."[31]

A complete unknown walking off the street into the chair's office and being given keys to a room of his own in arguably the world's greatest department of human genetics in a matter of minutes? It was the miracle George Price had been waiting for.

. . .

Tucked away on little Stephenson Way east of Euston Station, the Galton Laboratory at UCL was a storied home. Great names had been attached to it—Karl Pearson, R. A. Fisher, J. B. S. Haldane, Lionel Penrose—and now Harry Harris at the helm. "Like the chambers and corridors of some vast battleship," an observer remarked, "its rooms often seem to be below surface even if they are not. Its parquet and paneling are easily overlooked: its underlying grandeur is subordinated to the practical demands of intellectual inquiry."[32]

Cedric Austin Bardell Smith, Hamilton's old boss, was the Weldon Professor of Biometry into whose office George had walked. A Leicester-born Quaker five years George's senior, CABS was known for his gentle heart and mathematician's quirky sense of humor. What did Jesus mean when he said, "Heaven equals $ax^2 + bx + c$"? was an example (answer: It's a parabola); another was the invention of a new system of arithmetic based on counting from one to five and replacing all numbers greater than that with the same number subtracted by ten and printed upside down. As a student at Cambridge he and three equally offbeat friends formed the Trinity College Mathematical Society, publishing solutions to arcane problems under the pseudonym Blanche Descartes, a mythical Frenchwoman still referred to in the mathematical literature.[33]

One example was the problem of whether it is possible to cut a square into smaller squares each of which is different. The solution gave birth to the "square squared," a notion that proved highly useful in the design of electrical networks. Another example was the counterfeit coin problem: If you have twelve pennies, one of which is counterfeit and differs in that it is slightly heavier, and are given a pair of balance scales—what is the smallest number of weighings that will pick out the false coin? Cedric's solution ended up pioneering the field of search theory, a branch of mathematics commonly used in computing, economics, and in locating airplane crashes and lost mountaineers. (The answer to the problem is 3.) Most of all, though, Cedric Smith had been a protégé of J. B. S. Haldane, inventing powerful methods to map genes on chromosomes. In fact he had succeeded him. He was one of the world's leading biostatisticians.[34]

Meanwhile the skeletons of George's past continued to haunt him. He had left his wife, abandoned his daughters, been a lousy son to his aging mother. His behavior was self-destructive, people said, and deep inside him he knew it was true. He was still "daydreaming about torturing Ferguson." Back in America, Julia had come into her inheritance and could take care of the kids now. He was essential to no one; if he died not a soul in the world would be the worse for it. The UCL appointment was flattering but wouldn't pay the bills. Unless something extraordinary happened he planned to kill himself, he wrote to Annamarie, "since it isn't worth the bother of working just to stay alive."[35]

And yet . . .

Family, strangers, altruism, spite—the ideas swam in his head like drunken piranhas. To tame them he'd need to turn to science. In a way, he knew, they were his lifelines. Under the positively impressed, somewhat flabbergasted gaze of CABS he hunkered down once again and went to work.

His goal was clear: to fathom the mystery of family. Mate choice, fatherhood, individual interest versus common good—these were the issues he would tackle. It was to be a clean affair, and perfectly rational, nothing like the mess he had made of his own life. Developing mathematical tools for making evolutionary inferences would be the only way "to protect against biasing effects of emotional prejudice." Besides, quoting Haldane, with whose work he now began to become familiar, "an ounce of algebra is worth a ton of verbal argument."[36]

In a direct translation of the optimization work he had done for IBM on the "register problem," he set out to model human behavior. An "optimal" behavioral strategy was one that would maximize the frequency of an individual's genes in the next generations. Just as Ardrey and Morris had done, the method would be to consider a problem facing tribes of twenty to fifty hominids in the Middle and Upper Pleistocene, imagine a number of alternative behavioral strategies that might serve as solutions to the problem, and then to compare them with present behavior. Under the assumption that our ancestors were very likely to have developed genetically optimal behavior and to have maintained it for a long time—after all, *Homo sapiens* was a highly successful species—the strategy most similar to observed behavior today would likely have been the one that had evolved.

He started with basics. How, for example, did our ancestors allocate food? One optimal solution could have been complete sharing and cooperation: promiscuous, noncompetitive mating, cooperative rearing of the young with little or no recognition of individual motherhood and fatherhood, and retaliation against anyone out for himself. Just as with the antler model, an individual would increase his fitness by cooperating with others and thereby avoiding punishment, and by helping to punish others deficient in cooperation and thereby causing them to cooperate. It would have been a veritable Stone Age Plato's Republic.

But there was an alternative. What if the tribe chose cooperation in hunting by adult males, but individual and family action in all other areas? Hunting spoils would first be divided among the males, who would then distribute their share to women and children as a matter of personal choice. Such a system, George quickly saw, would favor monogamy, or at least something close to it. The reason was that if a man tended to keep the meat to himself when food was scarce so that his "wife" and children suffered severely while he ate comfortably, he would, on average, leave fewer descendants. Genes correlated with such behavior, therefore, would soon diminish in the tribe, and the behavior become less common. On the other hand, if a man was a wonderful provider and yet tended to swap wives every few years, most of the time he'd be providing food for the children of other men while neglecting his own older offspring; therefore, since gene frequency is a ratio rather than an absolute amount, the better he was at providing, the more effective he'd be in decreasing the frequency of his own genes in the group.[37]

True, there existed sexually promiscuous societies that deviated substantially from monogamy. But considering the enormous changes in living conditions that had occurred over the last twenty thousand years, it was remarkable how much the vast majority of humanity seemed to behave in rough accordance with the family model. Hypothesis 2 was a better bet than hypothesis 1.

Family, then, had developed under selective pressures related to food distribution at a time in human evolution when hunting by all-male bands became important—perhaps when hominids came down from the trees and began walking the savannas. Amazingly, exalted "fatherhood" might have been an optimal solution to the mundane challenge of securing daily grub. Even the heights of love were just an invention to oil optimality. After all, George wrote to an instrument-maker friend in the States from the days when he was thinking of building Skinner his Teaching Machine, love couldn't be an automatic consequence of "reinforcement" since it doesn't necessarily bring happiness. Something more powerful, like genetic evolution, had to be responsible.[38]

But this was all conjecture. To rise above it George would have to turn to math. Math would help make sense of mate choice and sexual selection, of nepotism and spite, of reciprocity and cooperation, of the interaction of cultural and genetic inheritance. Most of all, though, it might help solve the ultimate riddle: Where had evolution placed its eggs—in the individual or the group? In the gene or the family? Whose interest was it really trying to optimize?

With these thoughts in mind, he took another look at his equation.

. . .

It was the beginning of March 1969. Back in America, Kathleen and her new husband, Ronnie, were expecting a baby, and Alice's health was deteriorating. Clots in two toes had led to gangrene, Edison reported, and she was going to have to have her left leg amputated above the knee that week. The antlers paper had been provisionally accepted by *Nature* in February on the condition that it be substantially shortened. George was excited, but there was no time to celebrate. Once again the surgeons were having their way; he'd need to fly back immediately to see his mother for the last time.[39]

When he arrived at Midtown Hospital two days later, he found Alice in bed. The amputation had been a success but all was not well. "What happened in school today," she asked, taking his hand and looking up at him sweetly. "Did you wear your shoes. . . . Did you drink your warm milk?"[40]

As Alice declined, Kathleen delivered a baby boy in California, Dominique. She had no idea that her father was in America. Fear of Julia causing trouble over arrears in his support payments had overcome any parental, and grandparental, sentimentality. He was staying at Alice's home on West Ninety-third Street, going through all the old papers and photographs and closing it down. If she recovered, Alice would be going to the DeWitt Nursing Home. Meanwhile, George was selling furniture and getting rid of all her cats. Above all, he was battling Miss McCartney, the intransigent roomer, a seventy-five-year-old, two-hundred-pound, alcoholic former legal secretary, a "creature from a nightmare" who knew all tricks of the law and was unwilling to leave the apartment. Where, for goodness sake, had the days of the Japanese gentlemen gone?[41]

George had started the eviction process but was losing the battle. Short on cash, he had written to Al; Ludwig Luft, the instrument maker; and his elderly aunt Ethel in Michigan asking for loans. In his quirky way he even wrote to the president of Air Products & Chemical Corporation, based in Allentown, Pennsylvania, a man who in 1955 had contributed money to the University of Minnesota for George's research on ESP. To Bentley Glass, geneticist and president of the American Association for the Advancement of Science, he sent a fifty-two-page mathematical treatise on selection, asking whether Glass might help secure a fellowship for him, something, say, like a Guggenheim. None had replied yet, except Luft with $200. Edison had generously paid for George's airfare and expenses, but bitter over little help from him at the apartment, George was depressed and exhausted.[42]

To escape Miss McCartney he'd take the subway downtown to the Forty-second Street Public Library in the afternoons. He had failed to interest any magazines in articles but got $275 for helping to write the master's thesis in business administration of the uncle of a cute Yeshiva University grad student he'd met in the library. Sandy was more than twenty years his junior; going out with her made him feel young again. However wonderful the relationship, though, self-destructiveness, as usual, proved more comfortable a companion. "I am careful to keep my hate alive," he wrote to Tatiana, "since to let it abate would be giving in to the evilness of Ferguson. . . . He has beaten me physically but as long as I hate him and seek revenge, he has not beaten me mentally."[43]

Then, in mid-April, a third paper caught his eye.

. . .

Population growth, the biologist Garrett Hardin argued in "The Tragedy of the Commons" in *Science*, was a "no technical solution problem."[44] Like winning a game of tic-tac-toe against a competent opponent, or gaining more security in the Cold War by stockpiling weapons, maximizing population growth in a finite world was a technical impossibility. Malthus had been right, Adam Smith and his Chicago School followers mistaken. Limited resources render the Invisible Hand a farce; far from bringing about a collective paradise, individual interest will hasten the ruin of all. Borrowing a metaphor printed in a little-known pamphlet by an amateur mathematician in 1833,[45] Hardin explained:

> *Picture a pasture open to all. It is to be expected that each herdsman*
> *will try to keep as many cattle as possible on the commons. Such an*

arrangement may work reasonably satisfactorily for centuries because tribal wars, poaching, and disease keep the numbers of both man and beast well below the carrying capacity of the land. Finally, however, comes the day of reckoning, that is, the day when the long-desired goal of social stability becomes a reality. At this point, the inherent logic of the commons remorselessly generates tragedy.[46]

The reason was because each and every herdsman asking himself, What is the utility *to me* of adding one more animal to my herd? would answer in the same way: Since the benefit would all accrue to him while the cost (of depleting the commons) would be shared by everyone, adding one more animal would always be the thing to do. And another, and another. "Therein is the tragedy," Hardin bemoaned. "Each man is locked into a system that compels him to increase his herd without limit—in a world that is limited."

The ultimate result would be the destruction of the commons. Whether it was the use of national parks, radio frequencies, parking, fishing, foresting, or pollution, there was a true conflict between personal interest and the common good. John von Neumann and Oskar Morgenstern had already shown that maximizing for two variables is a mathematical impossibility; if the tragedy of the commons was going to be solved, nothing but "a fundamental extension in morality" would suffice.

For Hardin, a hardened realist, an appeal to conscience wouldn't work, though; it was "mutual coercion mutually agreed upon" that was humanity's hope for deliverance. The trick, far from expecting all men suddenly to become angels, was to devise clever mechanisms of regulation that, as far as possible, would allay the conflict between the common good and the pursuit of personal gain. In a way it was Skinner all over again. Freedom, the Universal Declaration of Human Rights notwithstanding, could not always be just a matter of personal choice. At times, as Hegel had realized, it amounted to nothing more than the "recognition of necessity."

Sitting in the New York Public Library, fretting over his return uptown to the "nightmare" Miss McCartney, George saw immediately how the tragedy of the commons applied to evolution. After all, in nature there was also often a conflict between the good of the group and the individual. And yet, as far as he could tell, group selection was dead: Hamilton, Maynard Smith, and Williams had seen to its demise.

George decided to keep a more open mind. For group selection to work there needed to be differences between groups in a population. Evolutionists had rejected such a possibility based on their rejection of Sewall Wright's 1945 model; migration between groups would swamp any differences created by random drift. But Wright himself, George had learned in correspondence, now saw his own model as a gross simplification; from his perspective its rejection was neither here nor there with respect to group selection.[47]

Group selection was theoretically possible; this much even Maynard Smith had willingly allowed. But it could be more than this, possibly a reality. In fact, Hamilton had suggested a perfect example of group selection in his paper on extraordinary departures from Fisher's 1:1 sex ratio, though no one seemed to notice, including Hamilton himself. Mammals, too, seemed to challenge the wisdom of the day. While both possessed well-developed social dominance systems, neither chimpanzees nor gorillas, for example, exhibited competition between males for mates. Still more puzzling were male wolves: The small amount of evidence available suggested that there is an inverse relation between dominance and mating success. True, Wynne-Edwards had been consigned to the back pages of history, but such behaviors were difficult to square purely on individual selection.[48]

When it came to the evolution of man, it seemed obvious to George that group selection must have played a role. Early humans had lived in groups, and cultural inheritance could have gone a long way in preserving the kinds of genetic behavioral differences that would otherwise be swamped by migration.

Huxley, Kropotkin, Allee, Wynne-Edwards, Emerson, Fisher, Wright, JBS, Maynard Smith—he read them all and more. Every one was occupied with the question of the unit of selection and every one seemed to have an answer. Still, was it conceivable that each held a portion of the truth, that none was entirely right but none entirely wrong, either? Could the same kinds of economic mechanisms Hardin argued were necessary to square the individual and common good exist, biologically, in nature? Could it be, in other words, as Darwin had noticed when contemplating the ants, that selection worked on different levels *simultaneously*?

At the end of April he said his good-byes to Alice, arranged for Miss McCartney's eviction, closed up the old apartment, and got on a plane. Back at UCL, Cedric Smith was pushing him to complete a grant proposal to the Science Research Council. Classical theory, CABS wrote in his report, assumed that each individual possesses a "fitness" independent of the "fitnesses" of others, and that by analyzing the situation mathematically, the course of the evolution of a population can be predicted. This was a good approximation of reality in some situations, as when an inherited disease shortens life, but where the interaction between individuals plays an important role in determining their fate it simply wasn't good enough. Sexual selection, parental care, formation of families and communities: All were situations in which conflict and cooperation were paramount. Interaction, not singularity, was the name of the game, and few besides Fisher and Hamilton had ever played it. "Dr. Price has come to the subject comparatively recently," Smith wrote almost apologetically, adding, "I have however been greatly impressed by his ability."[49]

For George, meanwhile, little had changed. "I continue to have the plan of limiting my life span to about 50 years," he wrote to Tatiana.[50] Soon after, the news came that the SRC had awarded him a three-year grant effective July 1, 1969. Walking home from UCL to the flat on Little Titchfield, he eyed the spire at All Souls. The only passion that remained in him was to crack the mystery of the evolution of family—that and exacting revenge on Ferguson.

A few days later Alice died peacefully in New York.

· · ·

Back in February, before leaving for New York, George heard a talk delivered at the Royal Society of Medicine by a psychiatrist from Maudsley Hospital named John Price. "The other Price" as he became known, had been interested, too, in ritualized animal combat: If there was no physical pain or incapacity, what made animals yield to the winner? Price's idea was that ritual yielding is subserved by mental incapacity and mental pain, and that human depression and anxiety—both painful and incapacitating—might have evolved from it. The notion was attractive to a clinical psychiatrist. It suggested a wealth of animal models for the study of human neurosis as well as a battery of new ideas for prophylaxis and treatment.[51]

Most important to George, though, was the gamelike logic behind the claim. Why, in fact, should there be any variation in yielding behavior? In terms of the group the answer was obvious: The greater the variation of yielding tendency in the population, the greater the chances that any two contestants are unevenly matched, and the shorter the duration of battles. If selection between groups had been important in evolution, groups with shorter spats would have surely done better than those with costly, protracted encounters.

But variation in yielding behavior could also favor the individual, as "the other Price" explained:

> *The disadvantage of being a yielder is counterbalanced by the likely mortality when two non-yielders meet each other. Thus it is advantageous to be a yielder when everyone else is a non-yielder, and to be a non-yielder when everyone else is a yielder. This dependence of the advantage of one's phenotype on the phenotypes of the rest of the population . . . tends towards the maintenance of variation in the population.*[52]

It was precisely the same notion George had come up with in his paper on antlers, and it hinged on the logic of games. Hardin had shown that there was often conflict between individual interest and the common good, and von Neumann that you couldn't maximize simultaneously for two variables. And yet even here, where the individual interest and common good seemed to correspond, George saw no reason why natural selection couldn't be working on both at the same time. An "outsider" untrained in evolutionary theory, he really had no reason to choose sides: Why limit the scope of Darwin's theory? The only relevant question, it seemed to him, was not whether selection was working on the individual or the group, but how, in each and every case, to tell which force was stronger.

Once more he turned to his equation. Covariance was a most simple relationship, a mathematical tautology. To make it even better, George now saw, he'd need to include a further component: something called "transmission bias." If a trait moved from a "mother" population to a derived "daughter" population or, in evolutionary terms, from generation to generation, it was important to know not only if it helped to increase fitness but also what the chances were for it being inherited. If the trait was genetic, for instance, it would be important to know whether it had undergone any kind of mutation or, equally important, whether the gene somehow biased the system so as to pass itself on more frequently than would be expected. And so, to get the equation just right, George added a second term

$$\Delta \bar{z} = \frac{Cov(w,z)}{\bar{w}} + \frac{E(w \Delta z)}{\bar{w}}$$

where $E(w\Delta z)$ is a measure of the extent to which the trait, z, will be passed on faithfully. Usually parents pass on their genes to children at random, meaning that genes don't affect which particular sperm will fertilize the egg. Since it almost always equaled zero, the transmission term could therefore usually be safely ignored. But now his equation was formally complete: Given a trait, z, its fitness, and the likelihood of its transmission, it could tell you precisely how it would evolve from one generation to the next.

The equation partitioned trait change in evolution into selection (Cov) and transmission (E). This was valuable enough,[53] except that the new term did much more. With a few simple substitutions it showed that selection could work at two levels simultaneously; in fact it could even partition them to see how much each contributed to the overall change. Instead of defining the two terms of the equation as the selection and transmission terms, corresponding to the individual and the genes in the sperm and egg respectively, they could be bumped up one notch in the rung and redefined as relating to the individual and the group. The key was for the far right-hand side part of the equation to be treated just like the left-hand side of the equation, only at a lower level.[54]

This was enormously useful. Imagine a group of people who

only know how to be altruistic to one another, and have never heard of selfishness. Imagine a second group whose members never heard of altruism but rather only know that each has to be out for himself. Clearly the first group would function better as a society, for it would revel in cooperation, whereas in the second everyone would be poking one another's eyes out. But what would happen if the groups were not entirely pure? An altruist would be easy fodder in the selfish group, whereas an egoist would quickly hoodwink all the members of the altruistic group. And so while the altruistic group is fitter than the selfish group, selfish individuals are fitter than altruists within each group. The question was: If two such groups live side by side, which trait will evolution select—altruism or egoism? Which is stronger—the interest of the group or the interest of the individual? The covariance equation could tell you the answer.

Without entirely meaning to, George had written an equation that had the power to do what generations since Darwin had failed to: watch natural selection work in all its glory at different levels at the very same time. In fact, since it was infinitely expansible, all the levels of life could be included: gene, cell, individual, family, group, species, even lineage. Like a giant eye perusing creation, selection could see everything; all one needed to do was to choose which two levels to compare. Finally, after all the evolutionists who came before George, not to mention Hardin's "Tragedy of the Commons," the equation could specify the exact conditions under which the good of the group would upstage the good of the individual. Crucially, contra Hamilton's kin-selection model, it needn't write off goodness as merely apparent: When selection worked more strongly between groups than within them, a genuine altruism could evolve.

Staring at his own creation, unbelieving, George thought of it as a "miracle."[55]

. . .

Hamilton had arrived back from Brazil with Christine, Romilda, and Godofredo in January 1969, and was invited in May to give a keynote address at the "Man and Beast" meeting at the Smithsonian Institution in Washington, DC. The conference had been organized with great hype and expense to impress on politicians and the public that biology and evolution were relevant to man, but the bushy-browed loner brought with him more than they had bargained for. Xenophobia, Hamilton stated—even relishing cruelty to others—was selected for in the evolution of man, since altruistic groups must expand at the expense of other groups and to do this they need to fight them. Reliance on the instincts of a supposed "noble savage" was no answer to the prisoner's dilemma or to Malthus: What produces altruism and kindness at one level only serves to produce hatred and violence at the other. Alas, group selection doesn't exorcise the harsher aspects of natural selection; actually it leads to fascism.[56]

The audience was shocked. At the aftertalk cocktail party a senator's wife pinned Hamilton with her "chin and fierce eyes." Could his theory help reduce crime in America's inner cities? she asked, desperate to eke out a measure of solace. Hamilton himself was deeply troubled by the thought: If not in human nature, where else could trust and hope in a peaceful and creative future be placed? And yet only honesty about the manner, he determined, stood the faintest chance of taming "the beast within."[57]

It was a terrible irony to come to terms with: With biology at the helm, goodness could be bought only at the price of cruelty. Worst of all, this tragic truth would be most pronounced in man; in all other creatures group selection was such an ill-defined abstraction, Hamilton thought, that it could be safely omitted from the evolutionist's toolkit.[58] The reason was that there was simply no plausible mechanism, like cultural inheritance, that would allow for it. Only in humans could the force of culture work to counteract selection at the genetic and individual levels; in the process it would allow selection to work at the level of the group. When it came to all other creatures, Hamilton remained a bitter enemy of the old notion of the greater good.

Altruism entailing malevolence had been precisely what George had written to him about a year ago. Hamilton had forgotten about it. But having worked out more of his equation, George was now ready to write to him again. "I must be one of the world's worst correspondents," his letter began, a year of silence having lapsed since Hamilton wrote his sweaty reply from Mato Grosso. Then he went on to retell the entire story of trying to rederive Hamilton's kin-selection math, coming up with the covariance equation, going into Cedric Smith's office, getting the keys and honorarium at UCL. Clearly, alongside altruism, spite was a possibility: What kin selection will do is increase the frequency of a gene that causes animals to give greater benefit to near relatives, but this always meant that lesser relatives had to be relatively harmed. Hamilton thought that kin selection could cause individuals to act altruistically to the group as a whole, but this was an ideal miscalculation. It was an oversight he would obviously need to fix. "But I did want," George wrote, "in view of your friendly correspondence, because I respected your work, and because every-one makes a mistake now and then—to publish in a way that would not embarrass you." Perhaps the best way, then, would be simply for Hamilton to publish his own correction. In any event George would be glad to speak with him on the phone.[59]

When Hamilton called the next day, he found the voice on the other end strange, rather guarded, "squeaky and condescending." His covariance equation, George said, was "surprising to me too—quite a miracle," really. But most surprising to Hamilton was what George had to say next, following a rather awkward silence "Have you seen how my formula works for group selection?" "I told him, of course, no," Hamilton later remembered. "and may have added something like: 'So you actually believe in that do you?' "[60]

After the phone conversation, each man returned to his own world. Hamilton was adjusting to family life and the two new adopted kids, as well as figuring out the outlines of a new theory of the "geometry of the selfish herd." George, for his part, had one more proof to complete the equation, and was using the FORMAC computers at UCL to help with the algebraic expressions. "I think it will be considered a very important piece of work," he wrote to his elderly aunt Ethel in Hiawatha, Michigan. "I wish mother might have lived to see it."[61]

. . .

October was the driest and warmest England had seen in centuries. Over the summer George had been working on a long paper, an ambitious attempt at a general theory of selection. Now, though, he had a change in publication plans. On November 11 he sent off a short letter to the editor of *Nature*, barely longer than a page and, reflecting his independent path, with not a single reference note. In it he derived his equation and explained how covariance could be applied to problems in genetic evolution, without dwelling on any in particular. It was a "preliminary communication," he wrote to Annamarie, just a tool that would ultimately allow him to crack the mystery of family.[62] A few months later he sent the longer paper to *Science*.

The rejection arrived in February; "It is too hard to understand," *Nature*'s editor, John Maddox, wrote. The letter did little to help George's mood: He was working as a teaching assistant in a statistics course at the Galton to earn a little extra cash now, and still thinking about possible magazine articles to sell in the United States. Obviously the readers were just "stupid"; after all, this was the miraculously simple equation that had earned him his room at UCL. The year 1970 had rolled in, and still no breakthrough.[63]

The next month, as he walked home to Little Titchfield beneath the ever-present All Souls spire, George discovered in the mailbox that his troubles hadn't ended there. *Science* had rejected the longer selection paper, too. It was too abstract, the note said, and besides that brimming with hubris, not obviously applicable to any particular scientific problem. George thought it "vicious . . . really breathing hatred toward me," but the rejection, perhaps, wasn't all that surprising. After all, in capital letters so that no one could miss it, he had stated that he was in search of a general "Mathematical Theory of Selection" analogous to what Claude Shannon had done with the theory of communication. It was quite a claim—especially coming from someone no one had heard of.[64]

Back in America, Al Somit had become the chairman of the Political Science Department and vice president elect of the State University of New York in Buffalo. He was planning a session at the International Political Science Association meeting that summer in Munich on "Biology and Politics" and wrote to ask his old buddy if he'd like to attend. "Our interests seem to be converging," he added amid their usual Chicago-day banter. George, for his part, complained that even in London he couldn't escape the "non-Aryans": his landlords, Basil and Howard Samuel; the kosher bakery across the street; the kosher restaurant at the corner; Sandy, the Yeshiva University fling who had visited in May; his instrument-making friend Ludwig Luft, who was lending him money; Harry Harris, the chairman of the department at Galton—all were Jews. Al shrugged it off as usual George.[65]

But not all was as usual. George was feeling the world slowly closing in on him. He contemplated Munich. He contemplated life. Down and out, dejected, he offered to speak about morality being nothing but a "masquerade."

His search for a miracle, it was patently clear, had ended in glorious failure. His mother was dead. Julia hated him. Edison and he had long drifted apart, but so now were his quickly maturing daughters, living out in California and starting families of their own. He was alone in a foreign country and had no friends. His arm and shoulder were partly paralyzed. His forays into human evolution were incomplete, dependent on a selection mathematics that had been humiliatingly rejected, twice. Swinging London was as much at odds with his grim lifestyle as the search for the origins of family was with his tattered personal affairs. His equation might say otherwise, but if there was any goodness in this world, George Price hadn't found it.

Selection and Covariance

THIS is a preliminary communication describing applications to genetical selection of a new mathematical treatment of selection in general.

Gene frequency change is the basic event in biological evolution. The following equation (notation to be explained), which gives frequency change under selection from one generation to the next for a single gene or for any linear function of any number of genes at any number of loci, holds for any sort of dominance or epistasis, for sexual or asexual reproduction, for random or nonrandom mating, for diploid, haploid or polyploid species, and even for imaginary species with more than two sexes

$$\Delta Q = \mathrm{Cov}(z, q)/\bar{z} \qquad (1)$$

The equation easily translates into regression coefficient (β_{zq}) or correlation coefficient (ρ_{zq}) form

$$\Delta Q = \beta_{zq}\sigma_q^2/\bar{z} = \rho_{zq}\sigma_z\sigma_q/\bar{z}$$

. . .

Fifth, it seems surprising that so simple a relation as equation 1 has not (to my knowledge) been recognized before. Probably this is because selection mathematics has largely been limited to genetical selection in diploid species, where covariance takes so simple a form that its implicit presence is hard to recognize (whereas if man were tetraploid, covariance would have been recognized long ago); and because, instead of using subscripts as "names" of individuals (as I have done), the usual practice in gene frequency equations is to use subscripts only as names of gene or genotype types, which makes the mathematics seem quite different. Recognition of covariance (or regression or correlation) is of no advantage for numerical calculation, but of much advantage for evolutionary reasoning and mathematical model building.

Some genetical selection cases (such as group selection) and many forms of non-genetical selection require more complex mathematics than that given here. I plan to discuss these and other matters in papers now in preparation.

I thank Professor C. A. B. Smith for help, and the Science Research Council for financial support.

GEORGE R. PRICE

Galton Laboratory,
University College,
Stephenson Way,
London NW1.

Received November 12, 1969.

Excerpts from George Price's seminal paper in Nature

10

"Coincidence" Conversion

*M*unich never materialized, but in May a turn arrived as unexpected as it was welcome.

"I am enchanted with your formula," the usually lugubrious Hamilton had written in a hurried script, barely able to contain his excitement. "I really have a clearer picture of the selection process as a result. In its general form I can see how we might use your formula to investigate " 'group selection.' "

It was quite an about face for a man who had judged group selection as so "wooly" an abstraction that it could be safely omitted from the tool kit of the evolutionist. Altruistic behavior was disadvantageous within groups; to evolve it would require groups fighting *among* each other. Kin selection, therefore, could not be an alternative to group selection: Both existed in nature's plan and needed to be related to each other. Accordingly Hamilton had taken George up on spite, and, in the process, devised a foolproof stratagem. It was a piece of trickery that made him especially proud: He would send his article on spite to *Nature*, wait for its acceptance, and then write to its editor, Maddox, to say that he would have to withdraw since it was entirely based on the covariance math of George Price, whose paper—in case Maddox forgot—had been rejected by him in February. Hamilton would, of course, be happy for "Selfishness and Spiteful Behaviour in an Evolutionary Model" to appear in *Nature* but not until the status of Dr. Price's article was finally resolved. A masterstroke of subterfuge, a coup of cunning and guile, amazingly, it worked. George's letter to the editor, *Nature* soon replied, would be published as an article, "Selection and Covariance."[2]

Walking home that day from UCL, as usual, George eyed the All Souls spire jutting up above the buildings. It was constructed of seventeen concave sides encircled by twelve Corinthian columns, the capitals Ionic in design, made from Coade stone, and the winged heads of the cherubs based on a design by Michelangelo. Three large clocks beneath the columns counted the minutes and hours with black and golden Hermle spade hands. The spire was a perfect cone, as sharp as it seemed foreboding, fearless in the way it pierced the overcast sky. As he turned from narrow Riding House Street into Little Titchfield, George began wondering about his luck, and, in particular, about coincidences.

Later he would write that the "coincidences" that forced him to convert were on the order of $1/10^{30}$, odds so fleetingly miniscule that he simply had no choice but to "give in and admit that God existed."[3] "About the beginning of June," he wrote to his brother, Edison, in the fall, trying to explain,

> *I happened to notice one surprising coincidence in my life, and this started me searching back through my calendar books and letters and other material, and noticing a long succession of other improbabilities, until the improbability level became astronomical. I listed a long series of independent events having improbabilities of the order of 1/100 or 1/1000, that fitted together into a meaningful pattern, and when I multiplied these together the product was something like i over i followed by twenty or thirty zeros.*[4]

What had been the initial "surprising coincidence"? It was bizarre and absurd, maniacal and eerie. But it had George entirely transfixed.

Back at Christmas, Bob and Margarite Sheffield, old Price family friends from New York City, had been visiting London with their two daughters, Anne and Sally. Almost immediately, though she was barely eighteen, George took a liking to Anne. Boo wasn't particularly happy about it, and hinted to George to back off. But Anne was continuing with a friend on a trip through Europe, and after visiting Finland, Sweden, and Germany, wrote to George innocently that she was scheduled to return to London on May 15, this time alone. "Since your visit played a critical role in this," he wrote to her later, explaining his conversion, "I know that you will be interested in hearing how this came about. Prepare yourself to hear some surprising things, for there are more things in heaven and earth than you, I presume, imagine."[5]

This is how it happened: When George first saw Anne over Christmas, he noticed her uncanny similarities to another Anne, the old girlfriend from the Midwest who had come to New York to see him just before his meeting with Emanuel Piore, director of research, on the twenty-third floor of the IBM Building in New York, on July 16, 1957. It was at that fateful meeting that George turned down Piore's offer to join IBM as a senior researcher, based on the draft of his Design Machine published a few months earlier in *Fortune* magazine. And it had been on the previous day, the fifteenth, that he met Anne and instead of offering to marry her, told her he'd think about things. When *he* had wanted to marry her around the time he had contracted polio, Anne had broken off their relationship for another man. Now, still jealous, he figured he could take his time.

Clearly, he came to believe, this had led to his downfall. For had he asked Anne, who was a Roman Catholic, to marry him that day, he would have been focused first and foremost on nailing down a stable job. And had he accepted Piore's offer, he would never have found himself in the drug-infested predicament in the Village, selling himself short on technical manuals for GE and Sperry-Marine, and trying and failing to write *No Easy Way*. In fact, had he taken Piore's offer, he would never have joined IBM on a lower rung, and might never have contracted thyroid cancer. If he hadn't been sick, he'd never have come under "butcher" Ferguson's knife, and his life might not have descended into misery.[6]

July 15, 1957, had been a fateful day, all right, of this he was certain. And so, meeting Anne Sheffield now, thirteen years later, couldn't just be a "coincidence." She not only looked just like the earlier Anne, she had the very same name, the very same inflections. It didn't seem to matter that he was forty-seven and she was eighteen and the daughter of close family friends. He wanted her, he wrote to her, "so very very much." On everything important to him—choosing to become a chemist, choosing to marry Julia, choosing to go to Ferguson, choosing not to marry the first Anne—he had always taken the wrong path. This time he was determined not to make another mistake.[7]

Innocent and spooked, Anne left London for home at the end of the week. Convinced that there must be more than just the hand of chance involved, alone again and lovestruck, George remembered a poem by Henry Constable that an old Harvard friend had once sent him. "To live in Hell, and Heaven to behold / To welcome life, and die a living death / . . . If this be love, if love in this be grounded . . ." He couldn't quite remember the rest, and started searching for the poem in his papers. When he couldn't find it, he ran over to the British Museum Library. Fingering through the index cards, he came upon another Henry Constable, not the sixteenth century poet but a twentieth century theologian and believer in conditional immortality whose titles—*Hades: or the Intermediate State of Man*; *Restitution of All Things*; *The Duration and the Nature of Future Punishment*— sent shivers down his spine. Once again—an identical name and a message![8]

Walking back to his flat on Little Titchfield beneath the spire, George contemplated his phone number. It was 580-2399, the last four digits signaling a minute before midnight. Of course 2359 was technically correct, but to him 2399 was meaningful, and that's what really mattered. Was this another message? Could someone be signaling to him that the clock of doom was about to strike, that only a moment remained to make the right choice in life, finally, *once and for all*?[9]

As he looked through his diaries and letters, he saw more and more "coincidences." Names, numbers, dates—they aligned in such ways that a pattern couldn't honestly be ruled out. Someone was speaking to him, of this he was sure. "It wasn't that I wanted to believe," he later wrote to Anne, "but there wasn't any alternative." Finally, forty-seven years old and a lifelong fanatical atheist, he gave in and bowed to the spire. On June 14, George Price walked out of his flat up the stairs and through the warm, honey-colored circular portico of All Souls Church.

· · ·

All Souls had been designed by John Nash, the favorite architect of King George IV, formerly the Prince Regent, and was consecrated by William Howell, the bishop of London, in 1824. A cartoon from those days depicting Nash impaled on its spire and referring to All Souls as "an extinguisher on a flat candlestick," made clear that the peculiar combination of Gothic spire and classical rotunda was not at first universally admired. But criticism soon died down. Since 1870, a history of the church explained, "All Souls seems to have been simple in worship and vigorous in missionary effort. Apart from two short periods it has always been evangelical in tradition." [10]

By "coincidence" again, George happened to walk in on the one Sunday of the month when services were intended for guests and novices rather than regular parishioners. With a program in hand (they weren't distributed on other Sundays), he took his seat on a spare wooden pew in the west gallery. In the handsome Spanish mahogany chancel above him, designed by Nash, the organ sang from newly gilded nonspeaking pipes restored after damage incurred in the war. It was a spacious hall, bright and airy. Behind the podium a large Richard Westall *Ecce Homo* painting of Christ in the hands of his enemies dominated. Reading the words above it, etched in gold in a half circle, George felt as though the whole service might have been planned especially for him: "God So Loved the World He Sacrificed His Only Begotten Son."

The next day, when he came to inquire about joining the Church of England, he was surprised to find that he had been baptized as an Episcopalian, which meant that he was effectively already a member. Since neither of his parents had had anything to do with Episcopalianism, and since it saved him a whole lot of hassle, this too seemed like another "coincidence." Going through his records later at home, a £120 covenant for the year in hand, he remembered that while living in Minneapolis, to assuage Julia, he had joined a Unitarian church whose minister was an agnostic and most of the congregation atheists. Now he really knew that nothing had happened by chance, that things had long ago been determined: The date, another "coincidence," had been June 14, 1955—fifteen years to the day before his first visit to All Souls.[11]

It was statistical evidence that had forced him to become a Christian, but now that it had happened new meanings soon arose. He had bought a secondhand Revised New Testament that Tuesday, and on Saturday came to listen to a sermon on "Christ on the Mount of Olives." He began reading his Bible furiously, making notes to himself. In his appointment diary "Scripture Center" and "Buy Book of Common Prayer" appeared side by side with "need guidance" and "Dr. Hawes—thyroid." In particular, he had begun intense conversations with John Stott, next to whose name in the diary appear the words, "Ask! Ask! Ask!"[12] The son of an agnostic Harley Street physician, Stott as a boy at Rugby in 1938 heard a sermon, "What Then Shall I Do with Jesus, Who Is Called the Christ?" and never looked back. With a double First in French and Theology from Cambridge he was appointed curate at All Souls in 1945 and rector in 1950, and quickly rose to become the leader of the Evangelical Church in England.[13]

George proved an opinionated novice. He was looking for hints in the Bible, a code. Stott told him to take things more slowly, to be less of a literalist.[14] George claimed that it was crucial to discover precisely what God wanted of man, and had opinions about Church misinterpretations. Conversations with Stott soon became acrimonious. More than anything now, George was concerned with the question of determinism: How much of fate was up to chance and free will, and how much had already been fixed firmly from above?

It was July 1970, a particularly warm summer. Earlier the previous year, short on cash, George had written to the old Harvard buddy who had sent him the Constable poem, Henry Noel, asking that he return half of a $62.70 shipping bill for furniture he had left in Henry's New York apartment when he was leaving for England in 1967. Henry had been a handsome, idealistic son of East Andover, New Hampshire, who left Harvard in 1942 to join the American Field Service in India and Italy, winning the Burma Star and Italian Theater Ribbon. But returning to the United States after the war and according to his own account, "heedlessly diving into the welter of contemporary American bourgeois decadence," he soon had a change of heart. Embarking as a purser on a Los Angeles Tanker Company ship, he sailed to France and voluntarily renounced his U.S. citizenship. "Harvard Alumnus Renounces U.S.," the *Boston Traveler*'s shocking headline ran, as did almost every newspaper in America from the *Plainfield Courier-News* to the *Dallas Times-Herald*. But Henry had already chosen his path: He was to be a "citizen of the world," never a national, and moved down to Kassel, Germany, to work as a bricklayer and "familiarize myself with the country and the people."[15]

Back in America now, having taken up residence as an alien immigrant admitted on the French immigration quota, Henry Noel was married with two children, a freelance editor of school textbooks working on the side on a book on UFOs as well as an autobiography. "I must say that I never thought of you as a likely convert," he wrote after hearing George's news.[16] He was delighted. He, too, had had a conversion some years before. In fact, besides himself, George was the only person "in all history, literature, and personal acquaintance" of whom Henry had ever heard who was drawn to Jesus by "the apparent inoperativeness of the laws of probability." His own experience in 1963 had involved the turning of the clock thermometer on the Union Square Savings Bank:

> *I was going to the Grand Central Station from our home on East 11th Street (where you once visited us) and one stage of my morning trip consisted in arriving at Union Square on an Avenue B bus which approached the square along a side street and reached its terminus, where I got out to take another bus just short of Park*

Avenue South. Every morning I would get out of the Avenue B bus and walk a few steps to the corner, where, in turning it, the first thing I would see would be the big turning clock, right by where I was to take my next bus. It quickly became my habit to check my watch against the turning clock, and to do so the instant I turned the corner and the clock came into view. But of course I could not do this if it was the thermometer side turned toward me in that instant that I turned the corner. It quickly developed that it was far more often the thermometer than the clock. . . . In the course of months—scores of times—of this refutation of the laws of probability (for surely there was no synchronization possible involving the arrival of the bus, the swinging of the clock, and the time it took me to reach the corner from the bus stop) I became progressively intrigued, then puzzled, then incredulous, then almost appalled.[17]

Henry's first son, Jack, had just been born, and he was feeling overwhelmingly grateful. But *whom* should he be grateful to? When he returned home one evening, flipped a penny that fell on heads nine times consecutively, called the next toss tails, and was right—he gave in and became a Roman Catholic right then and there. His old Harvard friend and George's Chess Club mate, Lloyd Shapley, now senior mathematician at the RAND Corporation, told Henry that if he had turned the Union Square corner and experienced the same thing, he would naturally have assumed that someone was playing games with him. Alas, Henry thought secretly, Lloyd had mistaken someone for Someone.

George was ecstatic. Could he show Henry's letter to Christians "as well as heathens I am trying to convert"? Henry's founding of the New York C. S. Lewis Society was a source of joy as well. Not yet familiar with all the writings of Britain's highest-profile convert (a hero to many an initiate), George had read *Mere Christianity* and *The Screwtape Letters* and borrowed *Miracles* from the library. In fact, after Henry sent him the contact information of an interested London fan, Kenneth Demain of Forest Gate, George began the business of creating a British chapter, even recruiting Lewis's own literary executor —his brother. Until a bigger space could be secured, meetings of the newly formed C. S. Lewis Society of Great Britain and Ireland would take place in George's living room on Little Titchfield Street.[18]

He was coming into his own as a Christian. He was "only 50 days old" he wrote to Henry, but he knew clearly, if he didn't mind him saying so, God preferred that Henry concentrate on spreading His Gospel through C. S. Lewis Society activity rather than writing a book about UFOs. And if Henry had yet to succeed in securing the cash needed to move his family to France, obviously it was because God wanted them to stay in New York a little longer. "I am not sure that we can so easily penetrate God's wishes by merely aligning them against the success or failure of our own pet projects," Henry replied. "All we really know of His wishes for us is that we should return to Him, knowing Him for our Creator."[19]

"How could you say such things?" George replied, indignant. "You tell me that God was concerned about whether you should flip heads or tails with a penny, and you doubt that God is concerned about whether you produce a UFO book or a C.S. Lewis book!" God was concerned about *everything*, and he had a plan for *everyone*. Matthew 28:19, Luke 9:60, Acts 10:42, Timothy 4:2—all made clear that He wanted man to preach the Gospel. Clearly Henry had been chosen to do this via Lewis. What he needed to do was to admit that "Lord" means exactly what it says. In case he didn't believe him, George wrote, he should go to a higher authority. "Henry, for God's sake, please ask God."[20]

In fact George already understood that God was constantly testing us. "You remember the story of Moses and the burning bush?" he wrote to young Anne Sheffield back in the United States:

> It says there (Exodus 3), "And Moses said, 'I will turn aside and see this great sight, why the bush is not burnt.' When the LORD saw that he turned aside to see, God called to him out of the bush, 'Moses, Moses! . . .' " Now the important point is that the bush that burned but was not consumed was something that Moses took notice of, that "he turned aside to see." And it was only after God saw that he turned aside to see that God spoke. So Moses, my New York friend, and I each noticed and "turned aside to see" something strange.

What we took notice of was quite different because we have different interests. We were each shown the sort of phenomena that we were interested in and that we would take notice of. Henry saw gentle intellectual "games" not because God is that sort but because that is what Henry would notice. And I saw especially coincidences involving girls because that was what I would especially notice. And, also, my coincidences included many involving names and dates because I tend to notice such things.[21]

It was clear then that Anne had been scheduled to be in London when she was in order to be George's "burning bush." Not that everything is predetermined; being able "to turn aside and see," was, after all, the key to every man and woman's fate. And yet, George now wrote, "it is obvious to me that I was allowed practically no freedom about the events of late May and early June." Christianity had been forced on him just as it had on Paul on the road to Damascus.[22]

His scientific work now assumed an altogether different purpose. Suddenly the fact that not one of the great minds since Darwin had come up with the simple covariance equation but rather he, a complete outsider, had, became illuminated: *George had been chosen.* He would abandon all thoughts of revenge against Ferguson. "You may later hear of some consequences from my conversion," he wrote to Anne, "since I evidently was converted in order to accomplish certain work, of which part should become widely known." Meanwhile, however, he had other duties to tend to. Annamarie and Kathleen and her baby, Dom, were scheduled to arrive on August 16 and planning to stay for a month. George was frantic. He was planning to convert them. If only he could find C. S. Lewis's "Difficulties in Presenting the Christian Faith to Unbelievers."[23]

. . .

"Selection and Covariance" was published in *Nature* on August 1, hardly noticed by anyone. Hamilton, it seemed, was the only person in the world who understood its significance. He'd been sending George drafts of his spite paper to allow him to check the math. Their meeting of minds had been so astoundingly familiar, he thought, that there was almost an "intellectual redundancy" in each other's presence. Like him George was a loner, Hamilton would later write to Edison; almost immediately he felt toward him "a very great kinship." Besides his obvious intellect, George's "rejection of compromise and his dauntless humour" were admirable. He was American (though the most "un-American American" Hamilton had met), well dressed, talkative, and clearly not a nature man. They were entirely different, Bill thought, and yet strangely almost identical.[24]

In July, Bill invited him over the weekend to the Berkshires, where he was living now with Christine, Romilda, and Godo near his sister Janet and her family.[25] Walking through the woods with dogs and children running beside them, stumbling through brambles and brackets, their shoes covered in mud, George told him about his conversion. Clearly George's soul was not in tune with quotidian human realities. Just like his own it was someplace else, inspired. But how had it ended up with Jesus? Years beforehand, Darwin had gotten a scare when he heard that Alfred Russel Wallace, his codiscoverer of evolution by natural selection, had turned to spiritualism and was now advocating the intervention of extrahuman intelligences in the evolution of man. "I hope you have not murdered too completely your own and my child," he'd written to him. A nonbeliever, gentle but firm, Bill now tried to grasp the scope of the affair, sending out his own feelers like Darwin before him:

> I read your ESP article with great enthusiasm. It is wonderfully clear and very cleverly composed. . . . No doubt my enthusiasm is partly based on the fact that I am hostile to ESP. Like you I find the phenomenon just too fickle and too "magic like" to fit comfortably into my picture of the world. I hope that your recent change of view about the "supernatural" hasn't changed the basic attitude you took in that article. Acceptance of ESP and religion seems to me to imply belief in a God who was omnipotent to fool and mislead people as well as being rather cruel, and that is repulsive.[26]

But George's conversion was only deepening. In fact, it had been strengthened by an unexpected discovery. Marie Lynch and her family were visiting London from America, and came by Little Titchfield on August 8 to say hello. Marie was the daughter of Caroline Doherety, a devout Irish Catholic who had worked as a maid in the home of the Price's neighbors in Scarsdale back in the 1920s and remained a close family friend, even becoming godmother to Annamarie and Kathleen. In fact Caroline had been George's first visitor the day he was born. Religious herself, her daughter Marie was thrilled by his conversion. But what would William Edison Price have thought, she wondered out loud jokingly, in an offhand comment. After all—he was a Jew.[27]

George was shocked. Could this be true? He wrote frantically to Caroline back in New York and received a positive reply. Mother certainly had succeeded in keeping it a secret! Suddenly a host of memories fell into place: how Alice spoke out against Jews but always added cryptically that there were "some very fine ones"; how most of his closest friends in high school were "Hebrews"; his anti-Semitic comments at the Chicago Co-Op coupled with his will to live and eat among its friends. Clearly George had always had a kind of strange, unexplained attraction to Jews; now, he understood, it had all been part of a plan. After all, "St. Paul tells us that God has by no means forgotten about the Jews and we can be sure that many are intended to become Christians." To Al Somit, the person who would be most tickled by the news, he wrote, wisecracking as usual: "I have the honour (note how I'm picking up British spelling) to inform you that I am both a Christian and a Jew. . . . Didn't you always sense, subconsciously, that I was too intelligent to be a Gentile? (I suppose what threw you off from guessing the secret was that you felt I was too handsome to be a Jew. But, you see, that came from my mother's side)."[28]

Annamarie, Kathleen, and Dom were arriving that Sunday. George was not planning on telling his girls the family secret yet; it would just be too confusing. Nor was he going to let them in on his plan. But to Caroline Doherety he laid out the truth:

> *The Lord is sending them over here at this time to give me an opportunity to try to correct the harm I did them in discouraging them*

from religion, and help lead them back to Christ . . . they won't real-
ize for a while what I'm up to. . . . I'm planning a big birthday party
for Kathy where most of the guests will be very advanced Christians
but who don't particularly look it to an outsider.[29]

And then they arrived. His conversion, the girls thought, was something of a joke, just another facet of their father's rather lovable quirkiness. Amused, Kathleen spent the night of her twenty-first-birthday
dinner party arguing about the (non) existence of God with Kenneth
Demain and Mr. and Mrs. Anand, all new recruits to the C. S. Lewis
Society.[30] It hadn't exactly gone according to the program, but overall
George was pleased. The Lord's plan was coming into effect. Everything would fall into place in time.

He'd already begun trying to convert other people: the theater
manager who ran *Oh! Calcutta!* in the West End, and the greengrocer
down the street, Mr. Angelou. Most of all, though, he had determined
to convert prominent scientists; that would be the most effective way
to combat atheism and spread the Gospel.[31] He wanted to invite
Hamilton to Kathy's birthday party but ultimately decided against it;
Bill's father-in-law was a Lutheran minister, but Bill wasn't yet ready
to be in a group secretly assembled to sway George's own family. No,
Hamilton himself still needed to be won over. "My change of view
about the supernatural," George now replied to him as Wallace had to
Darwin before him, "necessarily affected some of the views expressed
in 'Science and the Supernatural.' " Hamilton's remark had been very
perceptive, though the kind of "repulsive" violations of scientific law
that happen in tricklike ESP experiments sounded to George "more
like the work of the Devil than God." Just as altruism required spite,
it was impossible to construct Christianity without an "evil power."
For that reason, he explained,

I believe that ESP, if it occurs as Rhine and Soal have believed,
involves the supernatural; and that the supernatural, if it occurs in
the ways commonly understood, involves incorporeal intelligence(s).
I think these are two possible alternatives to the consequences you
suggest from belief both in ESP and religion.[32]

C. S. Lewis's *The Problem of Pain* and *Miracles* contained some interesting ideas bearing on these matters; why shouldn't Bill have a read? After all, he had followed George's covariance selection math "better than any one else I know," so maybe he could also follow his theology. Hamilton was unaware of it, but George had chosen him as his prime target for conversion. Part of the Lord's plan in "handing" him the covariance equation had been to help Hamilton see the light.

Back in New York City, Edison was deep into Iyengar yoga. "I suspect he is very far from Christ," George wrote to Caroline Doherety, "and heading straight toward destruction." Would she please pray for him? He was one of those who most needed it. Edison, for his part, remained characteristically staid. If Christ brought George happiness, so be it. Any attempts to convert him, however, would be useless. He recommended his brother a book on the lotus position instead.[33]

·   ·   ·

George's Christianity was growing stronger all the while, and increasingly adopting a tone of its own.

Back at the end of 1968, at a public talk at the London Zoological Society, he'd met a fetching young woman with shoulder-length dark hair. They spoke a bit, she smiled, he took her number, they had lunch; finally he decided against seeing her again. But now, almost two years later, sipping his tea across from her with the knowledge that although not particularly religious she'd been born a Roman Catholic—in fact her father had been England's leading lay Roman Catholic—George was convinced that Rosemarie was part of God's design. After all, she had been born on Ascension Day, and he had been born to a Jewish father. Clearly this couldn't have been a mere "coincidence." Everyone but her, it seemed, was in on the plan. To Julia he wrote asking for copies of their divorce decree, and to Al and his instrument-maker friend, Ludwig Luft, he disclosed his agenda: George was going to marry Rosemarie.[34]

He took her out to a few more teas, and then he laid it on. "I urge you and beg you to pray for guidance even if the Devil tries to make you think it's silly." Rosemarie wasn't amused. Her religious convictions were her own private affair, and George was insulting and presumptuous. "Be sober, be watchful," he replied, ignoring her firm request that he not contact her again. "Your adversary the Devil prowls around like a roaring lion, seeking someone to devour." Couldn't Rosemarie see that if the Devil could mislead Saint Peter when he was in the presence of Christ then he could mislead anyone? They were meant for each other, and being able to see this was her test. It was the Devil who was keeping her away from him.[35]

The Devil, too, could explain little Dominique's flu, back in California. It was He, after all, who had pointed Kathleen to the Women's Lib Cooperative Baby Sitting Establishment, where, George was certain, "you automatically get a selection of anti-Christian women." "Resist the Devil and he will flee from you," he quoted from James, explaining to his part-baffled, part-amused daughters the real cause behind Dom's fever. "Resist him!"[36]

He was sorry that he hadn't told the girls the family secret about William Edison being a Jew when they were visiting him in London, but, again, this was not just some "coincidence."

> *Dominique kept dragging me away from the table then, and since I had found him to be reliable in expressing or following God's wishes, I suspected that this meant that God didn't want me to tell you then. So after Dom had hauled me away from the table two or three times as I started to tell you, I asked the Lord to make Dom do it again if He didn't want me to tell you, and promptly Dom hauled me away again.*[37]

God was sending him messages, everywhere and always. Nothing in this world was without meaning, not even the tiniest, least significant detail.

In fact, the more he read the Bible, the more George found that he was seeing things that no one had seen before him. Generations of scholars, for example, in working out his ancestry, linked Jesus to King David via his father, Joseph. But Joseph had only been Jesus' legal father, not his biological one. And the figures in the New Testament who had referred to him as "Son of David" had been some blind men, a Canaanite woman (not a Jew), and the same Jerusalem crowds who later shouted, "Release Barabbas! Crucify Jesus!" This was not without significance. For hadn't Jesus himself warned against "blind guides" and "the blind leading the blind"? Clearly Scripture was hinting that what mattered was not Joseph's but *Mary's* ancestry, and with the help of hitherto unrecognized clues (the Bible stating three times that Rachel was buried on the way to Ephrat, among others), George was now able, Sherlock Holmes–like, to do what generation of exegetes had triumphantly failed to accomplish, linking the Virgin, and hence Jesus, to the Old Testament Joseph's son Ephraim. Among other things He had revealed to him the true meaning of 666 as well—a piece of information he now offered the increasingly freaked-out Rosemarie as a "bribe" if she'd be willing to meet with him. Clearly God was showing George things He had shown no mortal before.[38]

Most important, he now believed, He had helped him to solve "one of the great puzzles of all time, which during 18 centuries has received the attention of many famous and brilliant men beginning with Tatian of Assyria about 170 A.D." To Christians he offered it as the long-sought solution to the problem of the contradictions in the four Gospels, to Jews as a fascinating matter of ancient history, and to agnostics as an entertaining story of "puzzle-solving." He'd cracked it sometime in January 1971, and had now committed it to a fifty-two-page treatise that he sought to publish in one of the Easter Sunday papers. Nearly two thousand years of tradition and practice notwithstanding, George Price had news for the world: The conventional Passion chronology was wrong.[39]

Not only were the eight days of Holy Week plus Easter actually twelve; not only had Jesus been crucified for a full twenty hours overnight instead of just six or seven; not only had early Christian groups mistakenly followed a lunar calendar, against the expressed dictate of Christ—the very date of Easter was entirely mistaken. In actuality it had almost certainly been April 13, A.D. 27. "This means that the date for Easter that is favored by the majority of your member Churches (the Sunday following the second Saturday in April)," George wrote to the director of the Commission and Department on Faith and Order at the World Council of Churches, "will exactly agree with the Apostolic date about half the time, and be fairly close to the Apostolic date in other years—and thus this change would be an enormous improvement over the present tradition of dating Easter by the moon." In letters to countless clergymen and biblical and New Testament scholars, George had embarked on a mission. It made no difference that he was getting no positive feedback. Changing the date of Easter was the true will of God, for it followed the true meaning of the Gospels. The Devil would make everyone think it was blasphemy, but it simply had to be done.[40]

. . .

Once again he enlisted Hamilton to look over his theology; hadn't he been the only one to "get" covariance, when all was said and done? "After reading your absolutely fascinating article on the chronology of the First Holy Week," Hamilton replied, ". . . I thought I had better retaliate by sending you some Easter reading." There were two stories by Chekhov, a book of poems by A. E. Housman, and Samuel Beckett's *Waiting for Godot*. George's chronology seemed convincing, but Hamilton was really no kind of judge.[41]

George was thankful, but had some reservations about Housman's "Easter Hymn":

Suppose we use the number "70" to represent typical earthly life. Measured on the same scale, future life counts as 0 if atheism or pseudo-Christianity is correct, or counts as infinity if the teachings of Jesus are correct. Thus we have two different ratios of the importance of this present life [compared to the future life]:

$$70/0 = \infty \text{ or } 70/\infty = 0$$

As you see, these two views are radically different. What Jesus repeatedly taught was the equation on the right [that present life is worthless when compared to the afterlife]. . . . Housman's reasoning in that poem is in terms of the equation on the left [that present life is infinitely more important than the afterlife].[42]

No, a Christian who didn't believe in Heaven and Hell was no true Christian at all. But even though enjoyment and success in earthly life were of no consequence, what man did during his fleeting time on earth was still of great importance. Life was an examination, set for man by God. Far from being a strike against Christianity as Hamilton had surely wished to hint, Vladimir's hopeless confusion in *Waiting for Godot* over the events in the New Testament only went to show that figuring out what God meant in the Bible was another of His examinations. "If the Bible presented no difficulties," George explained, "but revealed clearly a superhuman logic and intelligence . . . then much of the examination would be negated."

What quality, above all, was God really testing for? Without a doubt it was *agape*, commonly translated as "love" or "charity." But there were other qualities that could enable one to pass the examination. Abraham's test when ordered to sacrifice his son, Isaac, had been a test of faith and obedience, rather than *agape*. And there was also meekness, the kind with which Moses had been blessed. George interpreted this to mean a kind of "intellectual receptiveness, approachableness, of being willing to accept evidence and listen to reason." It was a good thing that there was more than one way to pass God's examination, he wrote candidly to Hamilton, for "it doesn't look as though I can pass purely on the basis of *agape*." He was aware of his limitations: George was no kind of altruist. In obedience and intellectual openness, however, he was clearly proving his mettle.

Hamilton resisted. There was a story about an Irishman, he told George, who was asked whether he liked oysters and he replied, No, he didn't like oysters and he was glad he didn't like them because if he did he'd be eating them all the time when he hated the damned things. The message was clear: Hamilton was not about to embrace Jesus. "Is the story of the crossing of the Red Sea literal truth or myth?" he challenged George. "If it's literal truth why should we respect Moses if he was just a puppet carrying out maneuvers to foreshadow the crucifixion, and why respect Jesus for following a course that he was bound to follow anyway? A plot so elaborate would make life meaningless if we believed it—and ugly too."[43]

"What difference does it make whether you approve of it or not?" George fired back. "Do you think that it is something that I wanted to believe in?" Just like C. S. Lewis, he had been forced to become a

believer based on evidence. It had been a necessary scientific deduc-
tion, not a personal moral choice. Predictability and free will were not
alternatives; George's childhood had taught him that much:

> My widowed mother worked very hard during the Great Depression
> in the 1930s to support my brother and me. Though times were very
> difficult, she always succeeded in providing food and housing for us.
> Consequently, during that period I might have reasonably predicted
> one day that she would provide me with food the next day, and the
> following month, and the following year. And so she did. Should I
> feel no respect for her since she was just a "puppet" impelled by mother
> love or maternal instinct or whatever you want to call it?[44]

The same was true for Moses and Jesus. At any given moment they
could have exercised their own free will and departed from God's
design. But of course they didn't.

Minus the miracles, Hamilton thought George's Passion schedule mirrored his best work in evolutionary theory. In his irreverent
humor toward authority and the church, he offered that it was not at
all surprising that George had come up with a better Passion schedule
than anyone else in eighteen centuries; after all, he had the advantage
of not having gone to theological seminary. But George disagreed.
Theological training had not prevented the most extreme heretical
opinions from developing, as proved abundantly by the historical
record, so why should it prevent smaller deviations like the one he
was proposing? Nor could this be just a matter of enough time passing, as with scientific discoveries: After all, the ancients were better
versed in Aramaic and Greek, not to mention the intricacies of the
lunar calendar and dating. Moderns had no advantage over past
exegetes when it came to figuring out the last week of Christ.

Playing devil's advocate, George offered another possibility:

> But perhaps you will say that it is because I am so smart. Then why
> am I not a world-renowned scientist if I am that brilliant? After all
> I've worked far longer and harder on scientific problems . . . hence
> why such little success? Oh, true, it's a pretty equation, but the sur-

prising thing isn't that I found it but that someone else didn't find it before. Now, if population genetics had been a major concern of all Europeans for 18 centuries, and no one has seen this equation in that time, and I found it, why then that would be something.[45]

Intelligence couldn't be the reason; even Isaac Newton, for God's sake, had dabbled in Biblical exegesis and come up short. If anything, George had been at a disadvantage: He had been less familiar with the Bible before his conversion than with the poetry of Shelley and Byron, and was naturally appalling in foreign languages to boot. How could he compare, say, to William Whiston, Newton's successor as Lucasian Professor of Mathematics at Cambridge, a known geologist and biblical scholar versed in ancient tongues, who had translated Josephus and many early Christian documents?

The answer was clear. Sporting a silver cross on his lapel now, he wrote triumphantly to his daughter Annamarie: "Your father has surpassed many famous and brilliant men such as Sir Isaac Newton . . . and St. Augustine and a multitude of others and solved one of the great problems of all time." He had been chosen by God, there was no other explanation. For the Bible had been "supernaturally constructed so that various forms of sin would form various forms of blindness." Extreme hate, for example, would lead readers of Revelation 13:18 to view whomever they despise as the Antichrist, instead of the true meaning of 666. Misinterpretations about Jacob were caused by gluttony, misinterpretations of the Passion schedule by conceit. For some reason, at this time, George had been sent to reveal these truths to the world. He didn't know why but it was God's design.[46]

That Hamilton examine his Passion schedule carefully was therefore all the more crucial. It wouldn't take him all that long; it was much less technical and mathematical than most scientific papers he'd surely not think twice about reading. But it was important that *he*, more than anyone, have a serious look. "Even though you don't share my beliefs about Jesus and the Gospels," George explained,

you do have in high degree certain qualities of intellectual honesty, of Mosaic "meekness" . . . and of non-sneeringness, non-condescension.

*Therefore I expect that if you try this you will see what I have seen
. . . Why put it off?? Why not settle it now?*

It was a matter of great importance to humanity, George implored,
a challenge Hamilton could not refuse. "I think that if you try this
you will presently discover that more than human intelligence has
gone into the composition of the Gospels." Laying it out plainly,
George concluded with an astonishing revelation: "There is an extraordinary concealment cipher there that has remained largely unread
for 18+ centuries."[47]

. . .

If God had opened his eyes to great truths in the Gospels, he was
opening them too to great truths in science. Recently he had resent
his revised big paper "The Nature of Selection" to *Science*. But there
was something else he'd been working on too. R. A. Fisher's fundamental theorem of natural selection hadn't been a mystery for eighteen centuries, but nonetheless had baffled all students of evolution
for the better half of the century. Just as he was constructing the Passion schedule, George decided to crack Fisher's mathematics.

The fundamental theorem, Fisher had always claimed, was the
biological analog to the second law of thermodynamics: "The rate of
increase in fitness of any organism at any time is equal to its genetic
variance in fitness at that time." To the Anglican Fisher it was the
ultimate reply to the winding down of the universe, God's benevolent
reply to his own disconcerting law of entropy. What it seemed to
mean to students of Fisher trying to fathom the pride of place he had
granted it was that, left to its devices, natural selection would always
tend to increase the fitness of a population. Combating the tendency
to disorder, it was the ultimate device of progress.[48]

But how Fisher had derived this result mathematically no one
seemed to understand. Some called it "recondite," others "very difficult," still others "entirely obscure." Most thought the theorem only
held under very special, circumscribed conditions, like asexuality and
random pairing, and that the great Fisher had therefore been mistaken. Recently Fisher's own successor at Gonville and Caius College

in Cambridge had intimated that his teacher might have been exactly correct, if only we could understand what he meant.[49]

"Fisher's explanations of his theorem," George now wrote in drafts of "Fisher's 'Fundamental Theorem' Made Clear," "are afflicted by a truly astonishing number of obscurities, infelicities of expression, typographical errors, omissions of crucial explanations and contradictions." But Fisher, he thought, could not easily be accused of error. Carefully considering language, comparing words, and squinting at the mystifying mathematical notation, George now gradually began to see, as if looking through a crystal, precisely what Fisher had meant.[50]

He had just penned a letter to Henry Morris, the Texan founder of the Creation Research Science Center, who was known to Americans as "the father of modern creationism." A young earth creationist and a biblical literalist, Morris was the evangelical author of *The Genesis Flood,* and George was writing to congratulate him on his enterprise. Morris was surprised and delighted to hear such compliments, and from the author of "Science and the Supernatural" no less. Soon, however, even he was taken aback by George's fundamentalism.[51] "I am very much in sympathy with the claims of the society," George wrote to him,

> *and was expecting that I would be able to become a member—but I am afraid that I cannot. . . . You see, I try to be in everything a slave to the Lord, and bring large matters and small matters to Him for decision, asking in words similar to those of Saul in Acts 9:16. . . . I asked whether I should apply for membership. His answer was that I should read over the "statement of belief." As I read points 1 and 2 I commented to Him that I did agree. On No. 3 I commented that I thought the flood was worldwide but I was not absolutely certain. He instructed me to reread the account in Genesis. . . . I could not subscribe to the statement of belief unless He made me know it was correct. Then I asked Him again whether I should apply, and His command was that I should not. And of course I always obey.[52]*

Morris replied politely that he was sorry. Short of membership, though, he offered a subscription to the society's journal. "I am afraid that you did not fully understand the tenor of my letter," George's surprising retort soon followed:

> You write: "If you feel that you cannot join the Society for reasons of certain items in our statement of belief . . ." No, the reason that I do not apply for membership is that the Lord commanded me not to—as I stated in the next-to-last sentence in my next-to-last paragraph. It was not my choice but His, and this is an exceedingly fundamental distinction. I am sorry that you apparently did not understand that I really am a genuine slave, and not like the usual "evangelical Christian" who prays for "guidance" and then makes his own decisions. I do not ask for "guidance"; I ask for commands. And then I obey them. . . . I make no pretentions of being a prophet but boast only of being a slave.[53]

No reply was forthcoming to George's letter, the rest of which degenerated into a diatribe against pseudo-Christians like Morris and his like who ask only for God's guidance instead of his strict commands. Nor, for that matter, was there a reply from Rosemarie, to whom George continued writing increasingly disturbing letters, the last of which asked her to pray "Bless me, Father, for I am on the road to hell," and signed off, "sincerely His."[54]

But as George grew crazier and crazier to the world, his scientific insight only sharpened. At the precise time he was debating with the "father of creationism" whether the biblical flood had been worldwide or only local, he was penetrating the thought of the twentieth century's greatest evolutionary sage. He'd been sending drafts of "Fisher's 'Fundamental Theorem' Made Clear" to Cedric Smith for approval, and CABS had now decided to run it as the cover article in the *Annals of Human Genetics*. It was going to make a big splash. Once again this was no coincidence. "I've developed a considerable interest in trying to solve puzzles about Bible interpretation," George wrote to his friend Ludwig Luft in America. "It is possible that this involves some of the same abilities as are needed for interpreting what Fisher wrote on his theorem."[55]

All of Fisher's followers had tried their hand at it, but it was

George who cracked the mystery. Back in the early part of the century, Fisher had been the one to develop the notion of "variance." Variation in a population was due both to genes and to environment, for genetics might react differently in different environments, producing two distinct phenotypes based on the very same genes. What Fisher's mathematical definition of "variance" allowed scientists to do was to distinguish what part of the variation in a population was due to genes and what part was due to environment. Students of heredity with a determinist cast of mind would later use this concept to claim the supremacy of heredity over environment in the passing down of intelligence, but seeking out a general law, Fisher himself used it to construct the fundamental theorem. The principal misunderstanding had been that Fisher was talking about "total" fitness. In fact what he had claimed was that selection working on that portion of the variance that could be ascribed to discrete, "additive" (noninteracting) genes would always increase fitness, and that the rate of increase of fitness would be proportional to the rate of variation. The theorem classified everything outside of "additive" genetic variance—including epistasis (interaction between genes), dominance of genes over one another, as well as the truly nongenetic domain—as "environment." Fisher's fundamental theorem referred not to "total" fitness but only to the specifically additive, genetic part of it.

But if selection would always produce genes that were optimally adaptive, they could only be optimally adaptive to conditions that existed an instant earlier; the "environment" was constantly changing.[56] The total fitness of a population could therefore go down as well as up, it all depended on the "fit" of heredity and the outside world. In the battle between chance and necessity, then, genes were like agents of progress running blindly after whimsical masters. What George was able to show was that this statement was precisely true: Statistically speaking, genes would invariably become more adaptive over time, pushing their bearers to ever-growing heights. "Heights," however, could only be defined retroactively.[57]

There was a salutary lesson in this belated penetration. Fisher had believed that the true motor of evolution was selection working on discrete, "additive" genetic variance. Just as the theory of gases could predict outcomes based on the behavior of discrete molecules bumping about in a flask, his theory of selection could predict the fate of populations based on the behavior of their genes. He had yearned for a deep truth, an antidote in God's living world to the second law in His inanimate. What George now saw was that things were more complicated: Perhaps the fundamental theorem was the best we could yet describe, the most general law of life and its course. But it was fundamental only within its circumscribed boundary. Success or failure, elevation or denouement, ultimately depended on more than just the discrete particles within. It was how they interacted with the environment that mattered for overall fitness. Progress could be guaranteed only if inside and outside lived in harmony.[58]

. . .

It was the end of 1971, and Hamilton and his family were recovering from the loss of Romilda and Godofredo. "One does really develop parental feelings towards such fosterlings," Bill tried to explain in a melancholy philosophical letter to George, "and yet at the same time one doesn't completely." When Godo, the younger of the two, began to forget his true parents, and Romilda to engage in amorous affairs, there was no recourse but to send them back home to Brazil. Biology, in the end, was stronger than anything.[59]

Then Hamilton turned to the secret biblical cipher. If this really was some kind of "altruism code," the biblical analogue to George's covariance equation, Bill wasn't particularly interested in being convinced. Aesthetics were behind his avoidance of George's "homework" assignment, he explained; there was a "sense of beauty" he had grown up with, a kind of "sense of human freedom" in this case, that was whispering to him that the enterprise was wrong:

Why should I rely on a "sense of beauty"? Partly it is an intuitive feeling that this is what is deepest in me, and an equally intuitive disbelief that the creator who put it there intended me to put it [the creator] aside... It seems to me that my sense of beauty enables me to see genuine flaws

*and shortcomings in arguments that I have in no sense understood.
It even seemed to me at school and at university that I had a slight
advantage over my fellows in this respect, and I guess it was this kind
of aesthetic dissatisfaction with the elegance or completeness of the evolutionary explanations we were handed out that started me off on the
search for the basis of biological altruism.*[60]

If Hamilton's aesthetic sense had served him well, "like the far-
reaching antennae of a cockroach," then it was telling him now that
the extreme detail of George's unraveling of prophecy was unsat-
isfactory in some way. But his covariance equation, of course, was
something entirely different. Hamilton had just read an article by
the American mathematical population geneticist Richard Levins
who had protested (like George) against the idea that group selection
is an unproved and perhaps unworkable concept. And yet Levins,
Hamilton thought, had ended up showing that the conditions that
allow it to work are rather more stringent than he had anticipated.
Hamilton had high hopes that applying George's formula to nature
would allow scientists to do better. Really, he was much more excited
about following George's lead in this direction than he was his scrip-
tural exegesis. Meanwhile, he signed off kindly, "I had better get back
to the ants, bees and wasps."[61]

George replied that reading Hamilton's letter was like torture.
What was at stake was truth, not beauty:

*I recall one walk we had during the early period after my conversion
where I pointed out to you how closely Psalm 22 fitted the Crucifixion. Verse 16 reads ". . . they pierced my hands and my feet" and
I said to you that David's hands and feet were never pierced. You
replied that one should examine all the Psalms, note all the different
forms of injury described in them, and evaluate this statistically. In
short you were applying scientific reasoning to theology. Now you have
stopped applying scientific thinking and criteria to these questions,
and instead have shifted to aesthetics, philosophy, and undisguised
wishful thinking. . . . I think better of you as a scientific thinker than
as a wishful thinker, and wish you would drop this concern with*

aesthetics and concentrate on questions of scientific truth or error, as far as theological questions are concerned.[62]

That he should have accused his friend Bill Hamilton of ceasing to be scientific was amusing for the very same accusation was now being launched against George. In a fateful twist of irony, the man making the charge was none other than Joseph Rhine, the professor of parapsychology from Duke whom George had accused of fraud in 1955.

Shortly after his conversion, George had written to Rhine to apologize. He had assumed Rhine had cheated in his ESP experiments in order to promote Christian beliefs, and, being an anti-Christian in those days, had therefore sought to expose him to stem any possible tide. Now that he was a Christian himself he felt terrible about the matter, and had even sent a letter of apology to *Science.* In fact, if anyone had been a cheat, it was George himself. Back in Poughkeepsie in 1961 he'd performed a card experiment with a few IBM friends in which the odds against chance were beaten 300 to 1. "The odds then were such as to force a test upon me: not so low that I could shrug them off, not so high that I was forced to believe. I am afraid that I failed that test very badly." In fact, this had been George's first "burning bush"; his second, he reported thankfully, he finally "turned aside to see."[63]

Rhine was baffled, amused, mostly intrigued. He thanked George for his kind words. In his youth he himself had sought out an experience of conversion, but found the evidential basis inadequate to his questioning mind. Instead he had turned to science, where he came to question the dogmatic materialism he found, which in turn led to his curiosity about parapsychology. Far from being an evangelical Christian, he wasn't in the slightest religious. In truth, his interest had been purely scientific: Could laws of nature exist that had not been captured by conventional physics? George's apology was really unnecessary. Whatever merit the field had to offer had to rest on its own evidence, and couldn't claim any support from authority of any kind. In fact, he was glad to report, most scientists had come to feel the same way: The Parapsychological Association had recently become an official affiliate of the American Association for the Advancement of Science.[64]

Jokingly he recalled a story he had heard. Some years back a New York reporter had printed a request to his readers for descriptions of their experiences of a psychic nature. One reply came from a woman who, after describing her vivid conversations with the dead, wrote that her son had written a scathing review of the matter in *Science* back in '55, but that she knew more about this kind of stuff than he did. George thanked Rhine kindly, said it sounded like his mother, and told him that Alice had died two years earlier.[65]

But soon after "Apology to Rhine and Soal" appeared in *Science* at the beginning of 1972, what started as a friendly correspondence began to change in tone. While George was curious about why Rhine hadn't turned to religion, having found that his ESP results were incompatible with science, Rhine wondered how George planned to reconcile his newfound faith with science. He had never thought of ESP as incompatible with science; his only problem had been with the dogmatic assumption that known physics explains everything and that anything unexplained by it must therefore be unscientific. The more they wrote to each other, the more the differences between them became apparent. Rhine thought scientists sometimes make mistakes from being too dogmatic; George thought it was because the Lord "makes foolish the wisdom of the wise" in a world he constructed as a test for man. Rhine thought man was something more than matter, but that this didn't necessarily imply a godly spirit, as George believed, only scientific laws yet to be discovered.

Just as with Henry Noel, Henry Morris, and the Reverend John Stott before him, an amicable approach had turned nasty. Rhine was backing out of an offer that George write an article on the history of the 1955 paper for the *Journal of Parapsychology*; he wasn't about to have his name affiliated with him anymore. Scores of angry letters crossed the Atlantic. Alice was right, Rhine finally wrote, George really didn't know what he was talking about. "Why do you not just let parapsychology alone for now," he broke off, weary of the insults, "if only for your own peace of mind? I lectured at the University of Minnesota day before yesterday and nobody mentioned your name."[66]

It was less than kind, but it hurt because it was true. George had seen things that no scientist had seen before him. At the same time, retreating into a strange world of his own, he was ranting against family, antagonizing friends, pushing away acquaintances. Increasingly alone, he seemed almost willfully blind to the feelings of those around him. His equation would help solve the mystery of a truism, but his marriage to the Lord's commands had led to the darkest caverns of human selfishness.

Soon, however, all that would change dramatically.

March 24th, 1973

Dearest Kathy,

I called your mother again a little while ago, and she told me she had asked you about coming out to London this summer with Dom, and you said, "Absolutely not." Kathy, I wish and hope you'll change your mind, because this summer is likely to be the last opportunity we'll ever have to be together as a family. As you know, I went through a false conversion in 1970, and went to church for the first time as a partial (but false) Christian on June 14th, 1970. Last month I had a real conversion. My first time to Church after this real conversion was February 25th. I'm going to another church, not All Souls, but St. Mark's — Clerkenwell, which is about a forty-minute walk from here. On that day, February 25th, as I was getting close to St. Mark's, I began to have a faint, vague impression of seeing Jesus walking before me bearing the Cross. Two or three times since then I have felt that He gave me the message "Follow me." So this means that some suffering lies ahead for me. It won't begin very soon because I'm not ready for it nor

Letter to daughter Kathleen, March 24, 1973

11

"Love" Conversion

By the time he wrote to the Reverend Billy Graham, a large portion of the 58,148 American soldiers, 223,748 South Vietnamese soldiers, 1,100,000 North Vietnamese soldiers, and approximately 4,000,000 civilians on both sides of the divide had died in Vietnam. Richard Nixon had just recently appeared at one of Graham's revivals in east Tennessee, becoming the first president to give a speech from an evangelist's platform. George's was a heartfelt plea: The war had to be stopped. Could the reverend exercise his influence over the president to bring about the end of bloodshed?[1]

War was on his mind. He'd begun working again on the antlers paper *Nature* asked him to shorten back in 1969, plugging at the IBM 360/65 computers at UCL to come up with an unbeatable strategy for combat. He'd even gone up to the annual Animal Behavior Association conference in Birmingham in July 1970, returning to London encouraged by positive feedback.[2]

"Computer simulation of two animal conflict sounds dismally similar to the work in conflict theory which has been done by our students of international relations these past fifteen years" Al wrote to him that December, unconvinced that his friend's efforts were worth anything at all. But George plugged away all the same. "I don't know if my program will seem 'dismally similar' when you know more about it," he answered. "Aren't the conflict theory studies related to strategic advantages rather than psychological mechanisms? And if the latter, then it is interesting if animals and men go through similar reasoning in their conflicts (two main concepts in my treatment are 'who started it?' and 'get even')."[3]

Al didn't give up. "Pardon my skepticism about combat simulation programs. The difficulty at the most obvious level is that we have no clear evidence as to how people actually behave in real life combat situations. We can speculate all we want about this, but we can't get inside the participants' heads. Arguments by analogy, whether from animal behavior or from a computer simulation, seem to me to lack credibility."[4]

As usual George fought back with his quirky sense of humor. He'd just recently given his first sermon in church, and had begun sending a steady barrage of C. S. Lewis books to friends and family abroad. He was feeling good about himself. "In regard to your paragraph beginning, 'Pardon my skepticism about combat simulation programs,'" he replied, "it would have been more judicial for you to have written something along these lines:

> In general, I have much skepticism about combat simulation programs and about arguments by analogy from animal behavior or from a computer simulation. However, I have observed that you often take a somewhat original direction in approaching a problem. Why, in fact, you have sufficient originality so that it would not at all surprise me if you were to take some famous problem on which many thousands of scholars have worked during ten or twenty centuries, and come up with a full solution involving conclusions that no one ever thought of before, as for example to show that the crucifixion lasted over night or that "Palm Sunday" fell on a Wednesday. Of course I would expect that such novel results, coming from most scholars, would be total nonsense and full of obvious errors that even a non-specialist like myself could point out in a few minutes, but coming from you I would expect it to be so carefully thought out and so well

researched that no one (and certainly not any of the feeble scholars at this crummy university of which I am Executive Vice-President) could find a single way in which you have contradicted the Bible, nor any difficulty in relation to secular history that you have not at least discussed. Therefore, despite my general prejudices against the general type of approach to animal conflict theory that you have outlined in your letters, it may well be that you will come up with something new and useful. Especially since Nature accepted your original version, on the recommendation of John Maynard Smith (whom I know of as an outstanding mathematical biologist). . . .[5]

In truth, George was stuck with the computer simulations on animal conflict; by now he had practically abandoned them. His real intention in writing to Al had been to get him sufficiently riled up over the Passion schedule—part bemused, part challenged—that he'd send it to some of the biblical scholars at his university, if only to save face. But Al wasn't interested in what he took as his friend's new mumbo jumbo. He stuck to the science, where he remained unimpressed: "You apparently overlook one fact: no one has yet found a way to simulate a 'real life political situation' and I doubt that even a person of your own admitted talents can accomplish it." Then he added, "notwithstanding one John Maynard Smith."[6]

. . .

How had J. B. S. Haldane's beloved student John Maynard Smith become involved with George? It had all begun at UCL in the late 1940s, when he and his fellow students caught wind of Konrad Lorenz's recent discoveries in animal behavior. Lorenz was a bumptious, domineering, goateed Austrian, controversial for having been a member of the Nazi Party and yet recognized as the father of modern ethology and on his way to a Nobel Prize.[7] Ritualized fighting, Lorenz claimed, was the norm in nature, not the exception: When animals compete for territory or mates, they seldom use all the weapons at their disposal but rather settle for aggressive displays.[8] In line with the times, Lorenz thought that this was good for the species: If escalated fights were common, the abundance of injury might militate against the survival of the species. "Pangloss' Theorem," JBS furiously scribbled in the margins of a paper in which his student Maynard Smith adopted Lorenz's assumption; this was "group selection" nonsense and had no place in biology.[9]

A decade later, over in America, Richard Lewontin finished reading the current popular textbook on game theory, *Games and Decisions*, by Luce and Raiffa, two former members of the RAND Corporation. Lewontin was a quick-witted, politically engaged Harvard population geneticist, considered by many the most brilliant theoretician of his generation, and by others a dangerous, radical Marxist. Unable to cope with the idea of fitness being so dependent on context, he turned to game theory for answers. Maybe fitness could somehow be measured as a consequence of interactions; using von Neumann's approach, he set out to model a game between animals and nature.[10]

But the game Lewontin constructed could only measure the fitness of a species against a changing environment; constructing a game between the genes of animals interacting among themselves *and* the environment proved too complicated a task. Since modeling such interactions would entail making simplifications that to Lewontin seemed far removed from nature, he abandoned game theory as a tool useful to the biologist. Game theory might tell a coherent story, he thought, but that didn't mean that it was loyal to reality. And Lewontin wasn't interested in "interesting but not true."[11]

Meanwhile, in England, John Maynard Smith had spent the better part of the sixties working on the physiology of aging, the genetics of patterns, and, toward the end of the decade, on theoretical issues related to the evolution of sex. The question was: Why was there sex in the first place? If people were going to argue against Wynne-Edwards's group-selection explanations for his birds, he thought, it wouldn't do to put the evolution of sex down to the good of the species. And yet this was still the prevalent explanation: Sex had come about in nature because it was a wonderful way to create variation. And since more variation meant more gumption for a population facing a changing environment, even though individuals engaging in it would have to

give up passing down 50 percent of their genes to the next generation, not to mention expending all that energy in finding and bedding
a mate, groups with sex would outcompete groups that reproduced
asexually. By "inventing" sex, in its wisdom, natural selection had
overridden personal interest in favor of the common good, filling the
world with infinite variety. But this couldn't be right, Maynard Smith
thought, remembering Haldane's furious Panglossian scribbles. And so,
spurred to action, looking for an individual rather than group-selection
slant, he was working on cracking the mystery of sexuality.[12]

It was around that time that John Maddox, *Nature's* editor, sent him
a paper titled "Antlers, Intraspecific Combat, and Altruism," by George
R. Price. John hadn't heard of the guy, but boy, was this interesting!
First of all, here was a solid attack on the assumption of group selection: When all was said and done, it seemed to say, individual selection
was just as fine an explanation. But there was more to the paper than
an appeal to parsimony: Just as with two poker players staring each
other down, Price had considered combat to be a game in which each
animal's strategy is dependent on the other. This had led him to see that
if animals adopted a strategy of "retaliation" whereby they normally
fight conventionally but respond to escalated attacks by escalating in
turn, this would be favored by selection at the individual level. It was
a strikingly original insight. Of course all of von Neumann's games
assumed rationality on the part of the players, and doing the same for
animals was clearly not in the cards. Still, if one constructed a kind of
payoff matrix in units of fitness, natural selection could replace rational
thought. Lo and behold, short of drawing out an actual matrix, that
was precisely what George Price had done![13]

Excited about the paper, John recommended acceptance provided
it was shortened; *Nature* wasn't the place to publish articles fifty pages
long. About a year later he arranged to spend three months working
with the University of Chicago working with the Committee on
Mathematical Biology. It wasn't a bad place to choose if you were
interested in game theory, which, spurred by George's paper, John
Maynard Smith now certainly was. Although aware both of Fisher's
sex-ratio argument and Lewontin's game between species and nature,

John judged the field to be a blank slate as virginal as it was promising. After all, Fisher hadn't formalized his idea, nor had Lewontin applied it to the interactions between animals. Familiar with Hamilton's 1967 extraordinary sex-ratio paper, he initially hadn't seen its relevance to the combat of animals. The only existing bridge, therefore, was the paper on antlers. Working with what George had termed the deer's "genetic strategy . . . stable against evolutionary perturbation," he applied his mathematical skill in service of marrying game theory to evolutionary biology. The formal concept he came up with, closely based on George's, he called an evolutionary stable strategy, or ESS for short.[14]

John left Chicago toward the end of 1970, and back again in England was now writing a paper. An ESS was formally equivalent to a Nash equilibrium: It was a behavioral strategy such that, if a majority of the population adopts it, there is no "mutant" that would give higher reproductive fitness. It was a wonderfully enlightening concept: If one could find an ESS for combat behavior, for example, it would serve as a good argument that such behavior had actually evolved. Clearly he would have to reference George's paper as the major source behind the insight. Scouring through the literature he found no sign of it, though, not in *Nature* or anywhere else. "Antlers" had disappeared, as suddenly and fantastically as it had made its entrance. Having come across him briefly at a visit to the Galton in September, and wary of not giving credit where credit was due, Maynard Smith now sat down to pen a letter to George.[15]

Of course! He'd be glad to be thanked in reference to the antlers paper, George replied to John's kind request for the specifics of citation. Somewhat cryptically, though, he suggested a better way to express acknowledgment: "If one mentions an 'unpublished manuscript,'" he explained, "then someone may wonder about whether it was used with permission, but if you speak of 'discussion,' then no such suspicion arises." Maynard Smith didn't quite understand George's meaning and didn't press the point. He was glad simply to have found him, and besides, as the dean of Biological Sciences at Sussex University for close to a decade, he had plenty on his plate.[16]

. . .

Back in Little Titchfield Street, and once again provided a stage by CABS at the *Annals of Human Genetics*, George had just submitted another paper. In "Extension of Covariance Select on Mathematics" he showed more explicitly than in the short 1970 *Nature* paper how covariance could be extended to multiple levels of selection. This, after all, had been the true beauty of his equation, the special extra piece of the puzzle that showed that when selection worked at a higher level, genuine goodness between individuals could evolve.[17] Still, it was more of an outline of the approach than a direct application to any problem; having had the original flash of insight, it almost seemed, George had lost interest in fleshing out the rest. The winter of 1972 was nearing its end, and by now he had two utterly different distractions on his mind: Jesus and sex, in that order.

The first of these, he wrote to a friend, consisted in hearing God's commandments clearly and uncovering the Devil's designs.[18] The second had to do with the question of the rate of evolution, for sex could speed up evolution by providing more variation for natural selection to work with. He'd been following John Maynard Smith's papers on the matter, and was trying to work things out for himself.

Variation was on many people's minds. At the University of Chicago and now at Harvard, Richard Lewontin had recently shown in a set of experiments with *Drosophila* flies that there is much more variation in natural populations than previously suspected.[19] It was a bomb dropped right at the center of evolutionary population genetics: If natural selection was, in Darwin's words, ". . . daily and hourly scrutinizing, throughout the world, every variation, even the slightest; rejecting that which is bad and adding up all that is good . . . ," then shouldn't variation be at a minimum? Most mutations were injurious, not beneficial, so would be expected to be summarily culled. To explain the apparent paradox, a Japanese theoretical geneticist Motoo Kimura, posited a theory that soon became known as "neutralism": The reason so much variation existed in nature, the theory argued, was that most of the variation was neither here nor there with respect to fitness. Absent a reason to cull, natural selection would therefore simply remain indifferent, leaving high levels of genetic variation to accumulate and drift freely in natural populations.[20]

At the Galton, George's überboss, Harry Harris, was doing his best to fight the trend. If variation was neutral, what role then for natural selection? This was a serious matter: Unless a better explanation was found, Darwin's great theory would need to be thoroughly reconsidered. A Darwinist at heart, Harris was marshaling data from human hemoglobin to argue that Kimura and his gang were mistaken. To do so convincingly, though, he'd have to propose a superior mathematical theory, and who better than George for the job? In truth, George was more interested now in the Passion Schedule and the Bible. Still, he shrugged his shoulders and complied.[21] After all, this theoretical challenge on genetic polymorphism might help clarify some issues on sex, and both could serve as further scaffoldings for his continued search for the origin of family. Besides, all his attempts at simulating animal combat on the computers had by this time utterly failed.[22]

. . .

He was working on a model of an optimal mutation rate: It was "far superior to Kimura's," he wrote to the recent Nobel laureate MIT economist Paul Samuelson, with whom he'd resumed a short correspondence.[23] But then came a surprise. In the spring a letter from John Maynard Smith arrived in the mail. In it were computer printouts of John's latest attempts to model animal combat. To George it was practically a godsend.

"Fascinating. Congratulations," he replied, unable to hide his excitement, "it looks as though you've gotten well beyond the point I've reached." Combat, after all, had always been close to his heart. Even better was the offer of joint authorship, though George would only accept if John's name was listed first. At the moment he was trying to finish a paper of his own on "Sex and Rapid Evolution," so wouldn't have time to work on the combat paper just now; why didn't John write a first draft for *Nature* that he could look over and comment on? As for the different behavioral strategies of the model's animals in conflict, George suggested easily understood names like Dove, Hawk, and Prober.[24]

Once again the problem was how to explain why, in fighting over dominance rights, territory, or mates, animals often seem to stop short of actually hurting each other. George's original paper had concentrated on antlers in deer, but this was a general problem. The males of many snake species, for example, were known to fight each other by wrestling without using their fangs, a kind of "benevolence" not unknown even to praying mantises. The most absurd case was that of the Arabian oryx with the rhyming name, *Oryx leucoryx*: Its extremely long horns, pointed so absolutely in the wrong direction, forced males of this species to kneel down with their heads between their knees in order to direct their horns forward. How could kneeling oryxes, wrestling snakes, and deer that refused to strike "foul blows" be explained? How had Nature, in her wisdom so infinitely greater than man's, invented limited combat?

Maynard Smith and George took their cues from John von Neumann. For once again, as in poker, nuclear proliferation, or for that matter—Vietnam—animal conflict could be modeled as a game: the trick was to see that the strategy of each "player" depended on the other. Two contestants, for example, A and B, could adopt "conventional" tactics, C, that were unlikely to lead to injury or "dangerous" tactics, D, that were sure to lead to serious harm. And of course, they could simply retreat, R. A possible conflict between them could therefore look like this:

A's move   CCCCCCCCCCCDCCOCCCCD
B's move   CCCCCCCCCCCCDCOCCCCCR

with A probing on the twelfth and twentieth moves, B retaliating after the first probe, then retreating, and losing out, after the second. Each contest ended with particular "payoffs" to each contestant, measures of the contribution the contest has made to the reproductive fitness of the individual. Spurred on by Maynard Smith, George returned to the computer in the summer of 1972. Together they programmed five distinct strategies, sets of rules that ascribe probabilities to the C, D, and R plays as a function of what happened in the previous moves. There were five such strategies: "Dove," which never plays D; "Hawk," which always plays D; "Bully," which plays D if making the first move, D in response to C, C in response to D, and R following an opponent's second D; "Retaliator," which plays C on the first move, C in response to C, D in response to D, and R if the contest has lasted a preassigned number of moves; and finally "Prober-Retaliator," which if making the first move or in response to C plays C with high probability and D with low (but R if the contest has lasted a preassigned number of moves), following a probe reverts to C if the opponent retaliates but takes advantage by continuing to play D if the opponent plays C. There were fifteen types of two-opponent contests, and John and George simulated two thousand contests of each using pseudo-random numbers generated by an algorithm to vary the contests. With fixed payoffs and probabilities that seemed biologically sound,[25] the following matrix presented itself:

Average Payoffs in Simulated Intraspecific Contests for Five Different Strategies

		Opponent				
		"Dove"	"Hawk"	"Bully"	"Retaliator"	"Prober-Retaliator"
Contestant receiving the payoff	"Dove"	29.0	19.5	19.5	29.0	17.2
	"Hawk"	80.0	-19.5	74.6	-18.1	-18.9
	"Bully"	80.0	4.9	41.5	11.9	11.2
	"Retaliator"	29.0	-22.3	57.1	29.0	23.1
	"Prober-Retaliator"	56.7	-20.1	59.4	26.9	21.9

To see whether a strategy is an ESS against all others, all one had to do is examine the corresponding column. In a population full of "Hawks," for example, does "Hawk" do better than the alternatives?

The table showed clearly that it didn't: Both "Dove" (19.5) and "Bully" (4.9) did better against "Hawk" than "Hawk" against itself (-19.5). "Dove" too was not an ESS: "Hawk" (80.0), "Bully" (80.0), and "Prober-Retaliator" (56.7) averaged higher payoffs in a population almost entirely of "Dove." In fact "Retaliator" alone was an ESS, since no other strategy did better (though "Dove" ties), and "Prober-Retaliator" came in a close second.

How would such a population be expected to evolve? The answer was that "Retaliator" and "Prober-Retaliator" types would increase in frequency at the expense of "Hawk," "Dove," and "Bully." These last three types wouldn't become extinct, though, but rather remain in low numbers due to a constant flow of mutation but also because of senility, youthful inexperience, injury, and disease. The balance between "Retaliator" and "Prober-Retaliator" would depend on the frequency of "Dove," since probing was only an advantage against the meek: If the frequency of "Dove" was greater than 7 percent, "Prober-Retaliator" would replace "Retaliator" as the predominant type. The only way to make "total war" behavior advantageous would be to significantly alter probabilities for serious injury, or to give the same payoff penalty for retreating uninjured as for serious injury. Otherwise the simulations made abundantly clear that under individual selection "limited war" was superior to unbridled aggression.

Of course real animal conflicts in nature were infinitely more complicated. Besides the category distinction between "Conventional" and "Dangerous" behavior, which in all probability was more subtly graded, individuals varied widely in the intensity and skill with which each kind of tactic was employed. Still, by simplifying, the model allowed for certain predictions: The table, for example, showed that the best strategy against "Hawk" is "Dove," or in other words immediate retreat. This meant that it would be to the advantage of a given individual to simulate wild, uncontrollable rage, since those encountering him would do best to get out of his way and yield. If that was the case, a pretend "pseudo-Hawk" type would soon arise, leading in turn to selection for a second type with a special talent to "call a bluff." To counteract this dynamic, a true sign of mania-cal craziness would therefore be expected to evolve, one that could not be easily counterfeited. And this, it seemed, was precisely what sometimes happened in the wild. Elephants "on musth," for example, were often described by local peoples as invaded by wild spirits: As they rampaged uncontrollably, often with fatal results, a dark brown tarlike fluid secreted by the temporal glands ran distinctively down their faces. Combating deviousness, this was nature's way of signaling that here was no charade: With due respect to fakery, some things just couldn't be bought by guile.

These were important discoveries, George thought. Perhaps they'd prove useful to mankind.

· · ·

He was happy: not only had Maddox informed Maynard Smith that *Nature* was interested, but he'd have his name on a pathbreaking paper alongside a world-renowned authority. True, the insight had been his. But this was an infinitely better outcome than his first, original attempt alone.

As exciting as it was, however, George's heart remained elsewhere. He had no plans, he wrote to Kathy, to renew the SRC grant that was running out in July. Instead he was planning on spending the next few months working on a book about Shakespeare and the Bible.[26] Most of all, he was finally giving in to fate.

Recently he had preached a sermon on Matthew 6:25–34:

> *Therefore I say unto you, Take no thought for your life, what ye shall eat, or what ye shall drink; nor yet for your body, what ye shall put on. Is not the life more than meat, and the body than raiment? Behold the fowls of the air: for they sow not, neither do they reap, nor gather into barns; yet your heavenly Father feedeth them. Are ye not much better than they? Which of you by taking thought can add one cubit unto his stature? And why take ye thought for rai-ment? Consider the lilies of the field, how they grow; they toil not, neither do they spin. . . . Wherefore, if God so clothe the grass of the field, which to day is, and to morrow is cast into the oven, shall he*

not much more clothe you, O ye of little faith? Therefore take no
thought, saying, What shall we eat? or, What shall we drink? or,
Wherewithal shall we be clothed? . . . But seek ye first the kingdom
of God, and his righteousness; and all these things shall be added
unto you. Take therefore no thought for the morrow: for the morrow
shall take thought for the things of itself.

He couldn't stop thinking: He held an ordinary job, lived in
an ordinary flat, dressed conventionally, was paid a regular salary.
He'd *sort* of known, then, without *really* knowing it, that all this
would have to end. "Wishful thinking had kept me supposing that
it was not going to happen to me," he wrote to Maynard Smith in
October,

> *—or at least not in the extreme way that it seems always to happen in*
> *the accounts I had read by missionaries and others who had themselves*
> *lived "the life of faith." In those accounts the saving cheque always*
> *arrives at the last possible moment when disaster is at hand.*[27]

"I had optimistically calculated that deliverance had to arrive
around the 20th of September in order to avert disaster. However
it appears that God's standard of what constitutes 'disaster' are on a
different scale from mine," he continued.

> *Furthermore it appears that His standards are more accurate than*
> *mine, for in fact here I am almost a month later, still with food and*
> *other necessities and with all essential accounts paid. I don't know*
> *how much longer this is going to go on. The encouraging part is that*
> *I am now down to exactly 15p and my visitors permit for staying in*
> *the UK expires in less than a month. Thus I reassure myself by telling*
> *myself that God's standards of disaster will shortly be satisfied. I look*
> *forward eagerly to when the 15p will be gone.*[28]

John was very worried. "I have less faith than you do that the Lord
will provide. *Please* let me know at once if I can help." Hamilton too
was urging his American friend to renew his fellowship. If George

got evicted he was more than welcome at Berkshire. "If there is an Almighty I think you mistake his ways!"[29]

But neither Maynard Smith nor Hamilton understood the extent of the test George had put himself to, for it was far from just a matter of money. "It is now more than nine months since I last saw a doctor," George wrote proudly to a friend back in May the previous year; when he was feeling unwell he'd go to pray at the London Healing Mission. Soon after his conversion he had stopped his periodic thyroid checkups. After all, the coincidences, the penetration of the Bible, the equation—all were signs that he'd been chosen for some task. If the Lord so wanted, the Lord would provide. By now it had been two years since he'd been to a doctor. Pushing fate, he'd begun to eat just as little as necessary and had survived for the last week on barely a pint of milk a day. But there was something worse, darkly more ominous: George had completely stopped taking his thyroid medicine.[30]

· · ·

It was a neighbor from Little Titchfield, Mr. Wood, who found him, collapsed, in the stairway and rushed him to the hospital teetering on the brink of unconsciousness. Luckily an alert physician at the Middlesex ICU, Dr. Webb, noticed the deficiency in his thyroxine, and administered it intravenously while George lay asleep in bed. In most probability this was what saved him: With no corrective to his lack of thyroid hormone, his energy uptake and heart rate had reached such levels as to put his life in grave danger. George had been a heartbeat away from dying.

It was Christmas 1972, and he was recovering in the ward. His eyes were sore from the atropine, and when a patch was removed from one of them, it took awhile before he could stop seeing double. But most troubling at the moment was his imminently expiring visa. "If you say that I entered Middlesex Hospital in a state of 'starvation,'" he subsequently wrote to Dr. Webb, whose explanations would be necessary, "they are likely to say, 'Hah, he ran out of money, one of those American beatniks.'" George preferred that Dr. Webb write that he arrived in "a state of extreme bodily weakness" instead of "malnutrition." Also, he thought that it would be better not to mention "myxedema": Some immigration officer might know what that meant and wonder why he'd stopped taking his thyroid medication. Then he turned to the bigger picture: The reason he hadn't renewed his grant, he'd explain, was because he was in the process of switching his career "from science to Christianity."[31]

Finally, in January, release papers in hand, he was helped into a cab by Mr. Wood and was back in Little Titchfield. Weakened and shaky, he was doing his best to nurse himself back to health. "What is it that one does wrong in cooking rice that makes it stick together?" he wrote to Annamarie in sunny California.

> *Sometimes my rice sticks together and other times it doesn't. I don't cook it very often, so that it is difficult to remember how I did it when it worked well. The last time I cooked it it didn't come out at all well. The time before it was very good. Now maybe the difference was that it was a different brand last time, since I was opening a new package. Or maybe the trouble was that I didn't follow directions correctly. It said on the package to bring it to a boil, stir once, and turn it down immediately to a simmer. But mine boiled for a little while before I turned it down. Could that have caused the trouble? It stuck not only to itself, but also really stuck to the pan. So please tell me what are the critical points that make the difference between sticky and non-sticky rice?*[32]

He was fifty years old. Miraculously he'd survived the Lord's test. But more than ever now he grasped what seemed to have escaped him for quite some time. He was pale and thin, his fingernails brittle and blackening. His brand of strict Christianity had obviously gotten him nowhere. In this world at least, obedience to God notwithstanding, George was utterly alone.

. . .

Hamilton's rule, $rB>C$, was good for the family or at most, after George's correction, to those who shared similar genes. But what about altruism between nongenetically related organisms? This, after all, had been the moralist's concern throughout the ages, sacrifice among relations being a lesser riddle to spin. Recognizing the problem, the Judeo-Christian tradition exhorted its followers to "Love thy neighbor as thyself," but Aristotle was more of a cynic. "The friendly feelings that we bear for another," he wrote in his *Nichomachean Ethics*, "have arisen from the friendly feelings that we bear for ourselves."[33]

Pagan thought was not the only hardened philosophy on the moralism block: Saint Thomas Aquinas took a surprisingly utilitarian position, arguing in the *Summa Theologia* that we should love ourselves more than our neighbors. His interpretation of the Pauline phrase was that we should seek the common good more than the private good since the common good is a more desirable good for the individual. David Hume, in *A Treatise of Human Nature*, may have spelled out at the age of twenty-six what the venerable medieval theologian was thinking: "I learn to do service to another, without bearing him any real kindness: because I foresee, that he will return my service, in expectation of another of the same kind, and in order to maintain the same correspondence of good offices with me or others."[34]

In 1776, the year of Hume's death, his Edinburgh neighbor the economist Adam Smith put it even more directly in *The Wealth of Nations*: "It is not from the benevolence of the butcher, the brewer, or the baker, that we expect our dinner, but from their regard to their own interest. We address ourselves not to their humanity but to their self-love, and never talk to them of our own necessities but of their advantages. Nobody but the beggar chooses to depend chiefly on the benevolence of his fellow citizens." Most succinct of all, however, was the Dutch-born English satirist Bernard de Mandeville, who in a couplet from 1714 wrote: "Thus every Part was full of Vice, / Yet the whole Mass a Paradise."[35]

And so when a precocious son of Lithuanian Jewish immigrants started his education at Harvard in the early 1960s, he had quite a philosophical tradition to build on. But Robert Trivers was not interested in biology—he wanted to be a lawyer—and it would take a breakdown (his mania took the form of staying up all night, night after night, reading Wittgenstein and finally collapsing) to bring him closer to the animal world. Diagnosed with schizophrenia, Trivers took an illustrating course while recovering, and was hired to draw animals for a biology textbook. His mentor was Bill Drury, an Audubon ornithologist whom he learned to love and revere. "Bill and I were walking in the woods one day," Trivers once recounted to a reporter, "and I told him that my first breakdown had been so painful that I had resolved that if I ever felt another one coming on, I would kill myself. Lately, however, I had changed my mind, and drawn up a list of 10 people I would kill first in that event. I wanted to know if this was going forwards or backwards. He thought for a while, then he said 'Can I add three names to that list?' That was his only comment."[36]

With Drury's encouragement, Trivers read Wynne-Edwards and David Lack, and wondered about group and individual selection. Then he signed up for a doctorate in zoology armed with a sturdy plan to study monkeys. But his adviser was a herpetologist, and pointed him to Jamaica and to lizards instead. "When I flew to Jamaica," he remembered, "I took one look at the women and one look at the island and decided to become a lizard man if that's what it took to go back there."

Spending hours peering into the world of his lizards like Hamilton and Maynard Smith (and Darwin before them), Trivers came to believe that behavior was as much a product of evolution as were eyes and ears and fingers and tails. Looking into bird warning calls and cleaning symbioses in fish he came to see, like Aquinas and Hume and Adam Smith and the Dutch poet before him, that self-sacrifice could serve one's interest if the chance was better than decent that the good deed would someday be repaid. Following up on Hamilton, he wondered if altruism might evolve between nonkin. This depended on the rewards of cooperation outweighing the costs of conflict, on being able to remember encounters, and on coming into regular contact with one's neighbors. The theory of "reciprocal altruism," as he called his new invention, was modeled as a prisoner's dilemma, even though the actual proof of it as an ESS would have to wait until Hamilton got his hands on it. Still, Trivers was able to show that iterated encounters between nonrelated "players" would produce cooperative behavior since, in the long run, this provided more gain to the individual. On the principle "You scratch my back, I'll scratch yours," benevolence could be born of self-seeking: If promised in return, altruism would actually pay.[37]

As Trivers continued to work on his models, across the Atlantic, Maynard Smith and an ailing George were putting the last touches to "The Logic of Animal Conflict." Altruism was the opposite of conflict, just as spite was the negative of nepotism. But it bore a structural resemblance to pulling punches in a battle: The very same self-interested logic explained them both. On both sides of the Atlantic, game theory was being marshaled to bring home a century-old insight. Darwin (and Kropotkin) had been right, the "bulldog" Huxley notwithstanding: "Morality" (even if far from selfless) was an invention of Nature.

. . .

Meanwhile at UCL, Harry Harris and Cedric Smith were doing all they could to hold on to George. "The whole point of this application," CABS wrote in a new grant application to the MRC that spring, with minimal cooperation from George,

> is to take advantage of Dr. Price's originality, ingenuity, and ability in developing new ideas about the theory of natural selection. . . . As will be seen from his curriculum vitae, he has only been formally working in this field for about 4 years at the Galton Laboratory; his previous work was of quite a different nature. . . . Nevertheless, his interest and natural abilities do seem to lie very much in the direction of this kind of work, even though he has come to undertake it only comparatively late in life.[38]

Asking money from the government for someone like George took more than a share of explaining. He was unknown, had come from nowhere, and was now fifty years old. Besides, his requests and conditions were highly unusual. "I personally have been very impressed by the quality of Dr. Price's ideas," CABS continued. "But he tells me that for personal reasons he wishes to spend only about one year on the present project. The only way in which that would be practicable would seem to be through a grant to support him." All too well aware of his customer, he added: "—failing that, he would presumably be compelled to look elsewhere for some quite different occupation."[39]

CABS was nearer the truth than even he might have imagined. For soon after returning home from the hospital, spurred by a vision of Christ, George underwent a second conversion. Two years earlier he'd written to Hamilton that God's greatest test to man was that of *agape*, or charity and love. Luckily there were other qualities that counted in the Lord's eyes—like obedience and intellectual receptiveness—for, candidly admitting the truth to his friend, "it doesn't look as though I can pass purely on the basis of agape."[40]

But that was then. Now things were different. "Last February," he wrote to Hamilton, explaining,

> *I sort of "encountered" Jesus and found that I had never been a Christian at all but a Christian Pharisee and one of the world's best hypocrites. Now I'm trying to become a Christian. As you know one of the things that Jesus commanded his disciples was to give up things that were dear to them (e.g. money) and follow Him. One thing he has had me give up is the book. It doesn't really matter what the "true" date of Easter is—what matters is when people observe Easter. It's people who count in Jesus' eyes.*[41]

He was abandoning his biblical pursuits and reaching out to the world again. "I'd love to see you again, Bill," he ended his letter to Hamilton. "You're one of the people whose company I enjoy." Pulling out childhood photos of his daughters, he wept in silence, alone in his flat. "Poor Annamarie," he wrote to one of them, gazing at her picture:

> *It was probably taken not long before I moved out. I'm sorry I deserted you like that, and I'm sorry I was such a poor father to you. . . .*

Looking at your picture now makes me wish I could do it all over again. . . . You were a very sweet little girl, and I did wrong to leave you and then was very neglectful of you afterwards. I wish there were some way I could do a little toward making things up to you.[42]

He understood now, he wrote, that he had never really been a true Christian, believing in Christ intellectually but neglecting His teachings about mercy and humility. He was determined to make things right again. His first conversion had been brought about by coincidence and was "false." This time it would be "real" and was all about love.

Quickly he set about putting out old fires. "I write to you in very great shame," he apologized in a letter to Rosemarie. To his old Harvard friend Henry Noel, with whom his communication had soured, he wrote to make amends and explain his new path in life. In his vision Jesus had whispered to him: "Give to everyone who asks of you, and whoever takes away what is yours, do not demand it back." Henry was in France now with his family, living Luke 6:30 too, only from the receiving rather than the giving end, he wrote back with a smile. Still, he was glad about his friend's sudden change of tone. However much or little it might be like Dr. Skinner's rats or pigeons, he thought, George's choice to turn to *agape* was still a source of volition other than God's will.[43]

Whether George would agree was questionable, though reborn a new person, this time he'd keep his thoughts to himself. He was attending a new church now, Saint Mark's Clerkenwell at Myddelton Square, about a forty-minute walk from Little Titchfield. The first time he went there, he wrote to Kathleen, as he was getting close to the white stone West Tower,

I began to have a faint, vague impression of seeing Jesus walking before me bearing the cross. Two or three times since then I have felt that He gave me the message "Follow me."[44]

What this meant was clear to him: Suffering lay ahead. There was as much volition in this, he knew, as there was water in a desert—

Henry's sigh of relief notwithstanding. But George would accept his fate with open arms.

Then, with no words of introduction, he wrote to Julia in America, beginning dramatically: "Dear Julia, Would you be willing to marry me again?"

> After those terrible letters I sent you I can understand if you don't want to have anything to do with me. And if you think back over all the wrong things I did when we were married before, you will surely have doubts about marrying me again. Also I'm not in very good condition physically (a bit thin, tired), and I look quite a bit older now . . . and my financial condition is rather uncertain. And still another problem is that I am now hoping to live as a disciple of Christ, and this may force me to separate from you for long periods, and possibly even in some later year permanently. I don't know what the future will bring. . . . However, this would be separation, not divorce, and we should still be man and wife even if separated (see Matthew 19:27 on the possibility of separating at Christ's call). Thus there are many reasons why you might decide to say No, though I very much hope that you will say Yes. On the positive side, one thing I can tell you with much assurance is that you would find me much kinder than before.[45]

It was perhaps not the most attractive proposal, but George meant it from the bottom of his heart. His plan was for Julia to come to join him in the summer, he hoped with Annamarie, Kathleen, and Dom. The lease for the flat at Little Titchfield was ending on June 24, but until then they could live together, somewhat cramped, perhaps, but as one big happy family. Then he and Julia could finally go for a proper honeymoon in France, something they had had neither time nor means for back in the summer of 1947. Then they could return to Britain to travel with the kids to the "bonnie, bonnie banks of Loch Lomond" and to Tintagel in Cornwall, where King Arthur is supposed to have been born, and to Canterbury and "lots of other places." All the love and affection would help bury things past.

Julia wasn't sure about it all, she said to him in the first phone conversation they'd had since the sixties. But she agreed to come over in the summer, promises for a second wedding withheld. The pronouncements that Jesus wanted him to follow the path of suffering were worrying her. In truth, there was more to worry about than she knew of. For, opening his heart and his Little Titchfield apartment, George was already walking deep on the path of *agape*.[46]

. . .

It began in late March, around the beer- and piss-reeking, rat-infested corners of Euston Station and Soho Square. It was there that he was seeking out London's homeless and hapless dregs, men and women, young and old, down on their luck, those whom life had not smiled upon. "My name is George," he'd introduce himself. "Is there any way I can help you?"

Homelessness had been a part of London's landscape from its Roman beginnings. In modern times, made doubly visible by photography and statistics and sustained by often harsh Victorian poor-law institutions and a smattering of charities, vagrants became a celebrated if regretted part of the scenery. Disguised as a down-and-out sailor, the American writer Jack London painted a dismal portrait of life in the East End in the 1903 best seller *The People of the Abyss*. There was Poplar workhouse, or "the spike," as the shelter of last resort was known. There was "the Peg," the Salvation Army barracks on Blackfriars Road in Southwark, where the homeless could turn up on a Sunday morning and, if they were lucky, be given a free breakfast. And out on the streets there were knifings, and broken beer bottles, and the stench of death, and the silence of suffering, and the battle for a bed at night, and the constant mumbling under the breath, audible but opaque, punctuated by a barrage of expletives.[47]

In the winter of 1910 it was estimated that 2,747 people were sleeping rough in London. That number swelled after the Great War, when former servicemen, many of them disabled and disfigured, were reduced to begging on the streets. Air raids on the capital during World War II left an estimated one in six of Greater London's population homeless at some point, and squatting became a way of finding a home that would become part of the city in the years ahead. In 1966 the television drama *Cathy Come Home*, about a working-class single mother living in the streets, made a profound impact on a sentimental public, leading to the formation of the housing charities Shelter and Crisis, and the night shelter for homeless teenagers in Soho, Centrepoint, in 1969. No one knew precisely their number, but by the time George made his way to Euston Station from Little Titchfield in the spring of 1973, an estimated 40 percent of England and Wales's homeless—wet, grimy, and often inebriated—were living in and around the city.[48]

Soon he learned that there were many ways he could help them: A quid here, a sandwich there, a cup of hot cider, a word to a policeman. Most of all, though, he could offer room and board, and, beginning in April, they were flocking to his home. There was Peg Leg Pete, a temperamental redhead who lost his limb clambering over a wall while being pursued—falling and catching his leg on a hook, where he remained suspended until the police arrived. There was Smoky (real name: Trevor Russell), a hardened alcoholic who had been in and out of prison more than thirty times. Smoky "was tough and ready to fight any man of any size," George wrote to a friend; he couldn't stay at any hostel in London, tales of his disturbances having become legion. There were Bernardo and Chrissy, a sweet couple who hadn't had much luck and drank a lot. There was Aberdeen, another alcoholic with a mental hospital record, and a karate expert to boot. To all who stayed longer than overnight, George provided keys. At one point there were four men staying at Little Titchfield who had either done time or were wanted by the police and who were alcoholics or who had a record of insanity, each with his own keys to the place. One of the men had been in prison specifically for burglary.[49]

Friends and acquaintances were starting to worry. He had written to Henry Noel that before his love conversion he had been "vile" and was trying to make amends by following "the path of total depravity." Henry replied that it struck him that there might be a peculiar kind of pride in judging ourselves with more harshness than God Himself judges us. "Let God be the final judge—don't try to out-God God." The duty of charity, he reminded his friend, applied first and foremost to ourselves; George needed to snap out of this obvious "state of extreme." "The other Price" too, the psychiatrist John Price from Maudsley, wrote kindly to offer George to come stay with his family in Northumberland. This might help, he hinted gently, "to give you a wider perspective during a time of decision-making." Tatiana, for her part, wondered whether if he was feeding every man he himself was not growing hungry. Even Edison, all the way from the Village and still engrossed in Iyengar lotus-position yoga, implored his brother not to be so "damn self-critical. I really am fond of you," he wrote in a rare exhibition of brotherly affection.[50]

But as usual George Price was on a mission, fighting off any feelings of fear of the strange men in his apartment: "I told myself that if I was obeying Jesus, he would protect me against serious harm," he wrote. And, it soon became apparent, Jesus really was protecting him: These were rustlers and cheats, men, it seemed, with little moral compunction. And yet, from April through June, besides a bottle of wine that was later replaced, nothing was ever stolen. The place was filled with heavy smokers, most of whom were semi- or fully drunk much of the time, but there was never a fire. The neighbors in the flat below were not too happy about what was happening above, but following George's gentle explanations "soon became friends again."[51] Sure, he was giving out money freely now and leaving very little for himself. Sure, he was being taken advantage of. And sure, few of the men cared a hoot about him or would help him if he needed them. But this only strengthened his resolve.

Catching wind of the fact that George spent much of his time on the streets now, giving to the derelicts he encountered, the Reverend R. F. H. Howarth from All Souls wrote to say that experience teaches that giving money to down-and-outs "is seldom more than an easy way out for ourselves." But George's experience was different; this was no version, however sophisticated, of a Trivers tit-for-tat. To Henry Noel's concerned letters he replied that no, he had not deliberately put himself in a time and place where he would be asked for help. In fact he dreaded going down to Soho Square, where some of the fiercest and most violent hoboes held sway. But then he'd read 1 Thessalonians, and hear Jesus whispering to him: "Go back to Soho Square this evening."[52]

Soon he would have to leave his apartment. Suffering was on its way. Whether, as he'd written to Kathleen in March, there was ever really pleasure to be found in pain seemed beside the point now.[53] Aquinas, Hume, Adam Smith and the Vietnam War notwithstanding, after fifty-one years, finally, George was learning to love.

George Price in his University College, London, office, 1973

12

Reckonings

Toward the end of April 1973 the grant came through. Beginning May 1 George would be reappointed associate research fellow at the Galton, with a salary of £3,195 for a year of research on genetic polymorphism, "no fixed hours of attendance" stipulated. It was exceptional for the Medical Research Council to award one-year grants, not to mention for the Galton to demand so little; CABS's imploring had evidently done some swaying. But in truth genetic polymorphism in nature was rather far from his mind now. In May he wrote to Julia that he was planning to find a large derelict house once the lease ended on June 24, and to start a "Jesus people" commune with twenty to twenty-five people. Clearly her lack of response to his marriage proposal meant that she wasn't very interested. But she should still come along with the girls and Dom in the summer to stay with him there. He had no doubt that she'd be fond of Bernardo and Chrissy in particular.[1]

To Mr. Norman Ingram-Smith of Saint Martin-in-the-Fields he wrote in more detail about his wish to help solve the "bag-storage-place-to-wash-and-change-clothes problem" of London's homeless.

There were two explicitly Christian communes in the city: the Children of God commune on Walterton Road and the Jesus Family commune in South Norwood. Both were full-time evangelizing outfits, and both prohibited drinking and smoking: "That's fine for those who want that, but there are many people in London who enjoy helping others but who smoke and drink moderately and want to hold jobs. So there is a need for communes or co-ops with more permissive rules."[2]

The idea was to start such a commune by renting out a derelict house (not squatting; after all, the people living there would all have jobs, and it wouldn't be fair not to pay rent), and opening it up as a haven for homeless who needed help. Each member would decide what part of his or her possessions and income were turned over for common use to the commune. Particular effort would be made to get some members with skills useful for rehabilitating old houses: plumbers, electricians, and so on. It would be a happy and friendly place, "with a bit of style and swing to it," for advertising to help make communal life seem attractive. Members would be encouraged to invite guests for meals and to stay overnight or longer. But this would not be a one-off; it was part of a much bigger plan:

> *The hope would be that other communes would spring up that similarly had permissive and flexible rules, with members holding jobs, some of which would have more social service emphasis. One desirable feature would be to have young people and old people living together. Some communes might include bedridden or blind people. Others might make a special effort to help alcoholics. And so on. It would depend upon what people were interested in doing. If the idea took hold and many communes were founded, a lot of problems could be alleviated, including the problem of giving homeless people in some cases homes, and in other cases a place to leave a suitcase, wash up, and change clothes.*[3]

Back in Little Titchfield, he was learning lessons of love from his lodgers. In the living room he had built an altar, covered by a tablecloth and with a wooden cross standing atop it. It was rather simple,

George thought, but he had plans for a more splendid one, with a skirt of black velvet and a top cloth of white velvet, surrounded on the sides and back by drapes of blue velvet. Then one day Bernardo asked for some Vaseline for his hair, and, apparently not noticing that it was an altar, wiped his greasy hands on its cloth, and—to George's horror—hung his underpants on the cross. Just as he was about to give him a piece of his mind, George came to see that this was Jesus telling him—through Bernardo—that the rich velvet altar cloths were the wrong way, the old way, the Old Testament way, whereas giving good clothes that he himself wanted to needy strangers was the right way, the way of Jesus the Lord.[4]

As much as living with strange men was educating him, the lease came to an end on June 24, and, feeling utterly unworthy yet to follow Jesus' true path of suffering, the least George could do was not to renew it. The peculiar American had been a godsend to the homeless of Euston Station and Soho Square for the past three months, but he had completely failed to plan ahead for his own sake. Now George was going to be homeless himself.

. . .

The first few nights he slept in his office at the Galton, but clearly this was no kind of solution. Ursula Mittwoch, a colleague at the department, offered that George stay with her family and do some tutoring for their teenage daughter who was just then preparing for an English exam. She remembered his stay with fondness. Her daughter loved George's clarity and marveled at how he seemed to know all the poems she was supposed to learn. Everyone enjoyed his good humor at breakfast and dinner, his utter considerateness, tidiness, and gentle manner. Even George himself was making a good time of it. "Thus far I have enjoyed being homeless," he wrote to his old family friend Dr. Gilfillan back in the United States on July 3. "It is a good way to get acquainted with people."[5]

But the Mittwochs' was only a short-term solution, as were fleeting stays at other friends' houses from the Galton. He'd moved most of his books and papers to Wolfson House, the abode of the Department of Human Genetics and Biometry at 4 Stephenson Way. But soon George was beginning to see that this wasn't going to be all that easy. Before he could create his "Jesus people" commune, he would need to find a place to sleep.

Having to abandon George's apartment, too, Smoky was now once again behind bars. From Her Majesty's Prison, Pentonville, on Caledonian Road, he wrote on lined prison notepaper to thank George for the radio he had sent him, but also to explain his own philosophy. George might think that Jesus intervenes in people's lives, but Smoky was less salutary. "Lets be fair," he wrote, "if we do wrong we have to take the consequences." George's noble generosity was admirable, but in the end the price might be too steep:

> Come to think of it, you are better off keeping away from the square, those people there have no respect for you, all they want is money and cider off you, you have to consider yourself now and again, do they worry about you, when you are broke and hungry. I doubt it very much. . . . give them half the chance and they would squeeze you dry.[6]

Selfless friendship was difficult to come by.

> Talking about friends, where are all the ones who are supposed to be friends of mine (YOU EXCEPTED OF COURSE!) I haven't had even as much as a postcard off any of them. So I can assure you, I don't miss any single one of them, FRIENDS, people call them? They are or were only drinking acquaintances, I miss my drink and the privilege of walking the streets admittedly who doesn't in this place. In fact what purpose have these people got in LIFE? They live from one day to the next wondering where the next drink is coming from they hardly ever eat, have a bath, they won't work, honestly I think some of them would be better off in here, for a while.

He was praying that George and Peg Leg Pete find a place to rest their heads again. George was a rarity: a true and honest altruist.

He needed to watch out for himself. "It's not very nice in the Bruce House," Smoky warned him of the Centrepoint homeless hostel on the aptly onomatopoeic Drury Lane. "Pleased don't sleep that low!" Then he added, "I would suggest that you post the cash to me in your next letter, if you can manage say £10 to £15."[7]

Thinking little of his own problems, George had lately started to help old people around Myddleton Square near Saint Mark's Church, nursing them or running errands during the day and sleeping over nights when family found it difficult to stay. Muriel Challenger, a congregation member, acted as the go-between. There was the frail octogenarian Mrs. Rose on Chadwell Street, who wasn't doing all that well, and Mrs. Abercrombie, likewise, on Goswell Road. After explaining that he shouldn't worry if the old ladies sometimes seemed "changeable," Muriel wrote to George:

> *Mr. Eastop, 345 St. John Street, could do with someone to walk beside him for a very short walk. It would be good for him to go out and not sit all day but he has lost confidence since a severe illness. His little wife is housebound too & is not much of a companion as she finds her deafness difficult to cope with.*[8]

George had seen Julia briefly when she visited in the beginning of August. She'd come over, she made clear, to buy some antique jewelry and small collectibles to sell at weekend antique markets back in Michigan. It was a sad coda to the hopeful days all those years ago, when World War II was ending and the future lay ahead. If there had been some miniscule, wild glimmer of hope that they should get back together despite all their history, it had to be put to rest now. With George homeless and making radical selflessness his life's philosophy, it was clear to Julia that her relationship with that handsome man she had encountered at the Met Lab, the promising scientist who had become the father of her daughters twenty-five years before, was finally, irreversibly dead.[9]

George was staying at Bruce House now. Some nights a violent drunk would fight him over his cubicle, and always he would yield with a smile. During the day he'd walk to Euston Station and Piccadilly Circus to meet winos and beggars and see how he could help them. He was wearing a large aluminum cross against his chest, and twice already, he thought, it had come in handy. When he'd chanced upon two cases of police brutality to homeless men he confronted the coppers demanding that they stop. Each time he was told gruffly that it was none of his business, and each time he remarked that it was. Then, on both occasions, the policemen took a look at his cross and, silently if not entirely respectfully, retreated.[10]

He had testified at Smoky's trial, but the testimony failed to shorten the sentence. Never mind, George wrote encouragingly, this would give Smoky time to make a true promise to Christ. "You asked me for suggestions about what to do when you get out," he offered.

> *Well, Smoky, I may be totally wrong, but since you ask me for advice I'll tell you what I believe. Your ideas, from what you've written to me, are about getting a good job, working regularly, going to church regularly, and abstaining from drink. Well, I don't think you can manage it.*[11]

Instead George thought that he should take one or two drinks soon after he got out, that he should right away, even now in prison, stop attending church services, and that he should abandon ideas of getting a proper job and try to help the homeless instead. The reasons for all this ("by the way," George wrote, "this is very unconventional advice") were first, that not drinking entirely would only inevitably lead to a powerful urge for the bottle; second, that going to church was much less important than serving Jesus by loving and helping others; and third, that since he knew the streets better than anybody, helping homeless down-and-outs like himself would be the job he could accomplish with greatest skill.

> *If you try to manage a conventional, in-between life I think you'll quickly drift back to the way you were. So I think your only way out is to resolve that you're going to go to the other extreme and give most of your time and efforts to helping others, especially alcoholics. It's*

much the same way that some dangerous animals will attack a man if he tries to run away from them, but will run away from him if he goes directly toward them. So, in the same way, think of cider and wine, Soho Square and Bruce House and that whole way of life as a dangerous tiger that will hunt you down if you try to flee from it, but if you go directly toward it, armed with the "rifle" of intending to help people, it will flee from you.

Sealing the letter and addressing it to Pentonville, George might really have written the advice to himself.

. . .

Back in October when he'd expressed to Maynard Smith his joy over the outcome of their joint paper on the logic of animal conflict, he had paused to relate an embarrassing misgiving. "There is one matter that has to be brought up now and on which I need to hear from you before trying any re-writing," he wrote. "This is a rather unhappy matter to mention. It involves Bill Hamilton."[12]

What George was referring to was something Hamilton had once told him regarding the refereeing of his 1964 paper on the evolution of social behavior. Usually gentle and pacific, Hamilton felt that he'd been terribly wronged, and continued to harbor a burning grudge against Maynard Smith after all these years. For Maynard Smith had been the referee of his paper, Hamilton had learned, the one responsible for asking to split it into two separate papers—a request that eventually held up publication for close to nine months. This wouldn't have been so terrible if Maynard Smith himself had not hurried in the interim to publish the paper in *Nature* contra Wynne-Edwards, in which he coined the term "kin selection" for the very first time. After all, "kin selection" was a pilfering of Hamilton's idea of inclusive fitness; like King David with Uriah and Bathsheba, Maynard Smith had not only sent Hamilton away but absconded in the meantime with his beloved!

When George found out that Maynard Smith had been the *Nature* referee who read his *own* paper on antlers in 1963, he remembered Hamilton's ire and jumped to similar conclusions. Once again, it seemed, John had held up a new idea and in the meantime some-how found himself interested in the very same topic. But then George met Maynard Smith and saw what a kind and gentle person he was. Hamilton's anger, he now believed, had to be the result of some mis-understanding. This was the reason why, when Maynard Smith wrote to him in 1971 wanting to cite his antlers paper, George replied, "If one mentions an 'unpublished manuscript' then someone may won-der about whether it was used with permission, but if you speak of 'discussion' then no such suspicion arises." Having seen for himself that John was a good man, he wanted to protect him from a possible second accusation.

Hamilton, though, had never gotten over it. This was tricky business, George wrote to John in October, since in the draft of "The Logic of Animal Conflict" that Maynard Smith had just sent him there was no reference to Hamilton on kin selection. Clearly this needed to be fixed before George would put his name on the paper.

But something else needed to be fixed as well. For, following his love conversion, George saw that the current drafts had to be revisited. "This requires choosing some words with more care," he explained to John in February. "I think that I've found wordings that you won't object to and that won't shock Nature's readers by making them suspect what I believe."[13]

Maynard Smith replied that he was sorry about not citing Ham-ilton in the draft; oddly enough it was simply a glitch, and he'd fix it. He was sorry, too, about what had happened in 1964. In fact he'd quoted Hamilton's short 1963 paper then but could have done better by citing the as-yet-unpublished paper that he had just reviewed. He felt terrible about it, and would try to find the next opportunity to "set the record straight"; clearly, he wrote, "Bill certainly deserves any credit that is going."[14] As for the reworking of some of the language in their current joint paper, John would defer to George's sensitivities. He had no problem that the word "Dove" from "Hawk v. Dove" be scrapped; instead they'd just call a nonviolent, nonretaliating actor by the less theologically laden "Mouse."

. . .

"The Logic of Animal Conflict" made the cover of *Nature* in November 1973. "Game theory and computer analysis," the authors concluded triumphantly, "show . . . that a 'limited war' strategy benefits individuals as well as the species." With the newly introduced and formally defined concept of the ESS, it was a paper that would impact the study of the evolution of behavior dramatically.[15] Down at Soho Square most days, searching for Aberdeen and Peg Leg Pete, George had other things on his mind.

"As I need hardly tell you," Al wrote to him from Buffalo,

> *the moral and religious precepts of the Gospel reflect a profound understanding of human nature. I would think they are intended to identify goals toward which we should strive, given our frailties . . . to live up to them literally may be to attempt more than human nature can manage and, I suspect, more than we actually intended. Trying to live up completely by such principles might produce little in the way of peace of mind. And, it would seem to me, that where behavior based on religious precepts does not yield peace of mind, the eventual result will almost inevitably be the erosion of belief itself.*[16]

But George was deep in the forest on the path he had set for himself. "I can't remember whether I told you anything much about my way of life," he wrote to his brother, Edison.

> *I have no home, so use my business address as a mail address. . . . Usually I wear brown Levis, sneakers, and a colorful shirt. Many times each month I find myself reduced to one penny, a half penny, or zero. Most of my possessions have been given away, including my watch and coat (but I'll have to pick up a coat somewhere now with winter coming on). Everywhere I go I keep running into down-and-out alcoholics, to whom I give when I have anything, and with whom I sit and drink from their bottle if they offer me a drink. Increasingly I find myself on the opposite side from the police. Many of my friends have done time, and I've been in a house that was raided and had my things searched then, but I haven't yet been busted. . . . I do a lot of smoking, and also smoke cigarettes, though I haven't yet developed*

a fag habit. A substantial amount of my time is given to trying to help people in almost any way they ask me or seem to need help, whether it's by giving them money, cleaning a filthy kitchen, talking to a landlord, shopping for a housebound person, or trying to solve some mathematical problem for somebody here at work. A lot of this helping is of old people, especially women in their eighties. I live very cheaply and have been reducing my debts (which are large) fairly rapidly since I became homeless. . . . In spite of vast amounts of time missed from work, plus eccentric behavior such as sleeping here often and doing my laundry in the men's room and trying to borrow money from everyone in the department, my professor and the department chairman are friendly to me. (In fact, most people are friendly to me except the police, who seem to instinctively dislike me nowadays). I haven't gone to church for six or eight weeks, but I visit and try to help old people in connection with a church that I have often attended. I usually wear a cross of some aluminum-appearing or pewter-like metal around my neck, except that people keep asking me for it (especially old, sick people and down-and-out alcoholics) and I've given away seven of them and don't have any now and won't be able to buy another until pay-day). I generally try to say "Yes."[17]

Then he ended, "And now what's up with you?"

It was an amazing transformation from the prim, short-haired, gangling IBM worker he had been just a short seven years ago. Even Smoky was really starting to get worried about him, as was Paul Garvey, a homeless wreck serving time at Her Majesty's Remand Centre in Richmond, Surrey.[18]

But George was happy, perhaps the happiest he had ever been. A kind of peaceful quiet had finally descended on his soul. Lately he'd met "a bloke named Keith who is a follower of the Guru Maharishi," and the two enjoyed conversations on a park bench over chips and coffee.[19]

He was enclosing a picture of himself, he wrote to Kathleen, taken in his office by a photographer; the best faculty photo, people said, in all of the department. In it he is wearing a colorfully striped shirt and dark bow tie, sporting a wry smile above a scraggly red beard, counterbalanced by fine hair brushed back above a broad forehead. Only the eyes confuse an otherwise joyful portrait: Tucked behind dark-rimed glasses, one is small and kind, the other open wide and strangely empty.

He was feeling so good that he decided to send Kathleen a special surprise. His only likely source at the time was in Nottingham, where, he divulged to his daughter as if reading her a bedtime story, the sheriff who hunted Robin Hood had lived. But in the end he took out the quid's worth of pot, pressed between the pages of a C. S. Lewis book stamped for California. He had just moved to a new squat, and ended up smoking it himself.

Life among the destitute, as the "Monthly Message" of the London Healing Mission newsletter admitted, was "certainly never dull."[20] There was an alcoholic woman beater whose partner George had hid from him and who was demanding to know where she was. Increasingly he would come around the Galton, insisting to see George, and, when refused entry by the guard, would yell from the pavement up to his office. George refused to divulge the woman's whereabouts and soon was keeping his own whereabouts secret, too. "The reason for the secrecy," he explained to Kathleen after half apologizing for smoking the reefer,

> is mainly one very difficult man who has been coming around where I work to look for me and causing trouble. A week ago Monday he pissed publically on the front steps of the genetics building, smashed a bicycle lamp, scattered the contents of some student's satchel around, and shouted his best obscenities.[21]

The people he was living among and helping weren't always as friendly as he was, but, having been enveloped in a halo of serenity, George wouldn't allow this to dampen his mood. "I expect that one cover-illustrated article in Nature compensates for one urination at the front entrance to the building," he joked to Kathleen, his sense of humor still very much alive.

The administrators at the Galton weren't happy about the down-and-outs who were showing up, still less for their urinations and assaults on students. Late in the fall, George had met a young IBM programmer from America at the Russell Square Underground Station who turned out to share his liking for Proust plus his total lack of sense of direction and great inability in pronouncing unfamiliar names. Excited, George took him back to his office and they spent the evening talking about Proust and computer programming. As it happened, the young American was wearing blue jeans, leather boots, a green poncho, and had about a three-day stubble. Wary of George's exploits, the beadle immediately pegged him as an alcoholic and pressured Harry Harris into introducing a new rule against late-night and weekend working without special permission. CABS tried to help by sneaking George a key to the statistics library, but he was discovered. Soon he was no longer coming in to his office at UCL. Too quickly his colleagues at the department were losing contact with him. Most thought he had gone off the deep end. "He certainly flipped," one of them recalled.[22]

Then, in mid-November, Al Somit arrived in London for a visit. The UCL photo may have made George look healthy and chipper, but that was only a head shot and after a rare shower to boot; the professional photographer hired by the Galton had obviously done some magic. In reality things were very different. Al hadn't seen George for about eight years and was dismayed and appalled at what he now found: He'd first met him in the weight room at the University of Chicago, offbeat perhaps but handsome and muscular and hard; now George was as sinewy and gaunt as an old man, the spring in his step all but vanished. He was grungy and oily and shabbily dressed, his teeth were beginning to rot, his outgrown hair was as brittle as hay, his fingers yellow from smoking. They joked together like in the old times at the co-op before Al shifted, inevitably, to a more serious tone: "I'm not going to give you money for the new pair of shoes you obviously need unless you promise not to give it to these two leeches," he said to him in a coffee shop, eyeballing two alcoholics who had been on George's tail. George thanked him but said he couldn't make such a promise, and that no amount of convincing would help. It was the same old George, he thought, always contrary, always at the extreme. Holding his hand out with a smile to say good-bye, Al walked away from his old friend with the pound notes still deep in his pocket.[23]

"It was nice to see you again though, in all honesty, I think I would have preferred finding you in somewhat other circumstances," he wrote upon his return to Buffalo, suddenly feeling worried and regretful.

It occurred to me, as I reflected on our discussion, that you may be confusing the notion of serving your fellow man with loving your fellow man. If the former, surely there are more effective ways than the one which you have adopted.[24]

Then he added, with a candor far removed from their usual wise-cracking: "The latter may, in fact, be quite beyond your capacity—or mine."

. . .

When he left their meeting, George literally had nowhere to go. It was cold out. At Euston Station he met an alcoholic who told him about the headquarters of the Tolmers Village Association, just down the road. The address was 102 Drummond Street, and the man said, it was one of the few places around with a hot bath.

The Tolmers Village Association had been born in the summer of 1973, the fruit of community action to help save the neighborhood from Stock Conversion, a property developing company, and its owner, Mr. Joe Levy. Levy had become rich in the sixties, having obtained planning permission to build offices on a one-acre site on the corner of Euston and Hampstead roads in Camden. Demolition began in 1963, and within a short few years the run-down shops and houses west of Hampstead Road had all been bulldozed and replaced by "millions of pounds worth of windswept glass, concrete and steel, full to the brim with office workers." The first of the big property bonanzas, the Euston Centre whetted Levy's appetite. And as the conscientious members of the Tolmers Square Tenants Association were busy fighting in the Camden Council for the tenants down the road who were being displaced by the new massive high-rise tower, Levy and Stock Conversion were buying up their houses, literally under their noses, to do to Tolmers Square what they had done with Euston Centre.[25]

Levy hadn't gotten rich because he was lucky, and now the same acute developer's nose had led him once again to a practically assured gold mine. Tolmers Square had originally been built in the nineteenth century for the middle class, an almost complete, oblong circle of dark brick four-tiered Victorian terraced houses surrounding a church. Soon, however, it was taken over by the working classes, and when the main-line terminals of St. Pancras and Euston displaced thousands of residents toward the end of the nineteenth century, overcrowding and prostitution turned the area into one of London's worst residential slums. Heavy bombing during World War II, though, combined with rising health standards and—with the expansion of Euston Station—the encroachment of industry, commerce and offices, led to a population drop from an 1870s level of more than five thousand to just above one thousand in 1970. It was then that the Tory government attempted to stimulate the economy by letting the bank rate fall from 8 precent to less than 5 percent, resulting in a flood of capital into property. By the beginning of 1973 the market valuation of property companies in Britain was £3.6 billion, more than the entire gold and dollar reserves of the United Kingdom. With its run-down porticos and leaky pipes, Tolmers Square had little present value, but Levy's nose didn't fail him: Just half a mile to the north of Oxford Street, with Soho and the theater district just beyond, east of fashionable Primrose Hill, and only a few minutes walk from Regent's Park, Tolmers Square and its North Gower, Drummond, and Euston Street environs would fetch at least three times the rent for office rather than residential space, and considerably higher than for office space in other parts of London. It was a speculator's jackpot.[26]

But if the population had dwindled, there were still people living in the neighborhood. Attracted by the cheap rents and proximity of employment, besides a core of longtime English working-class homeowners, a constantly changing population from many nationalities had turned Tolmers Square into a colorful pageant. There were Cypriots and Maltese, Irish, Indians, Pakistanis, Bangladeshis, and Ceylonese. The Italians were especially proud that the former resident of 187 North Gower Street, Giuseppe "the Patriot" Mazzini had once lived there; the Muslim Asians were proud of the newly built mosque, and the Indians were proud of the five Indian restaurants and the neighborhood's new nickname, "Little India." There were chickens in some small backyards, pecking beside the many outhouses. There were old Georgian terraces and quaint Victorian pubs. Amid the tumult and hustle and bustle of Euston Road and the station, Tolmers had sewed the human patchwork into a living community "It was a place of atmosphere, always alive," one resident remembered. "Everyone knew each other."[27]

Except that no one was properly equipped to deal with the below-the-belt tactics of the wily new developers. By 1973 Stock Conversion had bought up almost five of the twelve-acre environs, a swath that came to be known as "Levy triangle." The overall strategy for future purchases, the boss determined, could be summed up in one word: neglect. With no particular reason to rush, since every passing day only served to raise the value, Levy enacted a deliberate policy of leaving his properties to decay. When a pipe burst or a terrace cracked, Stock Conversion–employed handymen and builders were instructed to botch repairs, and in any case not to spend more than £5, lest "the boss would kill me." That the company's net tangible assets stood at £62 million at the time was of no consequence; soon, enough demoralized tenants would leave to allow Levy to begin carrying out his plan. A few house collapses aided in the psychological warfare. Other abandoned properties were summarily boarded up, and before long many began to stud the square. What had once been a thriving piece of London was quickly turning into a decaying shell: More than fifty houses and a number of commercial buildings had been demolished, more than 10 percent of the land lay vacant, and a growing number of homes and shops were in varying stages of dereliction. Vandalism and rot soon followed; an empty shop on Drummond Street was nicknamed the "the pet shop" because people could watch rats playing behind the glass. The population had been cut by almost half; those remaining included a high proportion of pensioners, immigrants, and the poor. A council report from from the end of 1973 found that of the 213 tenancies in the development area, 30 percent had no access to a cooker with an oven, 70 percent had no access to a sink with hot and cold water, 57 percent had no access to a bath, and 79 percent shared a WC. Everything was going according to plan: "Anyone who observes a beatific smile playing over the features of Mr. Joe Levy of Stock Conversion," *Private Eye* reported, "may be sure that what is occupying his mind is not the recent disappointments over plans to knock down three streets near Piccadilly, but the two magic words 'Tolmers Square.'"[28]

But if private interest was trampling community living, if this was some kind of twisted revenge against an imagined tragedy of the commons, not everyone was going to stand idly by. Once again, as in T. H. Huxley's day, the winds of reform blew from "the godless institution of Gower Street." The School of Environmental Studies was the new name of the faculty housing the old architecture department, and student life at UCL was still spinning from May 1968 and the residual counterculture of the sixties. Passing Dan the beadle, uniformed bastion of the college traditions at the entrance in the corner of the quadrangle, one entered a wide hallway daubed with revolutionary graffiti:

> THE TYGERS OF WRATH ARE WISER THAN
> THE HORSES OF INSTRUCTION
> *(William Blake)*
>
> PROPERTY IS THEFT
> *(Pierre-Joseph Proudhon)*
>
> SOULS BROUGHT AND SOILED
> *(anon.)*

The first year curriculum included semiology, social history, Lévi-Strauss, Roland Barthes' *Mythologies* and John Berger's *Ways of Seeing*.

In short, though most of the students were only dimly aware of it, the dominant intellectual tone was antiestablishment neo-Marxism. And so, when a group of students in May 1973 embarked on a five-week planning project, it was only natural that they chose adjacent Tolmers Square. After all, they wanted to be involved in "reality," to break down the barriers between academe and the populace. The neighborhood was in flux, and yet the local council seemed to know next to nothing about the people who lived and worked there, or, in all honesty, to care. The students were outraged: "Unless Camden's planners can show that the redevelopment of Tolmers Square will have advantages for the people who live and work there at present, as well as those who might live there in the future, they should reconsider their plans," they wrote in their survey. "Merely satisfying property speculators is not good enough."[29]

Just before leaving for their summer vacation, they held a meeting in an office on North Gower Street. A community town planner and a councillor were invited, and joined nine neighborhood locals—a couple who ran a geological information center, two shopkeepers, and five tenants—in agreeing that some form of opposition had to be mounted against the council's pro-Levy proposals. Before long membership had grown to two hundred, and a group of students had started a community newspaper, the *Tolmers News*. The Tolmers Village Association (TVA) had come into being. Then, in October 1973, the association took over a shop on the first floor of 102 Drummond Street, turning it into the effective headquarters of the TVA. A local artist designed a letterhead, a rubber stamp, and painted a geranium on the window. It was there, four weeks before Christmas 1973, that George showed up asking for a bath.[30]

. . .

Whether it was his profoundly gentle sad eyes, his worn-out green cord jacket, straggly sandy hair and beard, or the fact that he mentioned that he worked on genetics at UCL, the good people at 102 Drummond invited him to stay. There was an empty room up two flights of stairs, the third-year architecture student, Ches Chesney, told him. Ches and his friends Tim Davies and Carla Drayton were squatting in the room below, and wouldn't at all mind George's company.[31]

He hardly spoke and was exceedingly retiring. They knew nothing of his background, and nothing of his present. At the very headquarters of a community battle against the interests of a selfish millionaire and his minions, no one had a clue that the strange middle-aged American had authored the equation that specified the exact mathematical conditions under which the interest of the group trumps the interests of individuals. What they did know was that he loved taking long, hot tubs, using dishwashing liquid to make bubble baths. Paul Nicholson, a first-year architecture student, remembered seeing George only on occasion, coming down from his room to the kitchen to make a cup of tea, "a flicker of a faint distracted smile passing over his pale features."[32]

Squatting, in fact, had become a major part of the battle against Mr. Levy, not to mention a national phenomenon. In London there were close to 30,000 squatters who had taken matters into their own hands: 12,000 people living in temporary hostels added to 200,000 on council waiting lists while 100,000 houses lay empty meant that the housing crisis was real. In Camden alone close to 5,000 dwellings lay unused, an increase of nearly 150 percent in a just a decade. The population in and around Tolmers Square had dwindled significantly owing to Stock Conversion's policy of neglect, and the remaining tenants were demoralized and weak. Squatters, the people of the TVA soon grasped, could bring with them new energy to fight the battle.[33]

In the beginning it was the winos who climbed through windows and broke through doors, finding refuge in often dilapidated abodes. But as the social movement grew, more and more activists found themselves doing the same: students, community organizers, Marxists, anarchists. There were artists, too, and drifters; young and old, men and women, families with toddlers, opportunists and ideologues. Of the 180 squatters, 16 were white-collar workers, 12 manual laborers, 12 skilled laborers, 13 artists, 11 children, 4 bakers, 3 travelers, 2 pensioners, 8 housewives, 8 teachers, 25 unemployed, 40 students,

and 6 "unknowns." Even the daughter of the celebrated pop art/
technophile–inflected professor of architectural history, Peter Reyner
Banham from UCL, had joined a squat on Euston Street, a fact that
made an impression on his reverential students at the TVA.[34]

The conditions in most squats were difficult. Water had to be carried from nearby taps in buckets, lighting was provided by paraffin
and candles, and open fires were the only form of heating. Besides
the outhouses the "superloos" at Euston Station or the ones at the
university provided necessary outlets. But a new spirit had come to
the neighborhood.

> *To Squat or not to Squat*
> *That is the question.*
> *The slings and arrows of outrageous speculation*
> *Or taking courage 'gainst a sea of troubles, end them;*
> *To sleep perchance to dream*
> *Of plans existent and plans yet to come*
> *Conversions dream'd of to improve our Stock*
> *And contracts social which with conscience limned*
> *Shall bring us peace that we may have a home*
> —From J.S. with apologies to W.S.
> (*Tolmers News*, No. 15)

Soon roofs were being repaired, windows were being painted,
bricks mended, even flower plants installed. A local furniture store
chipped in, providing free secondhand sofas, tables, and chairs. A
fruit and vegetable co-op was set up in the basement of Drummond
117–119; a print shop at Tolmers 19; a motorbike repair shop in the
adjacent house. A big carnival with inflatable stalls; a theater show
titled *Seeing It All Come Down*, about property speculation; a steel
band, Indian food, and a tug-o-war, enlivened the square in the summer. Derelict houses were turning into homes. And Tolmers Square,
once again, was becoming a community.

Which is why when George arrived one day at 102 Drummond
with a rather raucous and intimidating redhead walking on one leg,
no one objected that he join the squat as George's roommate. Soon

Ches, Paul, Tim, and Carla learned that the man's name was Peg Leg Pete; it was the same Peg Leg Pete who had stayed with George at Little Titchfield, and whom he now ran into again in Soho Square. Paul remembered the occasion:

> *They seemed from the first sight of them together the most unlikely companions—Peter being as loud and profane as George was quiet and studious. A battered wide brimmed hat hung down his back attached to a cord around his neck. He was wearing threadbare donkey jacket, greasy jumper and open necked shirt and a pair of dirty brown canvas trousers. A crutch lay propped against the wall behind him, and one of his legs was missing below the knee. He drank frequently from a bottle of cider, wiping his mouth with the back of his hand before squeezing some blues from a harmonica or speaking with some fervour about Leadbelly. From time to time he reached for a tin containing a quantity of stale tobacco accumulated from butts retrieved from the pavements, and rolled a cigarette. Hands and stubble beard orange streaked with the stain of nicotine and tar.[35]*

This was George's new roommate, and he was set on weaning him from his bottle. It was an impossible task. Often Peg Leg Pete would return from a night in the pub—the Exmouth or the Jolly Gardeners—drunk after closing time and having lost or unable to find his key. Shouldering the door, he'd break the lock and wake up all the gang, who, fearing the worst—police eviction—would run down the stairs only to find Pete sprawled on the floor in the hallway, cussing and oblivious. Sometimes George would discover him comatose halfway up the stairs in a pool of urine. At other times Pete would simply come crashing through the door in the middle of the day, collapsing in a heap. "Sorry, lads. Lost my balance. Got a snout?" Up in their shared room, he made a habit of pissing in bottles and throwing them out the window into the backyard; the bathroom was an outhouse and too tricky for a handicapped alcoholic. George tried to talk gently to him about changing his ways as they walked together in the neighborhood, or journeyed to the pub, or went to bed at night. But this only got Pete riled and abusive.[36]

He was doing his best to help anyone he could. Harry Robson was another tenant at 102 Drummond Street, a sixty-two-year-old former junk dealer whom Ches had found sleeping rough near the station. George was worried about him. He was suffering from chronic bronchitis and vomiting after every meal. "What I'm concerned about is diagnosis and treatment," George wrote in a letter addressed from the Department of Human Genetics to a Dr. Heaf at the University College Hospital. "I remember you as an excellent, conscientious and sympathetic doctor, so I wish that Harry could go to your clinic and receive examination and treatment under your supervision." Then he added: "I am concerned about Harry because he is kind and generous."[37]

He was still fighting for friends who had fallen on the wrong side of the law: There was Jim Crickett, behind bars at Leeds Prison, whom he was sending pound notes, and, as usual, there was Smoky. On behalf of the latter George wrote a long letter of surety to the official solicitor at the Royal Courts of Justice. "Yes, it is true that since June 25th I have not maintained a fixed abode," he attempted to justify his own qualifications.

> *But I have done this for reasons of Christianity in order to be able to give more money to the poor and as part of attempting to follow Jesus, who, as you will recall, said: "Foxes have hole, and the birds of the air nests, but the Son of Man has not where to lay his head." Thus I became homeless in order to be able to give more and take on larger responsibilities.*[38]

As for Smoky, why, sure, he was trustworthy. In fact he had often helped George with other homeless in the neighborhood. There was no reason to penalize him just because he made the street his home. Was Jesus unworthy of trust, George wondered, since He was a person of no fixed abode?

Often he would help anonymously. "Like an angel coming in," was how Elisabeth Mansell, the manager of an old people's home not far from Tolmers, remembered his entrance to SCS House on Christmas Eve, 1973. He had neither phoned nor announced his arrival in any way. But like his unbeckoned entrance to UCL with the equation, he came with special tidings. Elisabeth and George stayed up until well after midnight wrapping presents for the elderly tenants, hoping to surprise them in the morning. When Elisabeth woke up the next day she looked at her watch and panicked: She'd overslept and quickly charged down the stairs. But all was well. There, sitting calmly between the old folk, was George. He'd woken, dressed, and fed all twenty-one residents.[39]

Then, almost as miraculously as he had arrived, he disappeared.

. . .

Amid the chaos of Tolmers and Peg Leg Pete and the alcohol and drugs, George had not forgotten about science. Scribbled notes in an unsteady hand on scrap paper show that he was still working on dense mathematical calculations of genetic polymorphism for Harris.[40] Sexual selection, too, was not far from his mind. In fact in mid-February he went up to visit Hamilton at his home in Berkshire for a fortnight. It was a respite, to be sure, from his ragged life in London, but it was first and foremost an opportunity to begin a joint project on sexual selection. Bill and his wife were shocked at his appearance. He looked ill and starving, emaciated. They did their best to convince him to climb back full-time on the boat of genetics; perhaps he could apply for a job to work at Silwood. George said he'd think about. Population genetics wasn't his main work anymore, didn't they know?[41]

Back in town, two weeks later, he wrote to Bill to report on his progress. Things hadn't been easy. After finally leaving 102 Drummond following a string of violent attacks by Pete, George had slept for two nights at Elisabeth Mansell's old people's home. Then there were five merciful nights at a large luxury apartment of a student friend from Galton, with four other flatmates, though he'd caught some awful bug and ended up stretched out on the sofa most of the time. From there he left for a night at a B&B near Earl's Court, followed by two nights sleeping rough at Victoria Station. On one of these, a drunken down-and-out had attacked him, ripping the watch off his wrist as he fled. He was sorry, but he hadn't yet had time for sexual selection.[42]

Still, writing from what he dubbed "Hideaway No. 12," he had now found more conducive environs:

Address (to be kept secret): 45 Gordon Mansions, Huntley Street, Torrington Place, W.C.1. It's just a stone's throw away from the south-west corner of University College. It's the flat of an elderly lady who has had a stroke and uses one of those metal frames to get around, so that she doesn't like her guests to have many telephone calls or visitors. Needless to say, she will not permit me to take in homeless alcoholics. . . . So it is fabulous from the point of view of getting some work done.[43]

This past week he'd gone down to the computer room at UCL to run some programs modeling the effects of sexual selection in small populations. He was hopeful that before long the program would be working for medium-size populations and would include polygamous relatedness calculations, too. Bill was on his way to the United States for a few weeks, and implored George to feed himself properly. He, too, hadn't had much time to work on their project. Meanwhile, he sent a note to Imperial requesting that his collaborator be allowed to use his job number at the University of London Computing Centre. After all, George told him that he was avoiding going around the Galton. Besides, his MRC grant would be running out at the end of next month.[44]

When he was finally off the payroll beginning May 1, George had exactly £95 in his bank account. He was hopeful that this would get him through his next set of payments to Selfridge's and Barclays so that he could do his part for their paper before he'd need to take a job. No one at the Galton knew his whereabouts. He had just "crawled out of hiding," he wrote, to go to Wolfson House to collect mail for the past seven weeks.[45]

But George had been crawling out of hiding for another purpose, too. Back in October, before he'd moved into 102 Drummond, he'd met a woman outside the Swiss Center. It was raining and he was waiting for Smoky who had promised to come along with him to help some alcoholics. She was waiting for a colleague, too—a woman who ran her charity in Nigeria. Her name was Joan Jenkins, she said, and she'd started the Woman's Health Concern, focusing on estrogen replacement therapy after menopause. She had trained as a nurse, even worked as a nurse during the war, though for years now her day job was a reporter with the BBC. She was about George's age, handsome, busty, reddish-sandy-haired; a prim and proper, well-educated, upper-middle-class lady.[46]

Smoky wasn't showing up. We're both stuck here in the rain, George said, Why don't we go in and have a coffee? At first Joan pitied him: He was terribly unkempt, smelled bad, mumbled all kinds of nonsense about Jesus. But he was also clearly a special person: he knew a lot about hormones, and told her about his own work in genetics. She took him around the BBC to meet her colleagues that day; even took him to her home later to meet her husband, Cyril, and their son. Soon the relationship had turned into a sexual liaison. Every Saturday they'd meet at a set location, usually Finch's pub close to the Notting Hill Gate tube station, and make their way to the Hotel Commodore at Lancaster Gate, just north of Hyde Park. Joan would pull the grimy clothes off of her gaunt lover, run a hot tub for him, and clean him. Then they would lie down on the fluffed-up bed and make love.[47]

Joan was beginning to fall for him, and he was trying to hold her back. "Seeking to rush things," he explained, "you just bring trouble and unhappiness to yourself and others and make me increasingly desirous of avoiding your company." Jesus was guiding him that he was not yet ready to marry. He was thankful for all her presents: a sweet set of notes left for him at the Angel House hostel, a pair of shoes to replace the sneakers that were long past worn, the meals that she'd buy him. But he didn't appreciate her growing attempts to straighten him out and organize his life. "With your intelligence why can't you understand that the way to please me is to leave me alone and especially to give up trying to convert me to your way of doing things."[48]

It was not a problem that he was experiencing from Kathleen. Having "dropped out" to live the counterculture life with Dom in Hawaii, George's daughter thought it was actually pretty cool that he was giving away his possessions. George found it amazing that they'd both started similarly irregular lives at about the same time: staying a lot with friends, smoking dope, living with no thought of the morrow.[49] He was frail now, but in his letters tried to sound upbeat and quirky, just the way his daughters liked him:

> *I'm pretty much ok now healthwise, except that I probably have about 20 cavities. Had a lot of illness last three months, a long cough, bit of flu, and so on. Also had some sort of recurring fever that made me feel very week. Suspect it was infectious mononucleosis (didn't see a doctor) though don't know how I got it since I hadn't been doing much kissing.*[50]

Once, when George jumped into UCL and was spotted, CABS tried to pin him down to write a short report to the MRC on his polymorphism work. At present George was living in a Tolmers squat with a man who worked nights as a baker. Maybe he too could find a night job that would leave his days free. That way he'd have some time for a little bit of science.

. . .

Slowly, something inside him was beginning to stir. He couldn't quite get a grip on it, but it was growing, he was sure. It started off as a faint impression when his thoughts roamed and his eyes glazed over. Often, when he put his head to rest at night, when the lights went out and his eyelids rolled down, he could almost catch a glimpse of it. Then he could feel it in his stomach, still later, creeping up toward his brain. George had spent the last eighteen months selflessly giving to strangers. But now, perhaps, it was time to take care of himself again.

The reckonings came from all corners. From Great Depression Upper West Side to speculation-decrepit Tolmers Square; what, finally, had life taught him? Skinner, coincidences, the Bible, the Bomb: all had gotten him thinking about the limits of free will. In particular he wondered about the volatility of his own deeds and emotions.

Had he been a bad father and a neglectful son because he wanted to or because he'd been "programmed" differently? Had he left his wife and IBM and Little Titchfield due to chance or necessity? How much control did he, and people generally, really have over their own lives?

Since the day he walked down the gangway of the *Queen Elizabeth* back in 1967, evolution had taught him a world of wisdom about kindness. Hamilton's $rB > C$ made altruism dependent on genetic relatedness, whether as its author had first perceived, of the familial kind or, as George had pointed out to him, by simply sharing the trait with strangers in association. The logic of animal conflict, as well as Trivers's work on reciprocity, made clear that cooperation and mercy would evolve as long as they could be expected in return. A multilevel approach to selection taught that altruism could come about if the group had formidable competitors. And Garrett Hardin's tragedy of the commons was a warning cry that what was good for the individual was not always good for the community, and that clever ways needed to be imagined if the two were ever to live peacefully hand in hand.

But was science really enough? Could what was dearest to humans simply be the result of some empty evolutionary calculus?

He had found the Lord twice, and was pretty sure the second time that he was following His true calling. But if becoming selfless somehow trumped the logic of genes and reciprocity, then something was amiss. If true, pure giving, neither beholden to propagating more of its kind nor dependent on the promise of being paid back, nor even, as his own equation showed, made possible by groups having to triumph over others, was a heroic slap in the face to everything science had taught him, then obviously he had not cracked any riddle. After all, none of the down-and-outs he had tried to help had changed or would change, and he himself had fallen into a well of misery. He was no longer sure whether pure love and selflessness were possible; maybe, when all was said and done, the circumscribed kindness borne by evolution was really all that there was. If he was going to be honest with himself he would need to look the truth in the face: His attempts to make the world a better place had all utterly failed.

These were his thoughts as he hit rock bottom at the bank and set out to find a job. Luckily Hamilton, too, had not had much time for the work on sexual selection: Up until now all he had done was to collect a few rhinoceros beetles from a rotting log in Pembrokeshire, and most of those hadn't even mated. In the meantime George found work as a night office cleaner at a local branch of National Westminster Bank. It wasn't perhaps a dream job, but it would do for now. He liked the silence. It gave him time to think. In any case, he wrote to Bill, he was considered slow but dependable.[51]

Under conditions of starvation thousands upon thousands of social amoebas of the species Dictyostelium discoideum *come together to form a tiny fruiting body, a hairlike structure with an orb at its top. The orb is comprised of many fertile spores, while the stalk is made up of amoebas that died after producing strong cellulose walls. It is the self-sacrifice of these amoebas that allows the spores to climb up the stalk where, aided by a strong wind or the legs of an unsuspecting insect, they may be carried away to live to see another day.*

13

Altruism

wo things," the philosopher Immanuel Kant wrote, "fill the mind with ever new and increasing admiration and awe, the oftener and more steadily we reflect on them: the starry heavens above and the moral law within."[1] And, to be sure, from Darwin to Allee, Kropotkin to Fisher, Emerson to Haldane to Wynne-Edwards, the mystery of altruism, considered the highest form of morality, was attacked from all possible directions. Where did altruism come from: Could it have been borne by the invisible hand of natural selection working directly on genes, on individuals, perhaps, on communities, on groups? Each had a hunch, and each had an answer. Still in awe, still in admiration, no one came up with an entirely convincing solution.

Then came George C. Williams. "Group-related adaptations do not, in fact, exist," he ordained; while possible in principle, genes don't evolve via between-group selection.[2] Soon the entire world was let in on the secret: Using Williams's logic and building on Hamilton's inclusive fitness, Trivers's reciprocation, and Maynard Smith and George's ESS, a Kenyan-born Oxford biologist with a soft voice and a sharp rapier fashioned a gospel. Appearances to the contrary, it was genes running the show, Richard Dawkins explained in *The Selfish Gene* in 1976, a book that soon became one of the century's greatest best sellers. In the final reckoning individuals are just aggregations of elements that are shuffled and disbanded when the sexual gametes are made. It's the genes, not the soma, that persevere in evolution; DNA, not the body, which by sheer power of replicatory fidelity lives to fight another day. Even though, Huxley-style, Dawkins explicitly stated that humans were the only creatures in the world who could rebel against the tyranny of their selfish replicators, to many his message felt unusually deflating: When it comes to natural selection the best individuals could call themselves was "vehicles."[3]

But despite immediate criticism[4] and however weirdly counterintuitive, it was a robust theory: As biology marched ahead it seemed to unify a plethora of phenomena. Whether animals aid their kin (ants and wolves who help their sisters breed) or nonrelatives (vampire bats who share blood, mouth to mouth, at the end of a night of prey with members of the colony who were less successful in the hunt); whether they abandon their eggs (sharks and skates and stingrays) or goslings (eagle owls and leopard-faced vultures) or sacrifice themselves for the next generation (male praying mantises serve their heads during coitus to their avaricious ladies); whether they come together as a group (Siberian steeds forming rings against predators) or aid themselves at the expense of their hosts (from the common cold bug to proliferating cancers)—all living things are acting in the interest of their true masters: a cabal of genes whose sole imperative is replication. Volition and mind and "free will" notwithstanding, evolution fashioned genes that do whatever it takes to survive.

. . .

If Williams, aided by Dawkins, helped get rid of groups, kin selection had acted as a handmaiden. Scaling the eighties and nineties into the twenty-first century, family relatedness threw massive ropes down from Mount Modern-Evolutionary-Biology for others to safely climb. Haldane's mythological drunken insight and Hamilton's resulting rule, it transpires, hold up incredibly well in the face of winds, falling rocks, and negative slopes. From the naked African mole rat, the mammalian equivalent of the termite, sometimes called the "saber-toothed sausage," which forsakes procreation in order to help its chosen monarch, to the carnivorous spadefoot toad tadpole, which can actually "taste" relatedness and therefore spits out cousins and brothers—but not strangers—that find themselves in his mouth, relatedness has proven to be a robust predictor of altruistic behavior. Even cuckoos have figured out this metric: They take advantage of other birds' familial instincts by laying their eggs in complete strangers' nests, allowing the tricked parents to shoulder the burden of parenthood.[5] Beginning in the seventies, hundreds of biologists, ecologists, and evolutionary modelers have used Hamiltonian logic to make sense of many dramas of love and deception. With few exceptions, the general rule holds: The closer the kin, the greater the benevolence.[6]

Two very different examples help to show just how far kin selection has captured the imagination:

Moving through soil by extending its pseudopods, most of the time the cellular slime mold, *Dictyostelium discoideum*, is a loner. Usually it engulfs and eats bacteria, but when times are rough and bacteria are scarce, something amazing happens: The starving amoebas secrete a chemical, cAMP, which attracts the others along a concentration gradient, until chains of tens of thousands of them merge into a mound. Soon the mound elongates into a slug that begins to crawl, as one multicellular body, across the forest floor. When it reaches a place with some heat and light it stops, and the amoebas that formed the front 20 percent of the body arrange themselves into a stalk, laying down tough cells of cellulose, just like plants, to make it nice and hardy. Then the remaining 80 percent climb up the stalk. When they reach the top they reorganize themselves into spores, forming a round glistening orb. It is this 80 percent that will stand a chance to live another day, sticking perhaps to the wings or legs of some insect, or otherwise being taken by the wind. The 20 percent that formed the stalk, on the other hand, will have sacrificed themselves altruistically for all the rest.[7]

This is incredible, but what was discovered next is even more fascinating. In the wild most fruiting bodies form from a single clone: All the amoebas coming together to make the slug are virtually genetically identical. But when the husband-and-wife team Joan Strassman and David Queller mixed amoebas from different clones they uncovered the following: Able to recognize one another, members of one clone did their best to stick together at the backside of the slug. When the stalk was made, it was primarily they, and not the others, who shimmied up to become hopeful spores.[8]

If amoeba can recognize and aid kin, so too, of course, can humans; this shouldn't be all that surprising. What is surprising is that studies have shown that stepchildren are not only much less likely to be invested in than biological children, but also much more likely to be abused. Surprising, that is, if your names aren't Martin Daly and Margo Wilson. *This* husband-and-wife team has taken kin-selection logic to its end: Just like the slime mold, they claim, and the spitting toad and the cuckoo, humans are simply following Hamilton's rule.[9]

But if genetic relatedness was a handmaiden to the gene's-eye point of view, von Neumann games also proved a useful mountaineering partner. Soon its ropes, too, were being climbed by many a follower. The point of departure was George and Maynard Smith. Bolstered in *The Selfish Gene*, the concept of the ESS soon invaded the study of animal behavior. George and John, it transpired, had made an error in their paper: Retaliator, after all, was not an ESS. Since Dove did equally as well in a population of Retaliators, it could slowly drift into the population. When that happened, the true ESS would become a mixture of "Hawks" and "Bullies."[10] George, perhaps, might not have been glad to hear about it, nor to know that his "Mouse" had once again become a "Dove."[11] But considering that within a decade the application of game theory to evolution had revolutionized the field, perhaps he might have been assuaged nonetheless.

Once more, two illustrations from the many serve to make the point. Male dung flies, it transpires, are aptly named: Like fierce elephant seals or bucking red deer, they too defend their territory, even if in their case this is nothing but a patch of smelly excrement. The reason they do so is that females lay their eggs on the dung, and the fresher (and thus smellier) the patch, the more attractive it is to them. Having arrived earlier, males fight over the best patches; he who secures the most attractive dropping will win the right to mate with the female as she deposits her eggs. The question is: For how long should a male fly defend a patch of fresh shit before moving on to another? After all, the drier and crustier it becomes, the less chance that a female will choose to land on it. Clearly, just as in a von Neumann game, the answer depends on the actions of the other male flies. It turns out that, fashioning the minute fly a strategist, an optimal ESS can be worked out. On paper it is forty-one minutes, and incredibly, in nature it's just a few minutes away.[12]

But if an ESS is good for flies, once again it is not too good for humans. In fact, game theory analyses of animal, and even plant and bacteria, behavior have been so successful that the modifications made specifically to fit evolutionary problems are now being retranslated back into economics If neoclassical economic theory à la Milton Friedman assumed perfectly rational actors, it has since become clear that this is not really so: Risk aversion, status seeking, myopia, and other inbuilt cognitive biases are rampant in humans, and economic models of decision making need to take them into account. Introducing evolution-style games that assume minimal rationality, but whose dynamic depends on mutation, selection, and learning instead, has therefore become popular in economic theory. As an increasing number of theorists have found, this approach is helpful in figuring out problems like why firms don't always act to maximize their profits, or whether in a given competitive market investors should be aggressive or lazy. Darwin owed a debt to Malthus, and his followers are paying it back.[13]

· · ·

Alongside kin selection and game theory, Trivers's reciprocal altruism has also lowered a rope from Mount Modern-Evolutionary-Biology. One of the first to climb it, in fact, was Bill Hamilton himself. Trivers had sent him a draft of his 1971 paper, and, though Hamilton found the math flawed, he encouraged the young American to continue. It turned out that the two animal examples provided in that paper were not, in fact, good examples of reciprocal altruism: Cleaning fish not being swallowed by their larger hosts when danger came around was later repaid by hosts returning to the same cleaners, as were warning cries made by particular birds when predators were spotted lurking. These were more accurately instances of "return-effect" altruism rather than reciprocal altruism because the return benefit didn't come from the second party's *choice* to reciprocate but rather for other reasons. But despite the semantic imprecision and the weak math, it didn't really matter; Trivers had thrown down the rope. A decade later, together with the American political scientist Richard Axelrod, Hamilton proved mathematically that, alongside perpetual defection, the strategy of tit for tat is a Nash equilibrium: Through iterated encounters natural selection would favor social behaviors that exacted a fitness cost in the short run. Reciprocal altruism had been welded to the prisoner's dilemma. And while perpetual isolation was always an option, the rule of cooperation was simple enough for a bacterium. "The benefits of life are disproportionately available to cooperative creatures," Axelrod and Hamilton began, and Trivers, for one, thought it of "biblical proportions." "My heart soared," he wrote to Bill after sitting down one night with classical music to read the paper.[14]

Soon Mount Modern-Evolutionary-Biology was crowded with others climbing up the reciprocal altruism rope. The prisoner's dilemma, these researchers found, was too simplified a version of natural interactions. But allowing for the inclusion of punishment and forgiveness, delicate cheating, observer effects (when a third party looking on has an impact on the two-person game—something called "indirect reciprocity"), and many other subtleties eventually inched the fit between nature and such models closer together. As the years progress the laws of cooperation gain steadily: Theory and observation alike place them firmly as a powerful motor in the evolution of altruistic behavior.[15]

Pure direct reciprocal altruism between nonkin in nature, it must be said, has proved something of a rarity. For one thing the altruistic helpers might actually be related more often than Trivers and others suspected, rendering "reciprocal altruism" nothing but a version of Hamiltonian kin selection. Another problem seems to be that behaviors that were once interpreted as pure-cost assistance (baboons grooming each others' backs for fleas, for example), may actually just be a form of mutualism (the baboons gain valuable nourishment from eating their friends' fleas). Yet another impediment to the "you-scratch-my-back-I'll-scratch-yours" theory comes courtesy of Oscar Wilde. "I can resist everything except temptation," he quipped in *Lady Windermere's Fan*, and most animals, experiments show, are not all that different. Immediate gratification is the custom of even the most intelligent and social of mammals, a thorn in the side of establishing the courtly conventions that serve as requisites for social restraint.

Finally, a theory called "the handicap principle," espoused by the Israeli zoologist Amotz Zahavi, argues that animals that perform ostensible acts of sacrifice do so to prove that they are worthy of reciprocation. Thus, when a gazelle spots a lion lurking in the grass and begins to jump up and down in the air (a behavior called "stotting"), she is advertising to her friends that she is "willing" to pay a price for being part of the group. The problem with this solution to reciprocation's underbelly of deceit is that it is very difficult to falsify. What may seem like a selfless warning to her friends (or at the very least an act expecting reciprocation) might actually be a signal to the lion that he should focus his pursuit on a member of the troop less athletic and therefore more likely to end up on his palate. Likewise, a male peacock sporting a gigantic (and costly) colorful tail, or a bull elk showing off its large rack of antlers, may be signaling to potential mates that they need not look any further, that Numero Uno is stronger precisely because he carries a hindrance that would handicap a lesser fellow.[16]

Still, alongside kin selection, reciprocity and the games it engenders have been valuable spectacles through which to gaze at Nature and her ways. There is a downside, however, for what all three scaling ropes share in common is a rather dubious conquest: Whether a monkey scratches it's neighbor's back, or a bee fatally loses its entrails when it stings an invader at the hive, altruism à la Hamilton, Trivers, and von Neumann is never what it seems. In fact, when it comes to altruism, the gene's-eye view of evolution leads to positively uninspiring places. "The economy of nature is competitive from beginning to end," the biologist Michael Ghiselin wrote,

> . . . the impulses that lead one animal to sacrifice himself for another turn out to have their ultimate rationale in gaining advantage over a third. . . . Where it is in his own interest, every organism may reasonably be expected to aid his fellows. . . . Yet given a full chance to act in his own interest, nothing but expediency will restrain him from brutalizing, from maiming, from murdering—his brother, his mate, his parent, or his child. Scratch an "altruist" and watch a "hypocrite" bleed.[17]

The gene's-eye view can lead to strange places, too. "Consider a pride of lions gnawing at a kill," Dawkins asks us to imagine.

> An individual who eats less than her physiological requirement is, in effect behaving altruistically towards others who get more as a result. If these others were close kin, such restraint might be favored by kin selection. But the kind of mutation that could lead to such altruistic restraint could be ludicrously simple. A genetic propensity to bad teeth might slow down the rate at which an individual could chew at the meat. The gene for bad teeth would be, in the full sense of the technical term, a gene for altruism, and it might indeed be favored by kin selection.[18]

"Tooth decay as altruism?" biologist and philosopher Helena Cronin asked, amazed, in her 1991 book *The Ant and the Peacock*. "That's hardly how saintly self-sacrifice was originally envisaged." And yet, she added, "The logic is unassailable."[19]

After years of debate, it seemed, the genes, evolution's real scions, had finally provided the answer: Natural goodness was slowly being unmasked.[20]

. . .

Until group selection started making a comeback.

For both Williams and for Dawkins, the gene as replicator offered a strong argument against evolution working at the level of the group; if genes were the only permanent unit in nature, how could the eye of selection peruse anything else? As a bookkeeping device, scoring genes in populations helped keep good track of evolutionary change. But "selfish genes" were not fashioned merely as a metaphor. The point of view they engendered negated evolution at any other level.[21]

And yet, when reconsidered by theorists beginning in the eighties, chief among them David Sloan Wilson, it began to become clear that the unit of selection and the level of selection depended on entirely different criteria. Of course genes were replicators, and clear units of permanence. But whether a certain level of life could be viewed by selection depended not on *permanence* but on *where fitness differences resided in the biological hierarchy.* Here is why: If a population is viewed as a nested hierarchy of units, with genes existing within individuals, individuals existing within groups, groups existing within populations and so on, fitness differences can exist at any or all levels of the hierarchy because *heritable variation* can exist at all these levels. The gene's-eye point of view therefore says nothing about the possibility of group selection; genes, after all, can evolve by outcompeting other genes within an individual, or—via the traits they confer—by helping individuals outcompete other individuals within a group, or by helping one group outcompete another. Even if the genes are the replicators, it still needs to be determined whether they evolve by between-gene/individual selection, between-individual/group selection, or between-group/population selection. Despite all the history and hype, the gene's-eye view and group selection are not, and never should have been, antithetical.[22]

When it comes to altruism this multilevel selection approach is all the more important. For as Darwin himself perceived when contemplating the ants, altruism, like any other trait, is sure to have evolved. What makes altruism special is that it reduces individual fitness within the group while benefiting the group as a whole. The only real question, then, is whether between-group differences in fitness are ever strong enough for within-group altruism to evolve.

This is an empirical question. And it is why, after thinking it over following their first cryptic telephone conversation, Hamilton wrote to George that he was enchanted with his formula. The covariance equation, he suddenly saw, had elegantly wiped away years of confusion. Over his own head, and the heads of Haldane, Fisher, Wright, and Allee, of Emerson and Kropotkin and Huxley, it had returned to Darwin's own original insight: Evolution can occur at different levels simultaneously. The beauty of it was that this simple mathematical tautology could now help define where selection was acting most strongly for each and every trait.

Hamilton thought that he was the only person in the world who understood just how momentous George's formulation really was. For what covariance allowed to see was that while the different ropes scaling Mount Modern-Evolutionary-Biology seemed as though they had been thrown from the crest of an entirely different peak, in fact their climbers were simply making ascents up different faces of the very same mountain. Inclusive fitness made it seem as though "altruism" was always just apparent since by sacrificing himself an "altruist" might die, but his genes live on in the bodies of kin. Reciprocal altruism, on the other hand, also took the sting out of altruism, but the price in this case was always repaid to the very one who had made the sacrifice. Finally, variations on the prisoner's dilemma fashioned reciprocation a game, often making this ascent, too, seem as though it were progressing up a distinct slope.

The trick, of course, was to be able to see how each ascent related to the other, how scaling up one face could illuminate something about the mountain without having to exclude the lessons learned from alternative climbs. Inclusive fitness, for example, taught that genetic relatedness was important for the evolution of altruism. This *was* important. But what a multilevel frigate could do, armed with the ammunition of a covariance gunner, was to put it in perspective. Here are Sloan Wilson and the philosopher Elliott Sober:

> For all its insights, kin selection theory has played the role of a pow-erful spotlight that rivets our attention to genetic relatedness. In the center of the spotlight stand identical twins, who are expected to be totally altruistic toward each other. The light fades as genetic related-ness declines, with unrelated individuals standing in the darkness. How can a group of unrelated individuals behave as an adaptive unit when the members have no genetic interest in one another?
>
> Replacing kin selection theory with multilevel selection theory is like shutting off the spotlight and illuminating the entire stage. Genealogical relatedness is suddenly seen as only one of many factors that can influence the fundamental ingredients of natural selection—phenotypic variation, heritability, and fitness consequences. The ran-dom assortment of genes into individuals provides all the raw mate-rial that is needed to evolve individual-level adaptations; the random assortment of individuals into groups provides similar raw material for group-level adaptations. Mechanisms of nonrandom assortment exist that can allow strong altruism to evolve among nonrelatives. Nothing could be clearer from the standpoint of multilevel selection theory, and nothing could be more obscure from the standpoint of kin selection theory.[23]

If in a gene's-eye view individuals are "vehicles," the genes within the body saddled together as if on the very same boat, then a multilevel approach showing that groups are vehicles, too, can be translated into a gene's-eye view model.[24] Everything depends on whether the right conditions render individuals in a group akin to genes within a body, if "mechanisms of nonrandom assortment," in other words, do in fact really exist. Relatedness can be one such mechanism, but it needn't be the only one. Goodness depends on association, not necessarily family.[25]

It took him some time, but Hamilton began to grasp this after the cryptic phone conversation with George ("I thought you'd see that," George told him when he finally came around). Suddenly Maynard Smith's 1964 *Nature* article attacking Wynne-Edwards, the one that had made Hamilton so mad, didn't look so unequivocal: Sharply distinguishing between kin selection and group selection, after all, no longer seemed always to the point.[26] Hamilton made this clear in a paper he wrote in 1975 showing how the evolution of altruism between relatives is just an instance of group selection rather than an alternative explanation for an "apparent" altruism. It was the first application of George's full covariance equation to an evolutionary problem.[27]

Not only kin selection but games of reciprocation, too, could now be viewed from an entirely new angle. If two hamadryas baboons grooming each other are playing a kind of prisoner's dilemma they can easily be defined as a group. Clearly if one is an altruist and the other a free rider, the free rider will always win a benefit without paying any cost, and the altruist will always pay a cost without winning any benefit: Within the "group" it pays to be selfish. But what happens if the pair is compared with a second "group" made up of two altruistic baboons that groom each other loyally? Now, suddenly, selfishness becomes an impediment, for group B will be cleaner on average than group A, and therefore healthier and more likely to sire more offspring. The question then becomes whether between-individual/within-group forces are as strong as between-group/within-population ones. If association between the baboons is random, then they will be, and selfishness will beat out altruism. But if association is not random—if altruists, say, can choose to unite and stick together, leaving selfish types to pair vacuously with one another—altruistic genes enhancing grooming behavior will be able to evolve.[28]

Amazingly, Hamilton's paper fell on deaf ears. Even though George had succeeded in changing his mind about group selection, and even though, starting in the late seventies, Hamilton began to be referred to by many as the greatest Darwinian since Darwin and to win every possible accolade, group selection remained under the shadow of the gene's-eye point of view. If group selection exists in theory, most people said, it is too weak a force to play a role in nature. Into the eighties and nineties, and into the twenty-first century, theorists like Sloan Wilson continued to speak, frustrated, to those who would not listen.[29]

Slowly this is beginning to change.[30] In 1981 Hamilton's female-biased sex-ratio paper was exposed for what it really was—an instance of the test George Williams had asked for, showing group selection at work.[31] A further oft-cited example, which Lewontin caught on to very early, involves the virulence of pathogens: If evolution acts on each single pathogen within a host, a "race" will begin between the pathogen and the host's immune system, and the more hypervirulent the individual viruses the greater the chance they will all perish together with the host. On the other hand, if selection works on the entire virus population, attenuating it somewhat, there is a greater chance that the host will come into contact with other potential hosts (since not incapacitated or dead), and therefore a greater chance for the virus population to move on and survive. Just such an example was occurring in Australia: A virus, *Myxoma*, was introduced in the late sixties by its department of agriculture to help subdue a rabbit population wreaking havoc and growing dangerously out of control. At first the virus was devastating, but soon it began to attenuate. Lewontin explained:

> *When rabbits from the wild were tested against laboratory strains of virus, it was found that the rabbits had become resistant, as would be expected from simple individual selection. However, when virus recovered from the wild was tested against laboratory rabbits, it was discovered that the virus had become less virulent, which cannot be explained by individual selection.*[32]

In fact group selection even emerged as a clinical concept, since hospital procedures and public health practices can be employed to favor the evolution of low-virulence strains. This became known as "Darwinian medicine," and George Williams himself emerged as a pioneer late in his career.[33] Slowly but surely further examples began to surface.[34] And while group selection remains hotly contested, and good experimental examples are still few and far between, there's no arguing that it's making a comeback, at the very least in theory.[35]

Today even the entomologist Edward O. Wilson, whose 1975 book *Sociobiology* was a central pillar of the gene's-eye-view approach, has come to believe that selection works at different levels. "Superorganisms" like bee and ant communities are literally giant vehicles, and the altruism they foster, Allee and Emerson would have been glad to know, is in that sense genuine and real.[36]

The possibility of selection at many levels, of course, makes the covariance equation all the more important. And so, drawn from the closet, dusted, and sparkled, the "Price equation," as it has become known, is increasingly gaining new life. Theorists, among them notably Steven Frank, Alan Grafen, and the philosopher Samir Okasha, are finally joining in Hamilton's enchantment, showing just how powerful a tool George invented and building further on its foundations.[37] Ben Kerr and Peter Godfrey-Smith have recently even contributed a small improvent, adding a second covariance on the right hand side of the equation to make the causal structure more symmetrical and therefore more general, a development Price himself would undoubtedly very much have liked.[38] Still, like any tautology, it's sometimes unclear whether the Price equation is valuably invaluable or invaluably valuable: to really explain the natural world, it's clear, the pristine mathematical beauty of the equation needs to be filled with the messiness of biology.[39] But if the Price equation is better for framing a problem than for figuring out its messy biological details, it remains a crucial tool for understanding evolution in structured populations, where *inter*dependence is just as important as either dependence or independence. Its generality means that it can be applied to evolution not only of genes but also to other forms of nongenetic evolution, based on epigenetic or behavioral inheritance. Alongside its ability to represent selection working on different levels simultaneously, this is a major, if still untapped, attribute. Besides altruism, the Price equation has already proved useful in attacking problems as diverse as reproductive value, evolutionary epidemiology, genetic programs, human cultural evolution, rogue genes, meiotic drive, biodiversity, and ecology. And the more we think of it the more it stands a chance to help us understand many other problems.[40]

Hindsight is a wise master. Having thrown down the rope of game theory, it's now apparent, George Price helped others scale Mount Modern-Evolutionary-Biology with a stamina and strength that have proved enduring. But even more important, by providing the penetrating spectacles of covariance through which to measure all other ropes, he has helped us see more clearly than ever before that we are all climbing the very same mountain. It's a legacy that will continue to transform the view of what Darwin called the "entangled bank," a dramatically exciting window into the complexity and grandeur of life.[41]

They could hardly have imagined it when they met behind the curtain at the Belasco Theatre, but William Edison "Isak" Price and Alice "Clara" Avery would have much to be proud of in their son.

· · ·

What about altruism in humans: Is any of this relevant to us?

Return to the hamadryas baboons playing a prisoner's dilemma. Imagine first that each baboon is either altruistic or selfish by no fault of its own: Each is born the way it is, and can no more change its nature than a vegetable become a fruit. But imagine further that each baboon has both capacities and can begin to *choose* whether it wants to be a Good Samaritan or a cheat. Suddenly we are in a completely new game: We have entered the world of *mind*.

It was Trivers who saw where this was heading. If cooperation and ultimately altruism are games of reciprocal good faith, what better way to police the system than to be able to detect deceit? Benevolence, he reasoned, requires a strong sense of justice, because a sense of justice is necessary in order to appreciate dishonesty. Over evolutionary time, and beginning with animals, a genuine instinct of fairness would have been born out of the need to distinguish the trustworthy from the swindlers. This has been demonstrated in recent experiments with capuchin monkeys: Experimenters rewarded monkeys with sweet grapes when the monkey handed them a stone, but when the monkeys were divided into two groups, one of which was handed grapes in return for the rocks and the other much less attractive cucumbers, many of the monkeys in the second group would not accept the offer. In fact the sense of unfairness could even be magnified: When, in return for their rocky offering monkeys were given absolutely nothing, some began pelting the experimenters with stones in outrage. Both our human sense of right and wrong and our discretion over when and where to use it are, it seems, legacies of natural selection.[42]

This may help to explain why we are so good at detecting cheats. A number of neat experimental designs suggest that this betrays a human cognitive bias: We are better at solving logical riddles that have to do with exposing shysters in real-life situations than we are solving the very same puzzles presented in the abstract. Whether detecting cheats also requires a system of punishment is a question that is argued. While some find punishment requisite for cooperation, others claim that it's effective, but only for establishing dominance. True egalitarian cooperation, on the other hand, is best served by *withholding* recrimination: At least when it comes to peers (as opposed to the state), taking the law into your own hands is unprofitable.[43]

Whether punishment is necessary or not, it's clear that people often cooperate altruistically even when they have absolutely nothing to gain: The classical economic assumptions of self-interest cannot explain the richness of human behavior, Adam Smith's long shadow notwithstanding.[44] This may not be all that surprising if one considers how valuable regard, and its subtler cousin, self-regard, is to human beings. But *why* should this be so?

Following in Trivers's footsteps, the economist Robert Frank proposed that it all comes down to trust. To do well in life, our simian and hunter-gatherer ancestors needed to forsake the short-term temptation of self-interest, since they lived in small, mutually dependent groups. Emotions, or what used to be called "moral sentiments," helped to ease the retreat from egoism by establishing commitment ("I love you and won't leave you") and eliciting reciprocity ("I love you, love me back!"). Since in evolutionary terms life is a game in which one seeks trusted partners to play with, emotions that trigger loyalty and altruism (even irrational ones) are adaptations for regulating behavior. According to this view we have a genuine conscience

underlying our emotional repertoire in order to help us join hands
with other conscientious sorts to everyone's mutual advantage.[45]

But the logic is tricky. The possibility of deceit raises the probability of ever more subtle mechanisms for spotting deceit, which
raises the probability for mechanisms of *self-deceit*. Trick yourself
to trick another; what better way to conceal true intentions? This
in fact seems rampant in unsuspecting man: Ask a hundred people
whether they are above average in intelligence, or beauty or kindness
or virtue, and considerably more than fifty, hand on heart, will tell
you that they are. Still, if an arms race was put in place in evolution
between cheaters and detectors, perhaps the result has ultimately
been felicitous: After all, the best way to *seem* trustworthy and giving
is to actually *be* trustworthy and giving.[46]

Which is it, then: pure genuine selflessness or an egoism so cunning that even the trickster is tricked? Some argue that the genuine-
ness of goodness can be demonstrated scientifically, though their
arguments seem really quite feeble. The strongest case that can be
made for pure unsullied selflessness is that it is just as plausible as
unadulterated egoism. Proponents call this "pluralism," but in truth
it is more of an admonition: There are questions that, however much
we'd like it to, science simply cannot solve.[47]

Whether goodness is genuine or not, it was the father of the
theory of evolution by natural selection who instructed that the
moral sense has natural foundations. "Mr J.S. Mill speaks, in his
celebrated work, 'Utilitarianism,' " Darwin wrote in *The Descent of
Man* in 1871,

> of the social feelings as a "powerful natural sentiment," and as "the
> natural basis of sentiment for utilitarian morality." . . . But . . . he
> also remarks, ". . . the moral feelings are not innate, but acquired." It
> is with hesitation that I venture to differ from so profound a thinker,
> but it can hardly be disputed that the social feelings are instinctive or
> innate in the lower animals; and why should they not be so in man?
> . . . [Several thinkers] believe that the moral sense is acquired by each
> individual during its lifetime. On the general theory of evolution this
> is at least extremely improbable. The ignoring of all mental qualities

will, as it seems to me, be hereafter judged as a most serious blemish in the works of Mr. Mill.[48]

In private, rhetorical deference could duly be abandoned: "He who understands baboon," Darwin scribbled in his personal notebook years before, "would do more towards metaphysics than Locke." The mind had been crafted by millions of years of evolution. If man wanted to comprehend his own morality, he'd need to take a careful look back at the animals.

. . .

Everyone says humans are special, and, undoubtedly, this is so. Even if each and every species is idiosyncratic, if uniqueness is really not unique, however we like to spin it, our culture does set us apart.[49] Hamilton saw that this meant that group selection might play an even more central role in humans: The kind of cohesiveness borne by living in small hunter-gatherer groups and developing language might help solve the "free-rider" problem and would only have been strengthened by religion and lore. This left him cold: Group selection bumping whatever the opposite of altruism is to the next level was nothing to gain any solace from. For in order for true goodness to be born, a ruthless rival to the group was mandatory. This would both foster xenophobia between groups and necessitate a closing of the ranks within, a realization that led Trivers to dub Hamilton's 1975 study his "Fascist paper."[50]

It's a sobering thought, but it needn't always be true. A recent study in rats showed that the more a rat benefits from the altruism of a stranger rat, the more he will later act benevolently towards stranger rats himself.[51] This happened in a laboratory, in rats that were trained by humans to pull a lever that released food for a stranger rat in another cage—perhaps not the best approximation of what happens in nature. But we humans constantly find ourselves in situations where complete strangers need our help. In fact the notion that as egalitarian morality spreads through education and culture to an ever-greater number of people, the costs of altruism are reduced may mean that we have developed a unique capacity for altruism. It's true: History continues to serve up a litany of factious wars over religious and political influence.[52] But this doesn't mean things can't change. Cheap airfare, integrated markets, the Worldwide Web, and the extinction rate of languages are just a few of the harbingers of our increasing interdependence. And even if globalism fails or infinitely worse—the promise of universal tranquillity crashes—something else may end up uniting us. A classic Hardin "tragedy of the commons," the threat of a degrading environment, could graduate mankind to a higher level. And if Earth can serve as a genuine "vehicle," then the environment will be a proxy for a ruthless enemy. Short of recognizing the inherent goodness in universality and peace, coming together to protect our planet may prove a saving grace for a still-fractured humanity. In the end it may even lead to an expansion of our circle of morality to other creatures besides ourselves.[53]

These are grand affairs. What direction history takes is known only to prophets and fools, and from the dawn of time distinguishing between the two has been impossible. Increasingly, however, science is training its gaze inward, to the brain, in search of clues to the origin of altruism. The neuroscientists Jorge Moll and Jordan Grafman placed volunteers in a functional MRI unit and asked them to imagine either donating a sum of money to charity or keeping it to themselves. They found that thoughts of generosity activated a primitive part of the brain that lights up with reference to food and to sex, a result interpreted by them to mean that altruism is as hard wired as our most basic functions.[54] The psychologist Marc Hauser agrees. His research shows that all over the world people process ethical dilemmas in very similar ways, suggesting that the capacity for moral judgment, like language, should be thought of as intrinsic to our nature rather than a product of our culture. It is a well-documented fact that injuries to particular parts of the brain result in patients who seem to have lost their moral sense: Perfectly intelligent otherwise, they remain unable to tell right from wrong.[55]

But science is going further. In the most reductionist attempts to date to locate a "moral sense," researchers have begun searching for genes for goodness: A team led by the psychologist Ariel Knafo, the geneticist Richard Ebstein, and their graduate student Salomon Israel gave fifty shekels to two hundred students and asked each of them to play what is called a dictator game: The person with the money, the "dictator," must make a one-shot, nonreciprocated decision on how much of his pot to share with an anonymous "recipient." Not surprisingly there is quite a bit of variation: some split their loot judiciously down the middle, others selfishly keep it to themselves, still others give most of the money away. What the researchers found when they genotyped the students was that there is a significant correlation between the degree of giving and the length of a tiny DNA microsatellite in and around a gene encoding for a receptor of the neurotransmitter oxytocin in the brain. The same was true for a related neurotransmitter called vasopressin. According to these studies, and however incredible, altruism is literally in (and around) our genes.[56]

This kind of data is exciting to some, but leaves others scratching their heads. What do students divvying up fifty shekels in a controlled experiment on a college campus have to do with real-life altruism? If variation in the receptor gene also affects marital status, the chances for divorce, and autism, how specific is it really? Autistic people may find it difficult to marry, but aren't egoism and selflessness in a relationship relative and fluctuating? Finally, and perhaps most important, if the different DNA profiles account for just a few percentage points of the variation in altruistic behavior at best, in what sense do they account for the behavior at all? Causality is a tricky beast, especially when it comes to humans. Even the most sterling cancer research studies claim barely a 4 percent stake for their pet agents with regard to the occurrence of disease. And figuring out the relationship between altruism and genetics is infinitely more complex than figuring out the connection between lung cancer and tobacco.

What does all this mean?

Like many of us, Sally Adams of Hampstead, London, isn't sure what it all means, but she knows that she wants a baby. Having spent more than £15,000 on fertility treatments, and even traveled to Crete for her cause, Sally has not given up. Over fifty years old, a cat lover, and single, she has the requisite sperm in hand; only a tiny egg remains missing. Recently Sally advertised the fact that she is looking for an egg donor from among Oxford and Cambridge graduates. Her reason: Oxbridge graduates are not only intelligent but often also "very altruistic."[57]

Whether through brain imaging, animal behavior experiments, dictator games, genetic typing, gene-culture interaction models, or advertising in the *Daily Telegraph*, the timeless search for altruism continues, with "admiration and awe."

George Price, London, fall of 1974

14

Last Days

After World War II the old Victorian church in the middle of Tolmers Square was converted into a cinema. When it was being swept up after a showing, back to front, rats would come out the front door, run around the building, and go in the back door, every night exactly the same. Still the locals loved it. With stall tickets going for fifteen pence and circle ones for twenty-five pence, it was the cheapest theater in London, and queues of two hundred would gather for the Sunday-afternoon performance. When Joe Levy and Stock Conversion bought it out in 1972, a veil of sadness descended over the neighborhood. Appropriately *The Looters* and *Die Slowly, You'll Enjoy It More* were the last films ever screened. One pensioner wrote in rage to the local paper:

> *Dear Sir, Again a place of enjoyment . . . is to be closed and pulled down, although it had 1,000 regular patrons. Is this what we call progress? Who decides about this property? Do the 1,000 patrons have a say? Many age-old [sic] pensioners will miss it. We were told*

in 1918 and 1945 this was a land fit for heroes to live in; and take it from me, you need to be a hero to live in it.[1]

• • •

George moved into a squat in the square on July 15, 1974; the cinema had long been demolished, and a barbed-wire fence erected around the site had given it the look of a concentration camp. The entrance to the squat was through Tolmers number 19, but the walls between apartments 19–25 had been destroyed to make room for a large silk-screen-printing studio on the first floor. Together with two other squatters, Vince and Dave, George could share the entire second floor. Everything was extremely primitive: the toilet, the shower, the kitchen. But after what he'd been through, George thought it grand. The barbed-wire-surrounded outdoor courtyard notwithstanding, it was the nicest home he had lived in, he thought, since his babyhood home with his mommy and daddy in Scarsdale.[2]

The next day he sat on a bench to talk to Sylvia, the twenty-five-year-old American printmaking artist who worked together with Petal (a man) on the first floor. Sylvia was the daughter of a U.S. Air Force fighter pilot wing commander and had moved to Europe with an old boyfriend; now, together with a few friends, she had opened a gallery called Artists for Democracy on Whitfield Street. A group of them became the Poster Collective and had set up a squat studio at 19 Tolmers. She was living at number 68 Warren Street, just between Tolmers and the gallery. She had dark hair, a beautiful face and body, and said she came from Colorado.[3]

Even among the politicos, the artists, the eccentrics and the druggies, George cut an unusual figure. He was scraggly and often unclean. His teeth were rotting, and he suffered from a terrible raspy cough. And yet this stranger, his thin legs crossed, had the most amazing stories! He told her about his work at the Manhattan Project and at IBM, about Skinner, and Shannon, and Senator Humphrey. He spoke about Bell Labs and his Design Machine, about his attack on the Soviets in *Life* magazine. He told of his brother, Edison, back home, who had invented track lighting and worked with great architects like I. M Pei, Louis Kahn, and Buckminster Fuller.[4] He told her about his own journeys into genetics and evolution, about animal conflict and the equation that solved the problem of altruism. But mostly he spoke about the people he was trying to help, about their addictions, their sorrows, and hopes. George was a gentleman, she thought, a quiet, thinking man who talked in a high, squeaky voice, almost like a baby. There was something about him, she felt, that "just wanted to 'be.'"[5]

At the end of a day's work at the studio, she'd walk out into the square and set her sights back home. Increasingly, there was George, waiting to walk her across the Euston Road. He was lonely and she liked him; she was glad of his company. On rare occasions he'd come up to the collective kitchen for a cup of tea. But soon Sylvia began to see that for George this was more than just a friendship. Over in France, his old friend Henry Noel reproached him: Of course George might be an exception, but "when an ex-married man, over fifty, falls in love with a girl who could be his daughter, he is not on a Godward path."[6]

To Sylvia it was clear that his infatuation had nothing to do with her per se. The real reasons were mysterious, tucked deep inside his imagination. He was obsessive and intense but at the same time somehow not connected to the "real" world. When he asked her to marry him the first time, sitting together on the bench, she declined with a smile. When he asked again, she tried to quiet him with a chuckle. Really, George, she'd say. I like you. But marriage and romance are out of the question.

To Hamilton he divulged the true scope of his transformation: He had just recently left his job as a night office cleaner ("my supervisor was sorry to have me leave"). He was giving up his work on genetic polymorphism ("Harry, for reasons that I am not certain about, is not interested in having me finish"). He was giving up on their project on sexual selection ("I think maybe you'll be better off without me in your beetle theorizing"). He believed that a new phase had started in his life, "where Jesus wants me to do less about helping others and give more attention to sorting out my own problems," at least for a certain number of years. He was going to marry "someone named Sylvia (from Colorado)," and move to America to raise a family and maybe do some economics. Sylvia didn't quite share his feelings yet, he knew, but George was optimistic. "Let's wait and see what happens," he wrote, before reminding Bill that he had some books in the squat that he could pick up the next time he was in London.[7]

The following week George typed up a letter to the CPL Recruitment Division at 14 Old Park Lane. "Your advertisement in *Computer Weekly* for a programmer-analyst with IBM FORTRAN experience to work on 'a complex financial modeling system' at City Bank," he opened, "fits my background and present interests just about exactly." After describing the optimization work at IBM over which he had corresponded with Paul Samuelson back in the sixties, as well as his genetics work at the Galton, George explained that his current plan was to get in touch again with Samuelson and inquire about a research appointment at MIT. "That finishes describing the relevant parts of my rather odd occupational career," he wrote, explaining that his plan should not dissuade them from hiring him:

> . . . *even if an offer of an appointment should be immediately forthcoming, there is a matter much more important than economics research or jobs or anything like that that will keep me staying in London for an unknown length of time. To put matters briefly, I want to marry someone, and by "someone" I don't mean anyone, but one particular someone, who at the present time does not particularly seem to be close to agreeing with me in this matter. Hence the need for a temporary job.*[8]

Of course this would prevent him from being put in charge of any long-term project, not to mention a lower salary rung than advertised. But George wouldn't mind. He was a quick learner, and in view of his past experience, could be expected to start producing useful output during the first week. Then he ended: "I don't think money's all that important (tactless thing to say when seeking employment from a bank!), and so I would expect to receive substantially less than the 3375 pounds p.a. I was getting when I left University College, though substantially more than the 4.90 pounds per night I was earning in my last job as an office cleaner."[9]

He was feeling good about himself again, and no one could stop him. Recently two foot "fuzz" (policemen) had detained him for questioning on the street. He was "hairy, tired and decrepit"; soon two more cops appeared in a car, one of them six foot three or four, cold and tough and of high rank. The frisk job didn't take long to follow: pockets, shoes, socks, carrier bag—they even read his letters from Smoky in jail. The big cop radioed to the station to see if he had a record or was wanted. But George thought he had gotten the last word, he wrote to Kathleen, as is promised in Luke 21:15: "Because I will give you the ability to speak, along with wisdom, that none of your opponents will be able to resist or refute." He was following a "total honesty policy" now. What this meant Kathleen, back in Hawaii, wasn't certain, but George at least seemed to know where he was going. He had shaved his mustache and trimmed his beard. He had his hair in a neat ponytail instead of "floating in the wind." He had started acquiring more possessions instead of giving them away. "Am now apparently heading back upward," he promised.[10]

. . .

Finally scientific recognition was coming his way. In 1969 when he had come over to New York to see his mother for the last time, George had briefly met with Richard Lewontin. At the time it seemed to him that Lewontin had little interest in what he was doing. "Why don't you write your paper on Hamilton and not wait for me," Lewontin had replied in response to a follow-up letter from George; he hadn't had time yet to look at his covariance equation.[11]

But now, five years later, Lewontin's tone had changed. "I am sorry that I did not get to see you when I was in London," he wrote in a letter addressed to the Galton.

It has taken me a long time to come around to understanding the work you have been doing, which I was too stupid to appreciate when you first showed it to me. I hope that we might have some communication in the future.

Yours sincerely, R. C. Lewontin.[12]

James Crow, a leading population geneticist and longtime collaborator of the "neutral theory" apostle Motoo Kimura, wrote to CABS in the fall from Japan with similar tidings. "Much to my chagrin," he admitted, "I have only in the last few days 'discovered' the papers by him. . . . I haven't completely digested it yet, but want to make it a part of my thinking." He was especially enamored of the second covariance paper, the one in which George showed how selection could be partitioned at different levels. In fact, he had been thinking exactly along the same lines. As for George's brilliant explication of Fisher's fundamental theorem, he was at least in 95 percent agreement. This was a good thing because many of the statements made by Fisher were totally opaque: Probably only Fisher and God could understand them, "and probably God has trouble."[13]

CABS replied on November 1:

> With regard to George Price I do not suppose you have met him in person: he is a very intelligent and interesting man. He is a New Yorker who has come over to live in England because he likes the country. He seems to have a number of very bright ideas. He seems to have his own way of looking at the world and at problems which seems to differ sometimes from the way other people would think.
>
> He first came into contact with the Galton Lab when he walked in one day and said he had a new theory of selection which he would like to explain. I was very much taken with this and when his own money ran out I managed to get some money for him to work at the Galton.[14]

George's background was in chemistry and computers, he explained, and so, really he had come out of nowhere. As for cracking Fisher's meaning, it was CABS's opinion that George was "very like Fisher," and so, "on the principle that similar minds think along the same lines," it might just be the case that George had done it. Then he turned to explain the current situation:

> Since that time things have taken various interesting turns. For one thing George was suddenly converted to Christianity and by Chris-

tianity, being George he means accepting a very great deal. Includ-
ing in this to give all he had to the poor and to take no thought of
tomorrow extremely literally. This I must say is an extremely honest
and courageous thing to do but as you can imagine it also caused a
great deal of concern among his friends. Finally his funds ran out
completely. . . . He also decided to help some alcoholics. This again
seemed an extremely courageous thing to do and very well worth
while. Unfortunately things did not go straight. One of the alcoholics
found out somehow or other without George's permission that George
worked at the Galton and came round to see him. This particular
man was very strong and could be very violent and not very polite
and although he came without George's knowledge or permission,
George was accused of inviting him round and this caused a lot of
difficulty for a time. We hope this is over now. However the position
is that George's grant has now run out and officially he is no longer a
member of the Galton although he does actually look in from time to
time. I have left your message and I understand he has got it so you
may be hearing from him soon. In any case all this does certainly add
interest and excitement to life.[15]

In truth no one at the Galton really knew what was happening
with George, nor, they would later say, did they quite appreciate how
precarious was his existence. Lewontin's note had arrived in the spring
but only reached him in the summer; it had been that long since he
had popped in to check his mailbox. "I don't quite know what to say
in response to your extremely kind remarks about my genetics work,"
he replied in a handwritten letter.

I think the truth about me is that I have some intellectual traits
that tend to make a favorable impression upon very intelligent and
productive men, but my work repeatedly fails to measure up to the
impression given.[16]

The only genetics work he did which he thought of any value was figuring out the true meaning of Fisher's fundamental theorem. Sure, the
covariance was pretty, but he had yet to use it for any real evolutionary

explanation, and in six years of work he really should have accomplished more. Now he was thinking about moving into economics, where he hoped to finally achieve some "really useful results."

Lewontin's reply from August 27, just as George was writing to the recruitment office, was warm and friendly; evidently John Maynard Smith had filled him in on some of the details of George's life. He was sorry that genetics would be losing him, but was delighted that he was going into economics. "It is time that someone really original began to think about the important and relevant problems of economics," he wrote. He had no doubt that George would accomplish great things there.[17]

. . .

But it wasn't going to happen. As much as he wanted to head "back upward," George was finding it difficult. Back in late June he had written to his lover, Joan, that he could not be with her since it was necessary that he undergo some kind of "treatment" to learn to love properly—*really* love. Joan had come around to spend a night with him in the squat, and, not wanting to mix his two lives, George had ordered everyone out of his room. Now he felt terrible about it. He had been helping countless strangers, but to learn "real love, giving love, Jesus-style love" he would need to start again with just one person. It couldn't be her since that person had to be more innocent, less strong willed. He was sorry. But Jesus said: "Take my yoke upon you and learn from me," and that was what George was doing. Really, despite appearances, the only marriage permitted to him was to Jesus. If that happened, perhaps they could be together some day.[18]

Sylvia had woken up from tooth surgery at the dentist to find him sitting beside her, and she was beginning to get a little freaked out.[19] Clearly there was not going to be any relationship for a while. He was leaving Tolmers Square so as not to present any further problems. And the move down the road to 164 Drummond Street had gotten him thinking.

"I have indeed been serving the Devil most of my life," he wrote Joan in October. Trying to help all the homeless and old people had not been led by true love and had only done them harm. As with her, with them, too, he had just created false hope, and it was time for him to stop. To do this he would need to follow the way of Matthew 7:14: "Because strait is the gate, and narrow is the way, which leadeth unto life, and few there be that find it." The "strait gate," George now believed, was honesty about one's own self-recognition, "confession to one's self of one's deepest selfish desires." Before anything one had to be true to oneself. Then the "narrow way" was to follow Jesus' lead "along an easy, twisting, curved path of love"—love for others. George wasn't doing too well so far, he admitted. But he had stopped going around to help an old lady, Mrs. Veneham, two or three weeks before, and no longer popped in to Elisabeth Mansell's SOS House. He hadn't communicated with Chrissy, who had been in the hospital, and didn't even know if she was still there. His progress was slow, like that of the White Knight in *Through the Looking-Glass*, who kept falling off the chessboard to one side or the other. Still, it was a beginning. He was being more selfish and taking things easier. "Please return good for evil and pray for me," he ended. "God helps those who help themselves."[20]

His fifty-second birthday came along on October 16. "I pray that you will permit your soul to be restored," Joan wrote him, understanding finally that she would have to let go. "Dismiss your illusion . . . there are no such thoughts for you now." All the way from Grand Rapids, Michigan, his mother's cousin, Lewis Florman wrote kindly, too, to wish him the very best of everything: "As Paul said 'test the spirit—hold fast to that which is good.' Dig up that talent of yours and put it to good use. The world is sorely in need of scientists who are not afraid of looking at new truths. I feel that you could be a 'light unto the world' if you choose to do so."[21]

It was very good news, George wrote to Smoky, that the judge had decided to let his probation continue. But then, as kindly as he could, he changed his tone:

> *In regard to you and me, I have a feeling that we're sort of brothers, with me being the brother who has gotten all the good breaks, and you the one who has had all the bad ones. . . . But for the immediate*

future, I'd rather that we sorted out our problems separately. I'm right now getting the "treatment" from Jesus. . . . Every day a number of things go wrong for me as a result of my own fault. Also I'm unemployed, in debt, living on borrowed money.[22]

He would need to take a break from helping him, he explained. Meanwhile he recommended that whenever he found himself in an escalating fight he back down, and, no less important, to absolutely make up his mind not to touch anything distilled.

· · ·

He was living at 164 Drummond Street now, on the corner of Hampstead Road. Just two weeks earlier a couple of architecture students had broken into the building and changed the locks. It was a relatively new building, from the thirties or forties, and, though rather ugly and gray, as squats go not a particularly bad one. Behind the locked door there was an entrance hall and stairs up to the second floor with a kitchen and two rooms. A further set of narrow winding stairs led up to a third floor with a studio and a small adjacent room. There was electricity and, in the rooms, electric storage heaters. But there was no central heating, and as winter approached it was getting very cold.[23]

In the middle of November a pot of boiling water tipped over and burned his hands badly. He was finding it difficult to sleep at night, and couldn't grasp a pen to write. Finally, when the burns began to ease, he wrote to thank his brother, Edison, for the money he had sent him. He was leaving genetics, George told him, and was hoping to get a computer programming job at UCL with a professor who worked on the economics of developing countries in order to reacquaint himself with the field before trying his luck again with Paul Samuelson.[24] In the meantime he was contemplating writing up some genetics results from the past few years, and then perhaps returning to his work on the Passion schedule and the correct date of the death of Jesus. Earlier he had hoped to make a book of it, but now thought a pamphlet would do.

"In regard to religion," he explained,

*I've been heading back toward more conventional and conservative
Christianity, and have given up much of the amateur "social work"
I was doing. I'd like to get married again. Have about 4 kiddies, a
dog and a cat, house in the hills somewhere, about a five year old car
(to give a quick picture of the economic level I visualize), with lots of
time to hike around, read, write, maybe paint.*[25]

No doubt thinking of Sylvia, he envisioned "some small printing
business on the side" and perhaps putting out a local community
newspaper. But of course this was pretty long-range looking ahead.
"Wouldn't it be nice though?"

In reality George was in a much worse off state than he let on to
his brother. Shmulik Atia was a young Israeli, a former paratrooper
who had just gotten out of the army and come to join his artist friend
Asher Dahan who was living in the small room on the third floor of
the squat. Asher had turned the studio upstairs into his workroom,
painting into the wee hours of the night under the neon lights and
smoking a lot of pot. Once in a while George would come up with
a cup of tea, and join him for a smoke. Both remembered him as a
humble and reserved man, walking about like a ghost, down to skin
and bones, reeking, gaze downcast, mumbling about Jesus.[26]

"The Hounds of Heaven are closing in on me," he wrote Kathleen
that month, referring to the English poet Francis Thompson's work
from 1907.[27] Thompson had been born in Lancashire in November
1859, the very month that Darwin published *The Origin of Species.*
Moving to London after college to pursue writing, he soon became
addicted to opium and a street vagrant. Rescued years later by a
couple who chanced upon his poetry, he became a published poet
before dying, mentally unbalanced, of tuberculosis.[28]

> *I fled Him, down the nights and down the days;*
> *I fled Him, down the arches of the years;*
> *I fled Him, down the labyrinthine ways*
> *Of my own mind; and in the mist of tears*
> *I hid from Him, and under running laughter.*
> *Up vistaed hopes I sped;*

And shot, precipitated,
Adown Titanic glooms of chasmèd fears,
From those strong Feet that followed, followed after.
But with unhurrying chase,
And unperturbèd pace,
Deliberate speed, majestic instancy,
They beat—and a Voice beat
More instant than the Feet—
"All things betray thee, who betrayest Me."

"Like the hound follows the hare," J. R. R. Tolkien had written of it, "never ceasing in its running, ever drawing nearer in the chase, with unhurrying and imperturbed pace, so does God follow the fleeing soul by His Divine grace. And though in sin or in human love, away from God it seeks to hide itself, Divine grace follows after, unweary-ingly follows ever after, till the soul feels its pressure forcing it to turn to Him alone in that never ending pursuit."[29]

George was beginning to tire of the chase.

· · ·

His homeless friends were coming around looking for him. Bernardo and Chrissy arrived unannounced on Saturday night, November 30, only to find he wasn't there. "Bernardo is very upset that he was rude to you. He had a blackout through drinking," they wrote on the back of a torn Benson and Hedges Special Filter cigarette box, slipping it under the locked door. The next day they wrote a letter from their temporary abode on 40 Fitzroy Street. "We are very worried about you and we hope you are well. . . . We love you very much, Bernardo and Christina."[30]

George was falling fast. He had contacted Dr. O. W. Hill at the Middlesex Department of Psychiatry, and Hill set up an appointment for January 9, 1975. But that was more than a month away! Would he make it that far? To Bill and Chris Hamilton he apologized now for his vacillating behavior and refusal to accept good advice. They had always been such generous hosts to him. He should have heeded their words of counsel to stay in genetics. Did they think he might still get a research job at Silwood?[31]

Hamilton had seen George toward the end of November in London and was worried. He implored him to come stay with his family, if only to get some rest and talk about a possible joint scientific venture. George continued to vacillate, paralyzed almost, unable to make up his mind. Finally, after many delays and reversals, he climbed on a train at Victoria and arrived, weak and disheveled, in Berkshire. He told Chris and Bill about his unrequited love for Sylvia, about his debts, about all his help to the homeless having been for naught, about his plans to turn to economics where he might actually do something useful for humanity. Bill tried to keep him focused: Why not embark together on a theoretical project on altruism? They could begin to work right away on a grant.[32]

The Hamiltons were going to Chris's parents in Ireland for Christmas, so they couldn't invite him for the holiday. But by the end of the week George seemed to have been won over. Bill showed him the draft of a paper he was working on that applied the covariance to individual and group, and George was pleased.[33] He was determined to return to genetics, he told his hosts smiling, a flicker of clarity once again noticeable in his eye. After driving him back to the Maidenhead Station on the morning of December 19, Bill gave George a hug, watched him get on the train, and waved good-bye.

Then he rushed home, found a few sheets of lined computer paper, and hurriedly wrote out a letter by hand. It was addressed to a Dr. Kelly, the secretary of the British Teilhard Association, a nonsectarian educational society founded in 1963 to promote knowledge and understanding of the work of the French Jesuit evolutionist Pierre Teilhard de Chardin. Bill was sending a copy of George's "Twelve Days of Easter," and inquired if the society might be interested in publishing it. It would blow a welcome wind into George's sails, he was certain, and, in truth, the paper really was fascinating and original. "When he had his flat in central London and was at work alternately on the theory of natural selection and biblical exegesis," he explained, *he seemed to be an intellectual Sherlock Holmes in real life with a brilliant mind willing to work on any problem that appeared to him of being of permanent significance to man. Speaking of his achievements in my own field, I believe that his new presentation of natural selection effectively disposes of the problem dating back to Darwin of whether the individual or the group should be considered the unit of natural selection.*[34]

This was a matter of the gravest urgency; George was down and out, sickly and starving. His survival was at stake. If they couldn't print the Easter schedule, could they perhaps offer him a job as a janitor? Hamilton wasn't a religious man, but he would be eternally grateful.

*George Price's unmarked grave, covered by brush, left of the tombstone,
Saint Pancras Cemetery, London, 2009*

Epilogue

George Price killed himself sometime between January 5 and the morning of January 6, 1975. Shmulik Atia had been about to leave the squat to go to work washing dishes in a Hampstead restaurant when he saw an envelope someone had slipped beneath the padlocked door. Unable to read English, and figuring that it might be the dreaded eviction notice, he ran upstairs to show it to George. It was 7:00 a.m.

When he knocked on George's door, he felt it give in a little. Pushing it farther into the room, his eyes pointed downward, he saw a strange dark purple film, cracked like parched earth, coating the linoleum floor. A strong smell attacked his nostrils. He pushed the door still farther in, now feeling the weight behind it. When the aperture reached a certain angle, the full gruesomeness of the sight unfolded before him: A lone bare bulb lit the room from midceiling. The gray walls were peeling. Apart from a table, a chair, and an unmade bed, there were ammunition boxes serving as furniture. Hundreds of pages were scattered about. The entire floor was covered in blood. The window was cracked, and a cold wind was blowing through it.

There on the floor, with his back leaning against the slightly opened door, wearing black pants, a blue-and-white pinstriped shirt, and a worn-out black jacket—all too big for his skin-and-bones frame, was George. Shmulik looked down onto him: His legs were splayed, his arms fallen to his sides. His eyes were shut and his head slumped forward, leaning slightly to the left. His right hand clutched a pair of medium-size, sharp tailor's scissors, and the blood surrounding the gaping puncture wound in his neck pointed clearly to the site of injury. Shmulik wasn't sure when it had happened, but running up in a panic to wake Asher on the third floor, he screamed in his native Hebrew: "George is dead! He's dead! George is dead!!"[1]

Three days later Shmulik and Asher were called in by Scotland Yard. It was clear that George had taken his own life by puncturing his carotid, but, per protocol, there had to be a coroner's inquest. A professor named Hamilton stood up to speak, and someone whispered that he was considered one of the world's great evolutionary biologists. Hamilton said that George had been a guest in his house just a fortnight before the tragedy, and that he was perhaps the most brilliant thinker he had ever met. A handsome, well-dressed lady sobbed uncontrollably. Introduced as a BBC worker and a friend, she spoke of George as the most wondrous lover. *George?* Shmulik and Asher looked incredulously at each other. Could these people really be talking about *George?*[2]

He had stopped taking his thyroid meds, the inquest concluded. Hamilton explained that in the past George had felt that this was the best way to find out in which direction God wanted him to turn; probably he'd been seeking His guidance again. But a psychiatrist, Dr. Christopher Lucas, had apparently seen him just days before; George had come in off the street sunken and depressed. Lucas explained that with George's condition, erratic intake of thyroid medicine could have contributed to the despair. He was unsure, however, whether George might have suffered from the symptoms of schizophrenia.[3]

There had been a note, the inquest found, from January 1, at 4 a.m. "I came to England to a new part of the world," it read, "before I decided to kill myself." There was another, from January

2: "To Whom It May Concern: I guess I've had it. George." A third, addressed to Sylvia, was about how "pure" and "white" she was, about bringing her "shame and sorrow," about how he had mocked love by wanting her in a way that wasn't approved by the Lord. And then there was a fourth letter, one addressed "To my friends." George had felt himself to be a burden on them, he wrote, ever in debt, ever making the wrong decisions. "It seems the tim [*sic*] has come," he ended, "for me to go and meet my maker."[4]

On the day of the funeral at Saint Pancras Cemetery, the *Sennet*, London's student paper, announced the news to the world: "A prominent genetics researcher at UC Hospital gave up everything, including his life, for his religious beliefs. Dr. George Price gave away all his money, clothes and possessions to homeless alcoholics and left his flat in Bloomsbury to live as a squatter in Drummond Street, Kentish Town. It was there that he was found dead. A respected scientific researcher, Dr. Price was convinced that he had a 'hot line to Jesus.'"[5] Otherwise his death passed almost completely unnoticed.

That same day, after George was interred, Hamilton decided to visit the squat on his way home from the cemetery. The people from the American Consulate had already been there to pick up George's effects, but from a telephone discussion with a Mr. Cresty at the consulate he had a hunch that more papers were left behind. "I regard his ideas of such originality and such significance for evolutionary theory," Hamilton wrote to Edison, "that I believe that sometime someone may think it worthwhile to find out something more about him and wish to go through letters and papers with some care." Then he added, "and of course the strange life he has led for the past few years makes it quite a story."[6]

As Hamilton approached, the atmosphere around Tolmers Square was electric. The pavement outside the building had been ripped up, and that very day the tenants of 164 Drummond Street had had an altercation with the men of Stock Conversion. There was shoving and cursing, and the police came around. Sympathy aside, every one knew that the clock was ticking and couldn't be stopped. The building was due for demolition.

When he walked into George's room, Bill could still feel the blood crackling on the linoleum floor beneath his shoes. The mattress was gone, and so too the electric heater. His hunch had been right. Eyeing the piles of unmarked newspapers, magazines, cuttings, and old computer programs, he collected what was left of the papers. He found "The Nature of Selection," which had been rejected by *Science*, and some genetic polymorphism materials that CABS would soon help publish posthumously.[7]

A great sadness welled in his chest; in the silence of the room it was clear now that a friend "almost like a second self" was gone. "I felt a very great kinship with him," he wrote to Edison, "indeed apart from his rather private nature a feeling of intellectual redundancy in each other's presence may have been part of the reason we didn't meet more often." Racked by guilt for not being able to save him, he would later lower himself in comparison: "Where I had sympathized with flowers and bracken ferns partly because there wasn't much that I could be expected to do for them, making my sympathy cheap, he gave himself wholeheartedly to the crying children and homeless humans wandering the same streets." Bill was not sure whether any other evolutionary biologist in the world saw the interest in George's covariance equation in the way he did, but he wanted to believe that recognition for his achievement would come someday, however belatedly.[8]

. . .

A few days later 164 Drummond Street was demolished. On March 11, 1975, eighty-one people living in twenty-six houses were handed summonses to appear at the High Court in ten days' time. Stock Conversion's threats were finally materializing.

With the help of Camden Council, the day in court was postponed to four weeks. Immediately the community galvanized. The printing press churned out leaflets, and Sylvia and Petal's silk-printing studio churned out the posters: "Defend the Tolmers 81," "Stop Levy Fight Eviction"; there was even an invitation to a play at Leicester Square Theatre, beginning March 27: "Levy: He Will Tear Your Home Apart," it said, promising "your senses will never be the same."[9]

Finally the day in court arrived. Justice Croom-Johnson wasn't known in particular for sympathy toward squatters but, perhaps wary of the great local and national press interest, determined to follow the letter of the law to a T. And what the law said was clear: Obtaining a possession order had nothing to do with whether the owner had an alternative use for the property, or with whether the evicted tenants would become homeless. What it depended on was three very technical matters: (1) serving the summons correctly, (2) showing that the occupiers were there without a license or consent, and (3) convincing the judge that "reasonable steps" had been taken to find out the names of the occupiers

The vast majority of squatting cases are never contested; squatters don't bother going to court, and judges affirm possession orders almost automatically. But this was going to be different. On the morning of the court case the people of Tolmers, young and old, rich and poor, educated and illiterate, gathered in the square with banners and posters. With arms linked, singing songs, they marched in procession the twenty-minute walk to the High Court in the Strand.

Stock Conversion had hired the services of a renowned firm of solicitors, led by a hotshot property barrister who would normally have had nothing to do with such trifling affairs. The team included two barristers, four solicitors, two private investigators, and four representatives of the company. On the other side a young trainee solicitor who was squatting in Drummond Street managed to persuade his little firm to represent the squatters in defense. And although a lawyer experienced in squatting cases agreed to act as barrister for the defense, the matchup resembled that of David and Goliath. Nor did the proceedings start off well for the defense: Their motion to hear all the cases together was rejected; instead the judge decided that he would hear them case-by-case, household-by-household. All other parties were ordered out of the courtroom.[10]

The first case hinged on a doorbell. Had the squatters heard the ring of the private investigator hired by Stock Conversion to take "reasonable steps" in ascertaining their names? The investigator's report stated that he had "knocked loudly," but the squatters claimed that he should have rung the independent doorbell, for a knock would not be heard in the flat. After an hour the judge adjourned, pending precise information about the location of the bell and clarification from the private eye—who was absent—about the details of his knocking movements. The TVA knew that there was no chance of winning the battle in court against the speculators. But with the right tactics they might stall long enough to give the political opposition enough time to build up greater strength, or even for the development situation to change so that Stock Conversion would no longer have a reason for the evictions. The adjournment was therefore greeted with glee. But when the private investigator finally took his seat in the witness box the following week, he turned out to be a clean looking, decent ex-policeman, and his testimony that by "knocking" he had meant "knocking and ringing" was accepted by the judge, and the possession order was duly granted to the claimants.

One after the other the cases came before Justice Croom-Johnson, and in each case he found for the speculators. The Tolmers group waiting outside the courtroom was beginning to lose heart; no biblical miracle was going to fell this Goliath. But then came the case of the squat on 217 North Gower Street. The occupants—two medical students, a flute student at the Royal Academy, three local-authority social workers, and a housing surveyor—each swore on oath that they had been at the squat at the time the private investigator claimed he had come knocking at their door, and that they would have opened the door if they had heard a knock. They all seemed honest, and their conviction was strong. Croom-Johnson adjourned for ten minutes. Everyone in and around the courtroom held their breath. Finally, in a twenty-five-minute summation in which he spoke of both the private investigator and the squatters as reliable and trustworthy, Croom-Johnson announced that he had no option but to throw the case out and to award costs against Stock Conversion.

A roar of joyful laughter and clapping was heard outside the courtroom from the squatters in the hall. Immediately, a gofer was sent to give word to the press, and drinks were poured for all in celebration. The next day the Camden Council announced that it was going to buy all Stock Conversion land in the Tolmers area. It was "the victory of the year," the barrister for the defense was quoted, and the *Hampstead and Highgate Express* pronounced "A Square Deal for the People." The triumph, at least for now, was complete.[11]

It had been a true battle between collective and individual; the cooperative community of Tolmers against the selfish speculator Mr. Joe Levy. The selection and expectation terms of the equation had been applied to the individual and the group respectively, and the group-selection term had clearly made the difference. The anonymous man who had written the math explaining the structure of change better than anyone before him already lay deep beneath the ground at Saint Pancras Cemetery, but his brainchild had borne out its prediction.

. . .

The sweeping tale of attempts at an "altruism code" invites to the stage spitting tadpoles and "free-riding" cuckoos; sacrificing slime molds, naked blind mole rats, horn-locked oryxes, dung flies, rabbit viruses, and the overabused stepchildren of man. It spans the globe from the far reaches of the Siberian Afar to the South American tropics, from the prairies of Indiana to the African plains. It traverses from Aristotle and Jesus and Aquinas through Hume and Adam Smith and Malthus; from John D. Rockefeller to John von Neumann, from World War I to Vietnam. It takes in the Manhattan Project, the invention of the transistor and computer, Bell Labs and IBM; and it embraces the burlesque theaters of Great Depression New York as well as the subversive cinemas of squat-infested seventies London. From Darwin to Kropotkin and Huxley, from Warder Clyde Allee to Vero Cope Wynne-Edwards; from J. B. S. Haldane, the "last man to know all there is to know," to R. A. Fisher the Anglican apostle; from John Maynard Smith to Bill Hamilton to Bob Trivers; from Isak Preis to George Price; from nineteenth-century czars to mid-twentieth-century telepathists to biological mathematicians and brain imagers today.

But what, in the end, can the story of the attempts to crack the mystery of altruism teach us?

To begin with, many of the heroes of our tale had strong ideological commitments. From Kropotkin's anarchism to Huxley's statism, Allee's pacific Quakerism to von Neumann's obsession with war and deceit, Skinner's belief in a blank slate to Fisher's Anglicanism to Haldane's Marxism to Hamilton's naturism: Our tale teaches that the people doing science, their backgrounds, historical context, family histories, education, political views, religious affiliations, temperament—all play a role. Sometimes the role is straightforward; more often it is very complex. But whether direct or circuitous, subtle or brutish, to understand the science we should always look at more than *just* the science.

That real people do science is a salutary lesson we often learn from looking closely at the history of science, and, upon reflection, should never surprise us. But our story holds a deeper moral that may not be so easy to fathom: One of the pressing challenges of our times is defining the boundary between questions that can be addressed meaningfully by science and those that are outside its purview. Much of the acrimony between people who believe that science has all the answers and those who see science as a threat stems from a failure to rise to this challenge. At the poles of this debate, such as in the one between materialists who mock religion and creationists who lambast evolution, it often seems as though there is an almost constitutional inability to demarcate the lines—a deep confusion as to the territory of spiritual versus scientific pursuits. Ludwig Wittgenstein understood this problem well, which is why he wrote: "Even if all possible scientific questions be answered, the problems of life have still not been touched at all. Of course there is then no question left, and just this is the answer." He ended his famous *Tractatus* with the words: "What we can't speak about we must pass over in silence."[12]

In this context the riveting story of science's attempts to crack the mystery of altruism assumes a special role. For the search for a biological "altruism code" pits not only the individual against the group and the gene against the individual but also true goodness against masquerading self-interest, and heartless biological necessity against the transcendence of the soul. It is a story of chaos versus order, culture versus instinct, man versus animal. It is therefore a place where scientific fact and moral reasoning meet each other; where, face-to-face, nature and the species struggling to make sense of nature peer directly into each other's eyes. And so it is a place where the challenge is at its greatest.

At the outset our two nineteenth-century gladiators posed the question: How strong a grip should the biological approach to morality have on our imaginations? Kropotkin thought the answer was a lot, while Huxley rejected any kind of dependence on nature for moral instruction or solace. A century later we can see things more clearly: The Russian anarchist prince was too facile in attributing all our goodness to a "natural state," while the English bulldog was too strict in cordoning off humanity from its animal origins. Searching for reciprocity in Nature in order to urge it upon civilization is to get things backward: Reciprocal relations define the human condition, and real charity may in fact be a uniquely human invention. While Hobbes may have been right that we are out for ourselves, Rousseau was also right in believing that harmony and progress are possible without government. Self-seeking *can* produce genuine and true benevolence, and, it would seem, there are compelling reasons from evolution to believe that this all began in Nature. So Kropotkin and Huxley both held a piece of the truth.

But if the logic of selection in the prehuman world could have produced the seeds that were developed by civilization into what we call "morality," what does that say about this very dear invention? What kind of shadow, in other words, does an origin cast, even after its creation has spun far away and developed into something entirely different?

This is a sensitive question. The primatologist Frans de Waal, for instance, argues that "a viable moral system rarely lets its rules get out of touch with the biological imperatives of survival and reproduction," imperatives based on kin selection and reciprocation that first evolved in mammals.[13] This may or may not be true. But it also makes it not all that surprising that Hamiltonian and Trivers-like logic are sometimes attacked for implying genetic determinism. The thought that what is good in us, as well as what is bad, is somehow etched into us, and therefore fundamentally not of our making and outside our control, is abhorrent to most people. Such a reality could well strip life of much of its meaning.

Cynical biological determinists argue that this sentiment is just our self-importance talking; that to admit that we have binding constitutions is too difficult a psychological leap for a self-deluding, proud humanity. This may be true, but it is only half of the story. After all, the feeling that genes do not simply "run the show" comports not only with our vanity but also with our exceedingly healthy intuition about reality, as well as with what science is teaching us. Of course Haldane, Maynard Smith, Hamilton, and Price knew this, too. For when they spoke of genes for altruism, they were really only using a shorthand for genes that increase the probability that their bearers will behave altruistically. So long as such behaviors have a heritable component, evolutionary reasoning applies. Despite the incautious remarks of scientists and, more often, of science writers, this does not mean that a behavior is determined; culture and education are still acknowledged as playing a central, even exclusive role. There may be no behavior in humans, strictly speaking, that has no genetic component, but that's a world away from saying that our genes determine who we are and what we choose. As biologists and anthropologists and mathematicians and philosophers who study the subject today have come to see, natural selection based on *cultural* variation has produced behaviors that have nothing to do directly with genes.[14]

Most important of all, however, is the recognition that most human behaviors clearly have absolutely nothing to do with natural selection in the narrow, biological sense. For the biologist, the *consequences* of an action for reproductive fitness are what determine whether the action counts as altruistic, not the *intentions* with which the action is performed. Selfishness and altruism in human affairs have nothing to do with tadpoles spitting, mole rats digging, and cuckoos sneaking eggs into strangers' nests. Kindness is kindness only if it is meant to be so. Of course the brain is an evolved organ. But psychological altruism, the kind we talk about when we are thinking of human affairs, is entirely independent from biological altruism, or the kind that confers fitness on its bearer. Here is a quick illustration:

Helping an elderly lady across the street can hardly affect her fitness, since at her age she will not be having any more children. On the other hand, selfishly snubbing a young girl waiting in line to buy an ice cream might lead—who knows?—to her decision to have children who will exact revenge on you someday.

In *The Theory of Moral Sentiments*, Adam Smith softened up a little. "Man possesses the capacities," he wrote, "which interest him in the fortune of others, and render their happiness necessary to him, though he derives nothing from it, except the pleasure of seeing it."[15] Both biological determinists and those who think our evolved natures have nothing to do with our behavior would do well to pay careful attention to the words of the seventeenth-century Edinburgh economist: We are blessed with many gifts. What we do with them is our challenge, for together with kindness we were also given the capacities for cruelty, malice, nonsensicalness, and even boredom. Clearly, if our genetic endowments provide the foundation for both ethical and unethical action, then our moral lives are up to ourselves.

There can be no doubt that social life in Nature put into place the substrates that would one day allow the birth in humans of something we call our moral sense: the policing against cheaters, the mechanism for conflict resolution, and the capacities for empathy, shame, jealousy, sympathy, and rage are all stages on this wondrous evolutionary journey. This same journey also saw the laying down of the basic needs and compulsions of our species: the dependence of young on care, the interdependence of being part of a group, striving for status, sexual desire, motherly instinct, and of course the survival instinct itself. These basic compulsions and necessities are not infinitely pliable, nor should our moral sense ignore them. The fact that goodness may have natural origins—"the so-called moral sense is aboriginally derived from the social instincts," Darwin wrote—can be made to be confusing and coarsening. In the wrong hands it tends to produce a mentality that discounts what is not utilitarian in the biological sense.

But if the dramatic tale of attempts to crack the biological mystery of altruism teaches us anything it is that we need to do our best to resist this kind of scientistic "originalism." From Kropotkin, who thought men should return to the animal state, to Huxley, who implored that they shouldn't; from Emerson, to whom human life was akin to a termite mound, to Fisher's genes doing God's work on earth; from Wynne-Edwards's procreation-forsaking birds instructing about traffic jams to Haldane's genetics bolstering Soviet Marxism—everyone would have been better off for recognizing this.[16] Yes, by all means, natural dispositions play a part in our decision-making process. The brain really *is* an evolved organ, and in this sense there is a connection between biological and psychological altruism; both in the end can be studied from the point of evolutionary theory. But the fact that the brain has evolved is hardly its only interesting or distinguishing feature. Whether we want them to or not, the gifts bestowed upon us by evolution can never lead directly to the moral imperatives that help to shape our lives.

· · ·

But what about George? Beyond settling the meaning of Fisher's theorem, coming up with an equation that allows a multilevel approach to selection, and introducing game-theoretic logic into evolutionary biology, what further lesson can his unusual life confer? In a way trying to answer that question amounts to trying to understand his suicide. Why, in the end, did George Price kill himself?

The obvious answer is that George was unwell. Whether he was born that way or became that way, from early on he was different from the rest. His interest in numbers, his obsession with odds and ciphers, his awkwardness in social situations, his repeated behavioral patterns, his inability to hold a job or to stay in one place for long, his frequent insensitivity to the feelings of others—all point to the constitution of a highly gifted, highly unusual character. Before the days when Asperger's syndrome was routinely diagnosed by psychiatrists, George might have stood somewhere on the slippery spectrum between normal and autistic social behavior. And of course the thyroid problems couldn't have helped. In particular, failure to take thyroxine for the kind of problem George suffered from is known to cause depression. In George's case it might also have led to delusions. Joan, for one, felt that he kept getting things all wrong: Years after his death she would claim that she never had any intention of marrying him but had just been trying to save him, while he kept jumping to conclusions. While this is debatable—the letters from the period seem to suggest that she was entirely in love with him—increasingly, undoubtedly, he was becoming unhinged from reality. The last letters from him spoke of wanting to make love in threesomes, and of Jesus and Margot (a BBC colleague to whom she had introduced him) saying that it was not yet time for them to be together. But Margot had been dead for months, a victim of cancer; like his mother, George was talking to the dead. The combination of an unstable personality and the depression brought about by no longer taking his medication might well have pushed him over the deep end, and ultimately to suicide. This is what Joan and Hamilton and the people at UCL thought. It is also what the last doctor to see him, Christopher Lucas, seems to have thought, and there might very well be some truth to it.

But attributing George's suicide strictly to illness somehow doesn't feel entirely sufficient; there were life circumstances, too, after all. One of these was Sylvia. Unable to win her heart, George had become despondent. Whether he wanted to marry her because he thought Jesus wanted him to or not is, in a sense, beside the point. His heart was broken, and despair had descended. If the note that he left for her is to be taken at face value, George himself seemed to have decided that he would need to leave this world because of unrequited love.

And yet the suicide arrived just as, in his words to his daughter, George was "heading back upward." After a year and a half of radical, selfless giving to others, he had come to see that it was time to take care of himself once again. He had decided to hold on to more of his possessions, he had stopped working tirelessly for alcoholics in the square and station and courtrooms and squats. He had begun to try to do some economics, and was planning to write to Samuelson at MIT. Clearly conflicted about whether or not to stay in genetics, he nevertheless seemed to have determined that being a full-time selfless angel was going nowhere.

What might have gone through his head when he thought such thoughts? What, in the final analysis, had his giving been all about?

One way to understand George's unusual and courageous decision to take destitute and dangerously violent alcoholics off the street and bring them into his home is that it was a test. Pure and simple, this was a religious command. And since George himself came to believe that he was being led by Jesus on a path of suffering, he willingly accepted his fate. If giving to the needy meant having to leave his home, if having to leave his home meant living rough, if living rough meant being hungry and cold and losing all status—so be it. This was the Lord's wish, and George would follow it, period. And if the Lord had determined that it was time for George to look after himself, then he would follow that command as well.[17]

But is it possible to penetrate further beneath the religious injunction? Is there a deeper truth lurking behind George's own conscious explanation?

Widening the lens may help. Altruism, cooperation, and morality had in one form or another occupied George's thoughts from the beginning. Back in Chicago, working on uranium enrichment for the Manhattan Project, George wondered why nations should hate one another and fight. Struggling to complete *No Easy Way* in the Village between instruction manuals for Sperry-Marine and bouts of drug-induced incapacity, he had racked his brain for the solution to the threat of mutually assured destruction. And yet, as he wrote to his editor, the world was changing too fast to get any kind of grip on the problem; unable to see the answer, George finally gave up.

Then came Ferguson and the operation, and quitting IBM. Leaving one life, he acquired another; having crossed the Atlantic from America to England, he found himself swimming in the ocean of evolutionary theory. It was there, in search of the origins of family, that he became acquainted with the field of social behavior, with the dynamics of personal and collective interest, and most of all with the problem of altruism. Plunging into Hamilton's kin-selection mathematics and emerging with his own elegant covariance, George came to see that in nature, at least, goodness came about for a reason. Delving into game theory and surfacing with the logic of animal conflict, he understood that reciprocity was a utilitarian affair. Whether altruism came about at the altruist's own expense because it helped shuttle related genes into the next generation, or because it somehow ended up paying for the altruist later in his life, there was always an interested logic involved. Even when a "truer" altruism evolved under group selection it could only work if the good of one group was to triumph over the good of another. Whether conscious or brainless, intended or instinctual, altruism was never truly "pure."

But if science had painted a rather dour landscape of goodness, perhaps the spirit could transcend it after all. If ant workers helped their sisters only because it helped their genes; if a monkey helped another monkey only because he could cash in on the favor someday—perhaps man could do better. Perhaps *George* could do better. Inviting homeless strangers into his apartment was a beginning; in the vein of his constitutional extremism, giving them all he had and losing everything was the pushing of the envelope. If George's own mathematics described a world where selflessness was always selfish, perhaps in his own actions he could prove that in humans this wasn't necessarily so. If science could not provide the answer to the riddle of the origins of pure, universal goodness, if it could not even formulate the question necessary to fathom such mysterious depths, then perhaps George could find it elsewhere, in the fetid corners of Euston Station and the lonely benches of Soho Square. Perhaps pure selflessness resides in places science could never touch—in the unknown and unknowable recesses of the soaring human soul.

It was a courageous bid, a bold examination, and much was riding on it. But if this was indeed George's goal, then its outcome may best explain his demise. For in the final analysis selflessness only led to further despair: None of the homeless people he tried to help ever left the bottle, none returned to their families, none changed their ways. He himself had wallowed in misery, too sick and too hungry to be of any use even if he'd wanted to be. His own equation could reveal when altruism would evolve by benefiting the community despite being disadvantageous to the individual. But in his own life George had failed to find the balance.

This was bad enough, and yet it could always be chalked up to circumstance; coincidences, after all, were what defined the fundamentally incontrollable human experience. In truth a much more sinister realization was what had gotten to him, something that could not be a mere coincidence. It was a discovery infinitely more devastating. For no matter how much he wanted to forget himself, in the end he couldn't. How could he discern if his selflessness was not just a masquerade, the self fooling the self only to please the self and nothing more? How could he know, apart from the ant, the monkey, and all the other creatures that abound in nature, whether his goodness, *human* goodness, was really genuine and pure?[18]

He had been blessed with an unusual intelligence. He had seen things that the greatest minds had failed to see before him. And yet all his rational powers stood useless before the conundrum: It was impossible to know. Trying to transcend science he had found that he couldn't transcend biology: Despite the yearnings of his soul he was trapped in the prison of his brain. And, living in a squat with a broken window in the London winter, far away from his daughters, dejected, lonely, and weak, it may just have been a realization too difficult to bear. "Might go hay-wire but will never be humdrum," his Harvard interviewers had presciently divined back in 1940 when he arrived as a young hopeful—and they were right. In utter despair and utter anguish, George had finally, insolvably, hit "Wittgenstein's wall."

. . .

In 1996, Bill Hamilton wrote Kathleen Price a letter remembering his old friend. George's life, he explained, was like a novel, "kept exciting and unexpected right up to the last page."[19] Hamilton, who was to die at the height of his powers three years later from malaria-induced internal hemorrhaging following an ill-advised trip to the Congo to discover the origins of the virus responsible for AIDS, thought that George's life had been a "completed work of art." He may have been right. For, in a deep sense, what makes great works of art complete is that they remain forever incomplete. Explanations for events are at once myriad and mysterious; putting down a book, or walking away from a painting or a sculpture, or finishing listening to a piece of music, one always leaves with lingering thoughts that are neither questions nor answers. And so, whether George killed himself because of illness, unrequited love, confusion, or philosophical despair, his life and death continue to provide invaluable instruction. By crashing into "Wittgenstein's wall," George teaches us, like a great work of art, where the limits of our reason confront the depths of our soul.

Nothing makes this clearer than a sheet of paper Hamilton found among the discarded effects in the squat. Written by Richard W. De Haan, "teacher of the Radio Bible Classic, worldwide ministry through radio, television, literature," it was titled "Love is the Greatest!" and Hamilton paid little attention to it. But there was an important message there. "Men and women have always yearned for understanding, compassion, forgiveness, and deeds of loving-kindness from their fellowmen, but often they've been sadly disappointed," it read.

> *And today more than ever in a world torn by strife and dissension, the crying need is for a real demonstration of love. You see, love would pour the oil of quietness upon the troubled waters of human relationships, heal the ugly wounds of strife and contention, and bring together those separated by hatred, jealousy and selfishness. No wonder the apostle concludes the tremendous 13th chapter of 1 Corinthians by emphasizing that of all the gifts of the Spirit, including faith and hope, the greatest is love.*[20]

George Price lies in an unmarked grave in the Saint Pancras Cemetery in North London, flanked by a proud sycamore and a young weeping willow. Too weak or too ill or too heartbroken to follow this path peacefully, perhaps it was his ultimate and enduring message to the world he left behind.

Appendix 1:

Covariance and Kin Selection[1]

Price did not himself show how Hamilton's coefficient of relatedness in the narrow sense of common descent could be replaced with association. The first to do that was Hamilton in his 1970 paper. Later Queller expanded on this in a 1985 paper, "Kinship, Reciprocity and Synergism in the Evolution of Social Behaviour," clarifying further (if it had yet to be entirely clear) that association need not mean genetic relatedness in the narrow sense of descent via direct replication.[2]

To see how the covariance equation translates into Hamilton's kin-selection equation, we begin with the equation $w\Delta g = \text{Cov}\,(w,g)$ where g is the breeding value that determines the level of altruism. The least-squares multiple regression that predicts fitness, w, can be written as

$$w = \alpha + g\beta_{wg\acute{a}g'} + g'\beta_{wg\ddot{a}g+\varepsilon}$$

where g′ is the average g value of an individual's social neighbors, α is a constant, and ε is the residual which is uncorrelated with g and g′. The β are partial regression coefficients that summarize costs and benefits: $\beta_{wg\acute{a}g}$ is the effect an individual's breeding value has on its own fitness in the presence of neighbors' g′ (that is, the cost of altruism), and $\beta_{wg\ddot{a}g}$ is the effect of an individual's breeding value on the fitness of its neighbors (that is, the benefit of altruism). Substituting into the covariance equation and solving for the condition under which wΔg > 0 gives us Hamilton's inclusive fitness equation (rB > C) in the following form

$$\beta_{wg\acute{a}g'} + \beta_{g'g}\beta_{wg\ddot{a}g} > 0$$

where $\beta_{g'g}$ is the regression coefficient of relatedness.

This derivation, performed by Hamilton in 1970[3] following Price, clearly shows that it is natural to use statistical association instead of common descent. It is considered the first modern theoretical treatment of inclusive fitness.

Among other things, it allows us to see that spiteful behavior, where an organism acts in such a way as to harm itself in order to harm another organism even more, can evolve since the product of a negative relatedness and a negative benefit to the recipient (harm) is positive, meaning that benefit multiplied by relatedness can outweigh the cost.

Appendix 2:

The Full Price Equation

and Levels of Selection[1]

Unbeknownst to George Price in 1968, the simple covariance relationship

(1) $w\Delta z = Cov\ (w,z)$

had already been worked out and published independently by two other men, Robertson and Li, in 1966 and 1967 respectively.[2] But the uniqueness of the full Price equation stems from the inclusion of the further, expectation term. Here is the derivation:

Let there be a population in which each element is labeled by an index i. The frequency of elements with index i is q_i, and each element with index i has some character, z_i. Elements with a common index form a subpopulation that comprises a fraction q_i of the total population, and no restrictions are placed on how the elements are grouped.

Now imagine a second, descendant, population with frequencies q_i' and characters z_i'. The change in the average character value, $\bar{z}$, between the mother and descendant populations is

(2) $\Delta\bar{z} = \sum q_i' z_i' - \sum q_i z_i$

This equation applies to anything that evolves because z can be defined however one likes. For this reason, the equation applies not only to genetics, but to any selection process whatsoever.

What is special about the Price equation is the way in which it associates statistically between entities in groups, a "mother" and "daughter" population. Instead of the value of q_i obtaining from the frequency of elements with index i in the daughter population, it obtains from the proportion of the daughter population derived from the elements in with index i in the mother population. If we define the fitness of the element i as w_i, the contribution to the daughter population from type i in the mother population, then $q_i' = q_i w_i / \bar{w}$ where $\bar{w}$ is the mean fitness of the mother population.

The character values z_i' also use indices of the mother population. The value of z_i' is the average character value of the descendants of index i. The way it is obtained is by weighing the character value of each entity of the index i in the daughter population by the fraction of the total fitness of i that it represents. The change in character value for descendants of i is defined as $\Delta z_i = z_i' - z_i$.

Equation (2) holds with these definitions for q_i' and z_i'. With a few substitutions and rearrangements we derive:

$$\Delta \bar{z} = \sum q_i (w_i/\bar{w})(z_i + \Delta z_i) - \sum q_i z_i = \sum q_i (w_i/\bar{w}-1)z_i + \sum q_i (w_i/\bar{w})\Delta z_i$$

which, using standard definitions from statistics for covariance (Cov) and expectation (E) gives the full Price equation:

$$(3) \quad \bar{w}\Delta \bar{z} = \text{Cov}(w_i z_i) + \text{E}(w_i \Delta z_i)$$

The two terms on the right-hand side of the equation can be thought of as the selection and transmission terms, respectively. Covariance between fitness and character represents the change in the character due to differential reproductive success, whereas the expectation term is a fitness-weighted measure of the change in character valued between the mother and daughter populations. The full equation, therefore, describes both selective changes within a generation and the response to selection.

But the addition of the expectation terms also allows to expand the equation to show selection working at different levels. Here is how the equation can expand itself:

$$(4)\quad \bar{w}\Delta\bar{z} = \text{Cov}(u_iz_i) + E\{\text{Cov}_i(w_{ji}, z_{ji}) + E_j(w_{ji}\Delta z_{ji})\}$$

where E and Cov are taken over their subscripts where there is ambiguity, and $j \cdot i$ are subsets of the group i with members that have index j.

The recursive expansion of equation (3) shows that the transmission is itself an evolutionary event that can be partitioned into selection among subgroups and transmission of those subgroups. The expansion of the trailing expectation term can continue (say from gene, to individual, to group, to species) until no change occurs during transmission in the final level. Meanwhile, however, it will be possible to see how much of the change in trait is due to selection at each of the levels below.

Appendix 3:

Covariance and the

Fundamental Theorem

Consider the reduced form of the Price equation

$$w\Delta z = \text{Cov}\,(w,z) = \beta_{wz} V_z$$

where w is fitness and z is a quantitative character. The equation shows that the change in the average value of a character, Δz, depends on the covariance between the character and its fitness or, equivalently, the regression coefficient of fitness on the character multiplied by the variance of the character.

Since fitness itself is a quantitative character, z can be equal to fitness, w. Then the regression β_{wz} equals 1 and the variance, V_w is the variance in fitness. So the equation shows that the change in mean fitness, Δw, is proportional to the variance in fitness, V_w. This is what most people took Fisher's fundamental theorem to mean: The change in mean fitness of a population depends on the variance in fitness.

Price, however, showed that Fisher didn't mean this in a general but rather in a very specific way. Fisher had defined mean fitness in a way different than usually construed. For him it related only to that portion of fitness dependent on the additive genetic variance. All other components relevant to phenotypic variance—including epistasis and dominance—were categorized as "environment" and left out of the equation. Since it was known that epistatic and dominance effects can reduce fitness, along with the obvious fact that the environment can degrade, George interpreted Fisher's fundamental theorem as exactly true in its own terms, but not as biologically significant as Fisher had made it out to be.

Acknowledgments

I was sitting beneath a tree in a small park in Paris when I first read about George Price. It was 1999, and Andrew Brown's *The Darwin Wars: How Stupid Genes Became Selfish Gods* opened with a chapter called "The Deathbed of an Altruist." It was a mysterious story about a strange American who had come to London, written an equation that drove him first to Christianity, then to selfless aid to the homeless, and finally to suicide. There was some kind of connection between his equation and his life. What a story! I thought. This was the stuff of movies.

At the time I was in the middle of a doctorate and, returning to Oxford, forgot about Price. Seven years later in Jerusalem, while writing an essay review for the *New Republic* on Lee Dugatkin's book *The Altruism Equation: Seven Scientists' Search for the Origins of Goodness*, the memory came back to me, though the story had always remained in the back of my subconscious mind. Here was Price again, just as enigmatic as before, killing himself on the altar of altruism. But as with Brown, so too with Dugatkin: Price was afforded no more than a few pages. Was this all that was known of him?

Soon I learned that a few people had written about Price somewhat earlier. The source of all other accounts had been Bill Hamilton's autobiographical book *Narrow Roads of Gene Land*, in which the story of his meeting and subsequent dealings with Price was first revealed. Though he himself had not known Price, Steven Frank, a mathematical evolutionist, had also become interested in his life, and written an article about his scientific contributions to evolutionary genetics in the *Journal of Theoretical Biology* in 1995. Frank had even brought Price's "The Nature of Selection," apparently rejected for publication many years before, to posthumous print. So there were those who knew about him and cared.

But it wasn't until I came across Jim Schwartz's wonderful article "Death of an Altruist" in the now-defunct magazine *Lingua Franca* that I really became excited. The article was a beautiful exposition of the story, more detailed than anything else that had been written. Somewhat relieved that what had been referred to as "a biography" of Price was only ten pages long, I decided to contact Schwartz. Amazed at how little had been written of this incredible story, I held my breath and prayed that there were papers to work from. With any luck, I thought, there would be much more still to tell.

And so my first and heartfelt thanks go to Jim Schwartz. In preparing his *Lingua Franca* article Jim had contacted and met George's two daughters, Annamarie and Kathleen, and, as he put it to me over the phone, in the course of his research had "grown very close to George." Obviously George had been the kind of man who, even from the grave, could stir strong emotions, and it was clear that Jim had developed feelings for him. This makes it all the more special that in true gentlemanly fashion, and in an altruistic spirit of the common pursuit of history, Jim provided me with Kathleen and Annamarie's contact information, and, more important, with an introduction. Yes, there were papers, thank god. Jim wasn't certain that they'd be enough for a book, but I could try. Some early family material existed in boxes at the offices of the Edison Price Lighting Company in Queens. George's own personal papers had been rescued in January 1975 by representatives of the American Consulate in London. Others had been sent to Edison Price some weeks later by Hamilton, who'd fought his own way into the squat and come across more papers that the consulate men had overlooked. In 2003 Kathleen had donated a small portion of these papers to the British Library in

London, but the vast majority—notes, letters, diagrams, computer printouts, diaries—had remained in the homes of the two sisters. Over the years a substantial number of people had tried to reach the Price daughters to get their hands on his papers, and it wouldn't be easy gaining their confidence.

Soon I was conducting long transatlantic phone conversations from Tel Aviv with Kathleen in San Francisco and Annamarie in Solana Beach. Understandably they wanted to know who I was and what I wanted to do before opening up their father's treasure trove to a complete stranger. Gradually our conversations grew warmer, and in the spring of 2008 I traveled to California to meet Kathleen and Annamarie and finally consult the papers. That the papers provided a world of insight into the details of George's life, allowing me to reconstruct his story in its full richness, I hope the book demonstrates. This was a joyous discovery, and more than anything a huge relief. But what is more important to me is just how kind and inviting and warmly hospitable Kathleen and Annamarie proved. Sleeping in Kathleen's home and staying up late nights over wine discussing her memories, dining at Annamarie and her husband Ed's table and being trusted to share her deepest feelings and thoughts, rummaging through the papers together with both sisters in their respective dens, were more than any lucky historian can ever hope for. Like George, it was obvious, his daughters were special people, and it had been my good fortune to gain their trust. It would be my burden now to tell the story properly.

Precious little had been put on paper about George, but from the start I did not want this to be just a biography: This was going to be a history of a much larger quest. Already I had delved into the immense literature I would need to cover in order to write the book. Others had written about the problem of altruism and the evolution of cooperation, morality, and virtue. There were Richard Dawkins's classic *The Selfish Gene*, Robert Axelrod's *The Evolution of Co-operation*, and Matt Ridley's *The Origin of Virtue*. There were Marc Hauser's *Moral Minds* and Frans de Waal's *Good Natured: The Origins of Right and Wrong in Humans and Other Animals*. There was Helena Cronin's *The Ant and the Peacock*. And there were Elliott Sober and David

Sloan Wilson's *Unto Others: The Evolution and Psychology of Unselfish Behavior*, Richard Joyce's *The Evolution of Morality*, Samir Okasha's *Evolution and the Levels of Selection*, and many others, popular and academic. There were countless studies, articles, essays, and reviews. Some were philosophical accounts; others technical, theoretical, or empirical arguments; still others, like Cronin and Dugatkin, attempts at history. But the more I read the more I felt the need for a book that, in relation to altruism, would put history, science, biography, and philosophy all in one construction without the burden of blowing a particular horn. Most studies advocated one form or level of selection over another, one philosophy at the expense of its rivals, or were overimpressed with the ability of evolution, and a reductive genetics, to explain human goodness. It made it difficult to see the bigger picture, to appreciate the complexity and the nuances. Most acute was the lack of a good encompassing history. Certain pieces of the historical puzzle, like Adrian Desmond's magisterial *Huxley*, Gregg Mitman's wonderful *The State of Nature*, and Marek Kohn's stunningly beautiful *A Reason for Everything*, were in fact already in place. But that wasn't enough. The story of modern man's attempt to fathom kindness was a rich one, involving countless disciplines, characters, historical contexts, and scientific facts. It seemed to me that it had yet to be fully told.

Inspired by a novel I had read by the Peruvian author Mario Vargas Llosa, *The Storyteller*, I soon saw that the best way to tell the greater tale of attempts to crack the mystery of altruism, going back to Darwin, was to use George's life as a counterpoint. To get to the core of the problem, to really touch "Wittgenstein's wall," there was no better way than to let George's incredible story lead the way. By creating a double helix–like structure for the book, George's own personal tale and the greater problem of the evolution of altruism would reflect and resound off each other, finally becoming inextricably interwoven. Constructing the story this way, I could hope to show the extent to which scientific pursuits are embedded in the people and cultures that are responsible for them. Celebrating the majesty of science, I could also point to its limits; altruism—the meeting place of biology and culture—would be the perfect conduit. If, ultimately, there can be no scientific arbitration of the question of the existence of true, genuine selflessness in man, George's tragic life might be the closest one could come to understanding why.

Beyond my thanks to all the authors, scholars, and scientists on whose foundations I built to construct my story, I would also like to thank the many people who generously offered to answer my questions, whether in face-to-face interviews or via correspondence. Bill Hamilton's sister Janet and her husband, Rollin, generously hosted me in their Hampshire home, and I shall always remember the three of us by the fireplace in the evening listening to Tim Jackson's fictionalized BBC radio play about George, also titled *Death of an Altruist*, embers crackling in the fire. Earlier that day Janet had taken me to meet Tim, a professor of Sustainable Development at the University of Surrey and an award-winning radio playwright. I thank Tim, too, for a wonderful afternoon of discussion. His enthusiasm about George was contagious.

Thanks as well to Richard Lewontin, who devoted an afternoon and many subsequent emails to trying to explain Price's import and trying to figure out together where his scientific style or approach might have come from. The same goes for Warren Ewens, Danny Cohen, Ilan Eshel, Ariel Rubenstein, Joel Cohen, David Haig, and James Crow. I also want to thank Robert Trivers, Steven Frank, and Alan Grafen for answering important questions over the phone or in writing, and Mark Borrello, Ehud Lamm, Nathaniel Comfort, Marc Hauser, Jim Griesemer, Gar Allen, Michael Dietrich, Eva Jablonka, Marion Lamb, Stephen Stearns, David Kohn, Salit Kark, Alice Nicholls, Everett Mendelsohn, Razi Greenfield, and Ayelet Shavit for rich and informing conversations on the subject and, in some cases, for reading parts of the manuscript. Special thanks to Peter Godfrey Smith, who, following a lively conversation at a banquet in Brisbane, Australia, generously took the time to read the entire manuscript and provided invaluable comments. Daniel Kevles has been a warm critic and friend, and I thank him, together with Dudley Andrew and Howard Bloch, for the opportunity to discuss my ideas during a wonderful visit as the guest of the Whitney Humanities Center at Yale. Heartfelt thanks go to George C. Williams, sadly unwell but always kind spirited, for his friendly words of encouragement.

Those who knew George Price proved invaluable informants. M. Dan Morris and Richard A. Bader, Stuyvesant classmates of George, shared colorful memories of New York City in the 1930s. George's old friend from his Chicago days, Al Somit, met me in Solana Beach and shared wonderful stories before and after in correspondence; I thank his wife, Leyla, too, for her recollections. Former colleagues at Harvard's Department of Chemistry, Gilbert Stork and Leonard Nash, were kind enough to answer queries. At UCL, Ursula Mittwoch and Sam Berry knew George and kindly shared their memories, as did Newton E. Morton on C. A. B. Smith. Christine Hamilton was kind enough to share her memories of George's visits at her and Bill Hamilton's Berkshire home, including the very last one before his suicide. I thank Bill's close friend Peter Henderson for sending me a beautiful description of their time in the Brazilian jungles together, as well as Richard Dawkins for sharing a few memories of Bill and George.

No scholar can do without archives, and many people helped me in this respect. I've already mentioned Kathleen and Annamarie Price. A special further thanks goes to Emma Price, Edison Price's daughter, who welcomed me into her offices in Queens at the Edison Price Lighting Company and provided generous access to invaluable family letters and records. Jeremy Johns at the British Library proved most helpful in navigating through the Price, Hamilton, and Maynard Smith papers, and I thank him warmly for sending me photocopies and for his general kind spirit. Katherine T. Bendo of the Stuyvesant Alumni Association was immensely helpful to me in uncovering George's high school past. Barbara Meloni at Harvard University Archives, and David Pavelich and Allie Tichenor at Chicago University Archives were wonderful, as was Patricia Canaday at Argonne National Laboratory, Marcia Chapin at Harvard Chemistry, Dawn Stanford at IBM, and Karen Klinkenberg and Jake Williams at the University of Minnesota Medical School. In England the broad knowledge of Peter S. Harper helped me get a grip on the genetics scene at the time, and Brian Parson, funeral specialist, helped me understand how deaths were handled. Immensely helpful were the Tolmers gang, who with great candor and color shared memories both of George and of the times. I thank Nick Wates, Atalia Ten

Brink, Ches Chesney, Corrine Pearlman, Paul Nicholson, and Alon
Porat. Special thanks go to Asher Dahan and Shmulik Atia, with
whom I spoke at length about George's last days and suicide. Once
again I thank Jim Schwartz for generously sharing with me copies
of correspondence between George and Joan Jenkins that he himself had gotten from Jenkins before she died. Finally I thank Sylvia
Stevens who, when I tracked her down over the phone, after a long
silence, said, "You've rocked my world on a Friday afternoon!" Sylvia
shared sensitive memories of George toward the end of his life that
helped illuminate his mental state and gentle spirit

Life is no fun without friends, and luckily I have many of these
and really good ones. First of all, an enormous thank-you to Ben
Reis, my loyal and close friend, who was also throughout the writing
my sole and assiduous reader. This book is infinitely better due to his
extraordinary talents. Toda Ben Adam, and maks-ssim. Thanks too
to David Schisgall, who took the time to read the manuscript and
like a good friend told it to me like it is. My friend Noah Efron, a
prince among men, afforded me a sabbatical year from my teaching
duties at Bar Ilan University so that I could direct all my energies to
the writing, and, as always, was the most loving, smart, funny colleague anyone could ever have. I thank Eva Jablonka, too, for her
encouragement and wise words of counsel, as well as for her example
of how history, science and philosophy all need to be considered as
part of a whole.

The Halbans—Martine, Peter, Alexander, and Tania—were loving and fun hosts, always, on my frequent visits to London. Marty
Peretz has for years been a special friend—generous, caring, and hug-
gably irreverent—and via his introduction of me to Leon Wieseltier,
acted as something of a godfather to the project, which began in
miniature as a *New Republic* essay. Samantha Power not only encouraged me to write, among hippos on the Zambezi River, but also introduced me to her magnificent agent, Sarah Chalfant, who became my
friend and without whom none of this would have happened. Thank
you, Sam, and thank you, Sarah. I also am grateful to my old friends
Elizabeth Rubin, Maya Topf, and Gail'ad Zuckermann for early and
sustained conversations about what this book was going to be about,

and to my lifelong special New York families, Sam and Joann Silverstein and Hugh and Marilyn Nissenson, for always being in my heart. I am saddened that Erich Segal, whom I continue to love dearly along with Karen, Miranda, and Chessy, passed away before the book was published and long before he should have gone. Erich was blessed with a contagious lust for life, and a lovably mischievous twinkle. I always smile when I think of him.

Sue Llewellyn performed a masterful job copyediting, and I am extremely grateful for her hawk eyes and erudition, as well as for Don Rifkin's expert hand. In England, Kay Peddle at Random House was a fantastically perceptive editor and a friendly voice of encouragement. Thanks to Oren Dai, consummate professional photographer, ptitim-maker, and friend. My editor, Jack Repcheck, a wonderful writer in his own right, has been a constant source of wisdom and good cheer, and has also become a friend. Thank you, Jack—I couldn't have done it without you, and I look forward to many more years of work together at Norton.

Back home, the Organism continues to be a lifeline. Jeno ax yakar, my partner in crime: you are my source of sanity and insanity, shetavo alexa habraxa! My second family: the one and only Trulner Hakalil—shaderrrr; Nitzan Hakoks who did the unthinkable; Il Xamdelilah my animated chummus partner; Kata Ha'anak—friend, poet, intellectual; Fecht, the arak lover; Horror, Ace, Flotsenkranz, Salta and Shamna, and Ben Jaino ("aval lama bana'al?")—I love you all like brothers. To my smiling Taltal, thank you for your sweetness, for Purim, and for everything else. Abba and Imma: Thank you so much in everything, for everything, above everything—my most adorable, telephone enthusiast, deeply loving parents. Finally, to my sister, Danz, and brother, Mish, to whom this book is dedicated: You are the two most special, giving, and in a deep sense selfless people I know, and I love you very very much.

Notes

PROLOGUE

1. When Bill Hamilton wrote about the event twenty years after it happened, his memory played tricks on him. See *Narrow Roads of Gene Land* (Oxford: W. H. Freeman/Spektrum: 1996), 325; Saint Pancras Cemetery Registrar files, January 22, 1975; "Weather Report," *Daily Telegraph*, January 22, 1975.
2. I thank Ursula Mittwoch for her memories of the funeral and Martin Collier of the Camden Register Office for invaluable information.
3. The story about Saint Paul was recounted by John Maynard Smith to Ullica Segerstråle, *Defenders of the Truth: The Sociobiology Debate* (Oxford: Oxford University Press, 2000), 68.
4. Genesis 3:5; 4:9.
5. Charles Darwin, *Notebook M*, 1838.
6. Jack Repcheck, *The Man Who Found Time: James Hutton and the Discovery of the Earth's Antiquity* (Perseus, 2003), 6. Charles Darwin, *The Origin of Species*, 1st and 2nd eds. (London: John Murray, 1859).
7. For a description of the few partial accounts of George Price's life and science, see the acknowledgments.

CHAPTER 1: WAR OR PEACE?

1. Peter Kropotkin, *Memoirs of a Revolutionist* (London: Folio Society, 1978), 232.
2. Ibid., 232, 234.
3. Adrian Desmond, *Huxley: From Devil's Disciple to Evolution's High Priest* (Harmondsworth: Penguin, 1997), 9, 11; Leonard Huxley, ed., *Life and Letters of Thomas Henry Huxley*, 2 vols. (London: Macmillan, 1900); Julian Huxley, ed., *T. H. Huxley's Diary of the Voyage of H.M.S. Rattlesnake* (London: Chatto and Windus, 1935).
4. Kellow Chesney, *The Victorian Underworld* (London: Temple Smith, 1970); Thomas Henry Huxley, "The Struggle for Existence in Human Society: A Programme," *Nineteenth Century* 23 (February 1888),161–80, reprinted in T. H. Huxley, *Collected Essays* (London: Macmillan, 1883–84), 195–236, quote on 217; Desmond, *Huxley*, 11, 84.

5. Desmond, *Huxley*, 443–44.
6. Ibid., 361.
7. Kropotkin, *Memoirs*, 10, 238, 239.
8. Ibid., 237–40.
9. Quoted in Desmond, *Huxley*, xv. On the reception of Darwinism in England see Peter Bowler, *Evolution: The History of an Idea* (Berkeley: University of California Press, 1989), 187–245.
10. The "X Club" was a dining club convened by Huxley and eight other like-minded supporters of Darwin's theory of evolution and academic liberalism. It met regularly once a month between 1864 and 1893, and wielded a powerful influence over British science. See Roy M. MacLeod, "The X-Club: a Social Network of Science in Late-Victorian England," *Notes and Records of the Royal Society of London* 24, no. 2 (April 1970): 305–22.
11. Ibid. 424–25; V. Kovalevskii, "On the Osteology of the Hyopotamidae," *Philosophical Transactions of the Royal Society* 163 (1873), 19–94. See also Desmond Adrian, *The Politics of Evolution: Morphology, Medicine and Reform in Radical London* (Chicago: University of Chicago Press, 1989).
12. Besides Kropotkin's own *Memoirs* see George Woodcock and Ivan Avakumovich, *The Anarchist Prince: A Biographical Study of Peter Kropotkin* (London: T.V. Boardman and Co., 1950), and the more scholarly Martin A. Miller, *Kropotkin* (Chicago: University of Chicago Press, 1976).
13. Colin Ward, Introduction to *Memoirs of a Revolutionist*, by Peter Kropotkin (London: Folio Society, 1978), 8.
14. Kropotkin, *Memoirs*, 56; Ivan Sergeyevich Turgenev, *Mumu* (Moscow: Detgiz, 1959).
15. This was the title of an influential treatise by Chernyshevsky, and later of others by Tolstoy and Lenin.
16. Kropotkin, *Memoirs*, 35.
17. Ibid., 111, 80, 126.
18. Charles Darwin, *The Voyage of the Beagle* (Hertfordshire: Wordsworth, 1997).
19. E. Paley, ed., *The Works of William Paley*, 6 vols. (London: Rivington, 1830).
20. Charles Darwin, Zoological Notes, 1835. Much of Darwin's written corpus now exists on the Web at http://.darwin-online.org.uk, thanks to the labors of John van Whye and collaborators at Cambridge University.
21. Some historians have questioned Darwin's own claims about Malthus's influence on him. See in particular Silvan S. Schweber, "The Origin of the *Origin* Revisited," *Journal of the History of Biology* 10 (1977), 229–316. On England's leading evolutionists' appreciation of Malthus see Robert M. Young, "Malthus and the Evolutionists: The Common Context of Biological and Social Theory," in Robert M. Young, *Darwin's Metaphor: Nature's Place in Victorian Culture* (Cambridge: Cambridge University Press, 1985), 23–55.
22. Charles Darwin, *The Autobiography of Charles Darwin* (New York: W. W. Norton, 1993), 120; Charles Darwin, *The Origin of Species*, 2nd ed. (Oxford: Oxford University Press, 1996), 6.
23. Darwin, *Voyage of the Beagle*, 228–29; 471.
24. Darwin, *The Origin of Species*, 396; letter to J. D. Hooker (January 11, 1844), *The Correspondence of Charles Darwin*, vol. 3 (1844–46), ed. Frederick Burkhardt (Cambridge: Cambridge University Press, 1987), 2.
25. Kropotkin, *Memoirs*, 94
26. Ibid.
27. Ibid., 157.

28. Ibid., 201, 202. The best book on Russian anarchism is Paul Avrich, *The Russian Anarchists* (Princeton: Princeton University Press, 1967).

29. Helena Cronin, *The Ant and the Peacock* (Cambridge: Cambridge University Press, 1991).

30. *The Origin of Species*, 197.

31. A pre-*Origin* phrase coined by Alfred, Lord Tennyson, *In Memoriam A. H. H.* (1850).

32. Thomas Dickson has recently argued that "altruism," a term coined by the French positivist sociologist Auguste Comte in the early 1850s, filtered through Victorian culture in interesting ways. Contra most previous interpretations he claims that Darwin saw sympathy and love, not only selfishness and competition throughout the natural world. See Thomas Dickson, *The Invention of Altruism: Making Moral Meanings in Victorian Britain* (Oxford: British Academy Postdoctoral Fellowship Monographs, 2008).

33. Darwin also spoke of the "family" as a beneficiary, but "community" is the term he used more often.

34. *The Origin of Species*, 196, 392, 164.

35. Charles Darwin, *The Descent of Man and Selection in Relation to Sex*, 1st ed. (Princeton: Princeton University Press, 1871), 103.

36. Kropotkin, *Memoirs*, 254.

37. Ibid., 258.

38. Huxley letter to Frederick Dyster, November 25, 1887, quoted in Desmond, *Huxley*, 557; Ronald W. Clark, *The Huxleys* (London: Heinemann, 1968), 109. Huxley's first child, Noel, had died, aged four, in 1860.

39. He had won the Royal, the Wollaston, and the Clarke; the Copley, the Linnaean, and the Darwin still awaited him.

40. Desmond, *Huxley*, 572-73.

41. Huxley used this term in a lecture invited by the Prince of Wales at Mansion House, January 12, 1887; *The Times*, March 18, 1888; *Nature* 37 (1888). 337-38.

42. On social Darwinism see Richard Hofstadter, *Social Darwinism in American Thought* (Boston: Beacon Press, 1955); Mike Hawkins, *Social Darwinism in European and American Thought, 1860–1945* (Cambridge: Cambridge University Press, 1997); Peter Dickens, *Social Darwinism: Linking Evolutionary Thought to Social Theory* (Philadelphia: Open University Press, 2000); Paul Crook, *Darwin's Coat-Tails: Essays on Social Darwinism* (Peter Lang, 2007).

43. Herbert Spencer, "Progress: Its Law and Causes," *Westminster Review* 67 (April 1857).

44. Desmond, *Huxley*, 184. On Spencer see James G. Kennedy, *Herbert Spencer* (Boston: G. K. Hall & Co., 1978); Jonathan H. Turner, *Herbert Spencer: A Renewed Appreciation* (Sage Publications, Inc., 1985); Michael W. Taylor, *Men versus the State: Herbert Spencer and Late Victorian Individualism* (Oxford: Oxford University Press, 1992); James Elwick, "Herbert Spencer and the Disunity of the Social Organism," *History of Science* 41 (2003), 35–72; and Mark Francis, *Herbert Spencer and the Invention of Modern Life* (Newcastle: Acumen Publishing, 2007).

45. Coined in Spencer's *Principles of Biology* (London: Williams and Norgate, 1864).

46. Henry George, *Progress and Poverty: An Inquiry into the Cause of Industrial Depressions, and of Increase of Want with Increase of Wealth—The Remedy* (London: Kegan Paul, 1885).

47. Wallace even wrote a textbook titled *Darwinism: An Exposition of the Theory of Natural Selection with Some of Its Applications*, 2nd ed. (1889; reprint, New York: AMS Press, 1975).

48. Alfred Russel Wallace, "Human Selection," *Fortnightly Review* 48 (1890), 325–37. On Wallace see Martin Fichman, *An Elusive Victorian: The Evolution of Alfred Russel Wallace* (Chicago: University of Chicago Press, 2004); Ross A. Slotten, *The Heretic in Darwin's Court: The Life of Alfred Russel Wallace* (New York: Columbia University Press, 2004); *In Darwin's Shadow: The Life and Science of Alfred Russel Wallace* (Oxford: Oxford University Press, 2002).

49. For Darwin, humans had reversed the usual "female choice" in nature, making females the "chosen" and males the ones choosing.

50. Desmond, *Huxley*, 576.

51. Darwin used this image already in his essay from 1844, but it was made public in C. R. Darwin and A. R. Wallace, "On the Tendency of Species to Form Varieties; and on the Perpetuation of Varieties and Species by Natural Means of Selection," *Journal of the Proceedings of the Linnaean Society of London. Zoology* 3 (1858): 46–50.

52. Huxley notes for Manchester address, quoted in Desmond, *Huxley*, 558.

53. Huxley to Foster, January 8, 1888, in Leonard Huxley, *The Life and Letters of Thomas Henry Huxley* (London: Macmillan, 1900), 198. Quoted in Lee Dugatkin, *The Altruism Equation: Seven Scientists Search for the Origins of Goodness* (Princeton: Princeton University Press, 2006), 19.

54. "The Struggle," in Huxley, *Collected Essays*, 197, 198, 199, 200.

55. Ibid., 198, 199, 200.

56. Ibid., 204, 205. For a clear reading of Huxley on the evolution of sociality in man see Raphael Falk's review of Michael Ruse, ed., *Thomas Henry Huxley: Evolution and Ethics* (Princeton: Princeton University Press, 2009), in *Philosophia* 38, no. 2 (June 2010).

57. Ibid., 212, 235.

58. Daniel P. Todes, "Darwin's Malthusian Metaphor and Russian Evolutionary Thought, 1859–1917," *Isis* 87 (1987), 537–51, quotes on 539–40. See his broader treatment in *Darwin Without Malthus: The Struggle for Existence in Russian Evolutionary Thought* (Oxford: Oxford University Press, 1989). Also Stephen Jay Gould, "Kropotkin Was No Crackpot," in *Bully for Brontosaurus* (Harmondsworth: Penguin, 1991), 325–39.

59. Quoted in Todes, "Darwin's Malthusian Metaphor," 542.

60. On Malthus see Patricia James, *Population Malthus: His Life and Times* (London: Routledge and Kegan Paul, 1979); Samuel Hollander, *The Economics of Thomas Robert Malthus* (Toronto: University of Toronto Press, 1997); William Peterson, *Malthus, Founder of Modern Demography*, 2nd ed. (New Brunswick, NJ: Transaction, 1999); J. Dupâquier, "Malthus, Thomas Robert (1766–1834)," *International Encyclopedia of the Social and Behavioral Sciences*, 2001, 9151–56.

61. Todes, "Darwin's Malthusian Metaphor," 542, 540, 541–42. See also Thomas F. Glick, ed., *The Comparative Reception of Darwinism* (Chicago: the University of Chicago Press, 1988), 227–68. Also Engels letters to Lavrov, November 12–17, 1875, available at the Marx/Engels Internet Archive, http://marxists.org.

62. Darwin, *Origin of Species*, 53 [italics added].

63. Peter Kropotkin, "Aux Jeunes Gens," *Le Révolté*, June 25, July 10, August 7, 21.

64. Kropotkin, *Memoirs*, 261–338.

65. Peter Kropotkin, "Charles Darwin," *Le Révolté*, April 29, 1882.

66. "Without entering," he added, "the slippery route of mere analogies so often resorted to by Herbert Spencer." Peter Kropotkin, "The Scientific Basis of Anarchy," *Nineteenth Century* 22, no. 126 (1887), 149–64, quotes on 238.

67. Ibid., 239.

68. Desmond, *Huxley*, 599. The oft-quoted phrase is: "the ethical progress of society depends, not on imitating the cosmic process [evolution] still less in running away from it, but in combating it," T. H. Huxley, "Evolution and Ethics," in *Evolution and Ethics and Other Essays* (New York: D. Appleton and Company, 1898), 83; Desmond, *Huxley*, 598.

69. T. H. Huxley, "Evolution and Ethics: Prolegomena," in *Evolution and Ethics and Other Essays*, 25, 27, 30. Smith had spoken of the "man within" in his *Theory of the Moral Sentiments* (London: A. Miller, 1759; 1790), part 3, ch. 3; *Oxford Magazine* 11 (May 24, 1893), 380–81, quoted in Desmond, *Huxley*, 598.

70. Peter Kropotkin, *Mutual Aid: A Factor in Evolution* (1902; reprint, Boston: Extending Horizons Books, 1955) 4. Kropotkin thought Darwin, especially in his *Descent of Man*, had emphasized the role of cooperation, whereas his followers took to the narrower definition of the struggle for existence. "Those communities," Darwin wrote in the *Descent*, "which included the greatest number of the most sympathetic members would flourish best, and rear the greatest number of offspring" (2nd ed., 163). "The term," Kropotkin added, "thus lost its narrowness in the mind of one who knew Nature."

71. This was George Bernard Shaw's description of Kropotkin; *Mutual Aid*, xiv; Desmond, *Huxley*, 564.

72. Kropotkin, *Mutual Aid*, 14, ix. Professor Karl Fedorovich Kessler, rector of St. Petersburg University and chair of its Department of Zoology, had made this claim already in December 1879, at a talk before the St. Petersburg Society of Naturalists. Actually Kessler was to Kropotkin what Malthus had been to Darwin: his "law of mutual aid"—which Kropotkin read in exile in 1883—having struck him as "throwing new light on the subject"; Kropotkin, *Mutual Aid*, x. Other "mutual aid" supporters in Russia included M. N. Bogdanov, A. N. Beketov, A. F. Brandt, V. M. Bekhterev, V. V. Dokuchaev, and I. S. Poliakov; see Todes, "Darwin's Malthusian Metaphor," 545–46.

73. Kropotkin, *Mutual Aid*, 51, 40. 60–61; Kropotkin freely admitted that there was much competition in nature, and that this was important. But intraspecies conflict had been exaggerated by the likes of Huxley; it also often left all combatants bruised and reeling. True progressive evolution was due to the law of mutual aid.

74. Ludwig Büchner, *Liebe und Liebes-Leben in der Thierwelt* (Berlin: U. Hofmann, 1879); Henry Drummond, *The Ascent of Man* (New York: J. Pott and Company, 1894); Alexander Sutherland, *The Origin and Growth of the Moral Instinct* (London: Longmans Green and Company, 1898). Kessler, too, thought that mutual aid was predicated on "parental feeling," a position from which Kropotkin was careful to detach himself. See *Mutual Aid*, x.

75. Kropotkin marshaled evidence from varied sources, especially liberally interpreted archaeological evidence, to argue that man's "natural" state was in small, self-sustaining, communal groups.

76. Kropotkin, *Mutual Aid*, xiii.

77. Kropotkin, *Mutual Aid*, 75.

78. *Daily Telegraph*, July 5, 1895, quoted in Desmond *Huxley*, 612.

79. Kerensky was there, and offered him a ministry in the new government, which Kropotkin declined. Still, he did become active in party politics from the outside. See Miller, *Kropotkin*, 232–37.

80. P. A. Kropotkin, *Selected Writings on Anarchism and Revolution*, ed. Martin A. Miller (Cambridge, Mass.: MIT Press, 1970), 324–33, quote on 327. The meeting was recorded by Vladimir Bonch-Bruevich, an acquaintance of Kropotkin and close associate of Lenin.

81. An attempt to lay the foundations of a morality free of religion and based on nature, *Ethics* was published posthumously in 1922.
82. Kropotkin, *Selected Writings*, 336.

CHAPTER 2: NEW YORK

1. Intimations of this meeting exist in a letter from William Edison Price to Alie Avery, June 12, 1914, Edison Price Lighting Company Family Archive (EPFA).
2. *A Catalogue of Theatrical Lighting Equipment and Effects* (New York: Display Stage Lighting Co., Inc., 1923), quotes on 3, 50.
3. On Belasco see Craig Timberlake, *The Bishop of Broadway: The Life and Work of David Belasco* (New York: Library Publishers, 1954). *Madame Butterfly* was written by John Luther Long. A second Belasco play, his own *The Girl of the Golden West* (1905), was also adapted for opera by Puccini.
4. "An Old Friend Back at Belasco; Where Warfield in 'The Auctioneer' Charms with Blend of Comedy and Pathos," *New York Times,* October 1, 1913.
5. "Mrs. Emma Addale Gage Avery: Once An Esteemed Teacher in Big Rapids," *Bellevue Gazette*, March 26, 1925.
6. Oberlin College, the first to admit women, was founded by the progressive Reverend John J. Shipherd some years before he and a group of Congregational missionaries came north to create a college in the midst of the wilderness of southern Michigan. Recalling that the Mount of Olives had been a place of learning and contemplation, they named the college and the town they built around it "Olivet" and strove for a "harmonious Christian community." See Michael K. McLendon, "Olivet College: Reinventing a Liberal Arts Institution" (research paper, Center for the Study of Higher and Postsecondary Education at the University of Michigan, 2000).
7. Letter of recommendation for French teacher Clara Avery from W. N. Ferris To Whom It May Concern, July 18, 1904, EPFA; Letter of recommendation for Principal Clara Avery from G. A. Vail, May 17, 1905, EPFA.
8. The Preis family name is not a certainty, but both Annamarie Price and Kathleen Price, as well as their cousin, Edison's daughter, Emma Price, believe that this might have been the case.
9. The censuses of all major American cities can be found at ancestry.com. W. E. Price's information is in the Chicago 1900 and 1910 census, Wards 8 and 12, respectively, and in the New York 1920 census, ward 13. I am grateful to Annamarie Price for this information.
10. Wedding invitation, Clara to Mr. William Edison Price, EPFA.
11. *Morning Telegraph*, June 4, 1914.
12. The names of the businesses with which Display had contracts are detailed in the report of Saul Levy, Certified Public Accountant, February 29, 1928, EPFA.
13. Belle Hecht letter to Alice and William Edison Price, October 20, 1922, EPFA.
14. "List of Patents Granted to W.E. Price & D.S.L.Co," EPFA.
15. John Frederick Cone, *Oscar Hammerstein's Manhattan Opera House* (Norman: University of Oklahoma Press, 1966).
16. "Ordo Ab Chao: Suprema Council" (certification of William Edison Price joining the Masons as "Prince of the Royal Secret 32nd"), March 24, 1925, EPFA.
17. Emma Avery letter to Clara Avery (her family continued to call Alice by her given name), undated, EPFA.
18. "William E. Price Dies: South Broadway Resident Had Revolutionized Theatrical

Lighting With His Inventions," February 22, 1927, unidentified newspaper clipping, EPFA.

19. "Brokers Believe Worst Is Over and Recommend Buying of Real Bargains," *New York Herald Tribune*, October 27, 1929; "Very Prosperous Year Is Forecast," *World*, December 15, 1929.

20. Keynes was quoted in the *New York Evening Post*, October 25, 1929; Philip Snowden was quoted in the *Wall Street Journal*, October 4, 1929. On the crash see Harold Bierman, Jr., *The Causes of the 1929 Stock Market Crash* (Westport, CT: Greenwood Press, 1998), John Kenneth Galbraith, *The Great Crash, 1929* (Boston: Houghton Mifflin, 1961) and Irving Fisher, *The Stock Market Crash and After* (New York: Macmillan, 1930) for different, time-lapsed perspectives.

21. See Suzanne R. Wasserman, "The Good Old Days of Poverty: The Battle Over the Fate of New York City's Lower East Side During the Depression" (Ph.D. dissertation, New York University, 1990); also Thomas Kessner, *Fiorello H. La Guardia and the Making of Modern New York* (New York: McGraw-Hill, 1989).

22. Alice Avery Price letter to Mr. Zagat, Sunlight Realty Co., August 16, 1933, EPFA; Clara Price letter to Remco Real Estate, April 28, 1933, EPFA; Alice Avery letter to Mr. Zagat, February 14, 1933, EPFA; Rosendale and Cohen Counselors at Law letter to Mrs. Price, April 13, 1932, EPFA; Alice Price letter to Dr. Ralph Singer, June 18, 1934, EPFA; "Final Notice Before Suit," *New York Times* letter to Mrs. Alice Avery, November 13, 1934, EPFA.

23. Alice Avery Price letter to Remco Real Estate, Dec. 26, 1934, EPFA.

24. Alice Avery Price letter to Mr. Kelly, Transfer Tax Commission, Jan 30, 1935, EPFA.

25. Edison Price letter to Mr. Elias Aaronson, December 9, 1936, EPFA; Alice Avery Price letter to Edison, undated, EPFA.

26. Wasserman, "The Good Old Days of Poverty."

27. Alice Avery Price letter to Chief Inspector George G. Henry, September 14, 1934, EPFA; Alice letter to Transfer Tax Commission.

28. Gage H. Avery letter to Clara Avery, January 27, 1932, EPFA.

29. Alice Avery Price letter to Mr. Barzo, Nov. 16, 1932, EPFA.

30. See Robert C. Allen, *Horrible Prettiness: Burlesque and American Culture* (Chapel Hill: University of North Carolina Press, 1991); Michelle Baldwin, *Burlesque and the New Bump-and-Grind* (Golden, CO: Speck Press, 2004), especially 1–16.

31. "Moss Weighs Ban on 14 Burlesques," *New York Times*, April 30, 1937; "La Guardia Backs Ban," *New York Times*, May 3, 1937.

32. Alice Avery Price letter to Edison, July 9, 1936, EPFA.

33. See Robert A. Caro, *The Power Broker: Robert Moses and the Fall of New York* (New York: Knopf, 1974), and Kenneth T. Jackson and Hillary Ballon, eds., *Robert Moses and the Modern City: The Transformation of New York* (New York: W. W. Norton, 2007).

34. A good description of New York City in the 1930s can be found in Alan Greenspan, *The Age of Turbulence* (New York: Penguin Press, 2007), 19–37.

35. For an endearing fictional account of this development, see Michael Chabon, *The Amazing Adventures of Kavalier & Clay* (New York: Picador, 2000), 74–77.

36. Alice Avery Price letter to Edison, undated, circa 1936, EPFA.

37. Numerous letters between Alice Avery Price and Edison Price, 1934–38, EPFA.

38. "Louise Birch, 97, Dead; Co-Founder of School," *New York Times*, September 29, 1976; Birch Wathen yearbook (1935), 16. Two of the school's most celebrated alumni in later years would be Barbara Walters and Brooke Shields.

39. "The History of Greek Temple Architecture," by George Price, George Price Papers (GPP).

40. Alice Avery Price letter to Edison Price, July 5, 1938, EPFA; Alice Avery Price letter to Miss Louise Birch, September 6, 1938, EPFA.

41. Susan E. Meyer, *Stuyvesant High School: The First 100 Years* (New York: Campaign for Stuyvesant/Alumni and Friends Endowment Inc., 2005), 10.

42. See Diane Ravitch, *The Great School Wars: New York City 1805–1973* (New York: Basic Books, 1974).

43. Meyer, *Stuyvesant*, 14, 11, 15.

44. *New York Times*, October 27, 1907.

45. Interview with Richard A. Bader, Stuyvesant class of June 1940, May 7, 2008; *Stuyvesant High School Yearbook*, June 1940; Meyer, *Stuyvesant*, 24, 16–17.

46. Interview with Dan Morris, Stuyvesant class of June 1940, May 2, 2008.

47. Ibid.

48. Lewis Mumford, *Sketches from Life: The Early Years* (New York: Dial Press, 1982), quoted in Meyer, *Stuyvesant*, 45.

49. "'Til We Meet Again," *Indicator* (school yearbook), June 1940, 102; interview with Richard A. Bader, May 7, 2008.

50. W. H. Bradshaw, "The Citadel," *Indicator*, June 1940, 7.

51. Sinclair J. Wilson, "Lost Horizon," *Indicator*, June 1940, 13; interviews with Dan Morris, May 2, 2008, and with Richard A. Bader, May 7, 2008.

52. George Price letter to Abe Shlemewitz, August 15, 1940, GPP; George Price letter to Mr. Sternberg, August 17, 1940, GPP; "Harvard College, Principal's Report on Applicant," March 15, 1940, Harvard University Archive (HUC).

53. Greenspan, *The Age of Turbulence*, 24, 25.

54. David Gelernter, *1939: The Lost World of the Fair* (New York: HarperPerennial, 1996); "Stuyvesant High School Permanent Record—George R. Price," June 30, 1940, Stuyvesant Archives; "Harvard College Freshman Scholarship Application," March 15, 1940, HUA; "Harvard College Freshman Scholarship Personal Interview Report," May 24, 1940, HUA.

CHAPTER 3: SELECTIONS

1. Joan Fisher Box, *Ronald A. Fisher: The Life of a Scientist* (New York: John Wiley & Sons, 1978).

2. Charles Darwin, *The Autobiography of Charles Darwin*, ed. Nora Barlow (New York: W. W. Norton, 1993), 58.

3. Box, *Ronald A. Fisher*, 13.

4. A. W. F. Edwards, "The Genetical Theory of Natural Selection," *Genetics* 154 (2000), 1419–26, quote on 1420.

5. "Regression to the mean" was Galton's terminology, not Jenkin's. Fleeming Jenkin, "Review of Darwin's Origin of Species," *North British Review* 46 (1867), 277–318. Darwin's solution was to fall back increasingly on the inheritance of acquired traits, and to posit unseen "pangenes" whose very existence even he described as "provisional."

6. Peter Bowler, *The Eclipse of Darwinism: Anti-Darwinian Evolution Theories in the Decades around 1900* (Baltimore: Johns Hopkins University Press, 1992); Charles Darwin, *The Origin of Species* (Oxford: Oxford University Press, 1996), 70; William Provine, *The Origins of Theoretical Population Genetics* (Chicago: University of Chicago Press, 1971).

7. Quoted in Marek Kohn, *A Reason for Everything: Natural Selection and the English Imagination* (London: Faber and Faber, 2004), 95. The portraits of Fisher, Haldane, Maynard Smith, and Hamilton in this chapter and chapter 7 borrow generously from Kohn's wonderful book.

8. R. A. Fisher, "Some Hopes of a Eugenist," *Eugenics Review* 5 (1914), 309–15. Many contemporary eugenicists focused on "negative eugenics," meaning steps to be taken to contain the procreation of the weak and unsocial elements in society, such as sterilization. Fisher preferred "positive eugenics," namely steps to be taken to encourage the procreation of the fitter elements of society, such as family allowances and tax-cut incentives.

9. Kohn, *A Reason for Everything*, 101.

10. R. A. Fisher, "The Correlation Between Relatives on the Supposition of Mendelian Inheritance," *Transactions of the Royal Society of Edinburgh* 52 (1918), 399–433.

11. "Correlation Between Relatives" was not strictly about evolution, but it showed that trait similarity could be explained by recourse to Mendelian genetics, rendering clear the implications for evolution, at least for Fisher if not at first for others.

12. Udney G. Yule had made the same claim back in 1902 but went unheeded. See his "Mendel's Laws and Their Probable Relations to Intraracial Heredity," *New Phytologist* 1 (1902), 193–207, 222–38.

13. Kohn, *A Reason for Everything*, 142. See also Ronald W. Clark, *J. B. S.: The Life and Work of J. B. S. Haldane* (Oxford: Oxford University Press, 1963).

14. See Martin Goodman, *Suffer and Survive: Gas Attacks, Miners' Canaries, Spacesuits and the Bends: The Extreme Life of J. S. Haldane* (London: Pocket Books, 2007).

15. On J. S. Haldane's philosophy see S. Sturdy, "Biology as Social Theory: John Scott Haldane and Physiological Regulation," *British Journal of the History of Science* 21 (1988), 315–40. For J. S. Haldane's own writings see *The Philosophical Basis of Biology* (Garden City, NY: Doubleday, 1931), and *The Philosophy of a Biologist*, 2nd ed. (Oxford: Clarendon Press, 1935).

16. "I am enjoying life here very much," he wrote to his father in February 1915. "I have got a most ripping job as a bomb officer." To his mother he wrote: "I find this sort of fighting very enjoyable."

17. Haldane's wartime paper showing the genetic linkage between albinism and pink eyes in mice was written with his sister Naomi and his friend A. D. Sprunt, who was killed in battle before it was published: J. B. S. Haldane, A. D. Sprunt, and N. H. Haldane, "Reduplication in Mice," *Journal of Genetics* 5 (1915), 133–35; J. S. Haldane is credited with inventing the gas mask that saved countless Allied lives in World War I.

18. Kohn, *A Reason for Everything*, 149.

19. "J.B.S. was against—against authority, and against the government, any authority and any government; if possible in the cause of reason; if not as a matter of principle," Clark, *The Life and Work of J. B. S. Haldane*, 20.

20. John Herschel, *Physical Geography* (Edinburgh: Adam and Charles Black, 1861), 12.

21. For a perspective on Fisher's statistical accomplishments and legacy see Anders Hald, *A History of Mathematical Statistics 1750 to 1930* (New York: Wiley, 1998).

22. Edwards, "The Genetical Theory," 1423.

23. Interesting appreciations of *The Genetical Theory of Natural Selection* can be found in Edwards, "The Genetical Theory"; James F. Crow, "R. A. Fisher, a Centennial View," *Genetics* 124 (1990), 207–11, and "Fisher's Contributions to Genetics and

Evolution," *Journal of Theoretical Biology* 38 (1990), 263–75; E. G. Leigh, Jr., "Ronald Fisher and the Development of Evolutionary Theory," in *Oxford Surveys in Evolutionary Biology*, vol. 3, ed. Richard Dawkins and Mark Ridley (London: Oxford University Press, 1986), 187–223; Richard Lewontin, "Theoretical Population Genetics in the Evolutionary Synthesis," in *The Evolutionary Synthesis: Perspectives on the Unification of Biology*, ed. Ernst Mayr and William B. Provine (Cambridge, MA: Harvard University Press, 1980), 58–68; and in the foreword to the Variorum Edition of *Fisher's Genetical Theory of Natural Selection*, J. H. Bennett (Oxford: Oxford University Press, 1999). Quotation in Edwards, "The Genetical Theory," 1422.

24. Attempts to understand the fundamental theorem, and George Price's solution, will be discussed in chapter 10.

25. On Fisher's notion of progress see Michael Ruse, *Monad to Man: The Concept of Progress in Evolutionary Biology* (Cambridge, MA: Harvard University Press, 1996), 295–303.

26. Quoted in Kohn, *A Reason for Everything*, 96.

27. Friedrich Nietzsche, *On the Genealogy of Morals* (1887), trans. Douglas Smith (Oxford: Oxford University Press, 1996); Charles Darwin, *M Notebook*, 1838.

28. R. A. Fisher, "The Evolution of the Conscience in Civilized Communities," *Eugenics Review* 14 (1922), 190–93.

29. R. A. Fisher, "The Renaissance of Darwinism," *The Listener* 37 (1947), 1001, 1009, quoted in Kohn, *A Reason for Everything*, 108; R. A. Fisher, "Indeterminism and Natural Selection," *Philosophy of Science* 1 (1934), 99–117.

30. Cyril Darlington, "Recollections of Haldane" (draft), Darlington Papers (DP): C. 108: J. 86.

31. Quoted in Clark, *The Life and Work of J. B. S. Haldane*, 160.

32. Ibid., 115.

33. J. B. S. Haldane, *The Inequality of Man and Other Essays* (London: Chatto and Windus, 1932), quote on 121. On the culture and politics of British science at the time see William McGuken, *Scientists, Society, and State: The Social Relations of Science Movement in Great Britain 1931–1947* (Columbus: Ohio State University Press, 1984).

34. J. B. S. Haldane, *Possible Worlds and Other Essays* (London: Harper and Brothers, 1928), 220–21. See also Charlotte Haldane's *Truth Will Out* (London: Weidenfeld and Nicolson, 1949) for a colorful account of the couple's life together.

35. Haldane vacillated between attributing actual novel scientific discovery, as opposed to understanding, or knowing "what to look for," as opposed to telling you what you "are going to find," to the dialectical method. Compare "A Dialectical Account of Evolution," *Science & Society* 1 (1937), 473–86, to *The Marxist Philosophy and the Sciences* (London: Ayer 1939; reprint, Freeport, NY: Books for Libraries Press, 1969), 43. As for how he got to Marxism, Haldane cited English capitulation to fascism alongside "recent developments in physics and biology."

36. J. B. S. Haldane, *The Causes of Evolution* (London: Longmans, Green, 1932). The series of nine papers can be found in Mark Ridley, *Evolution*, CD-ROM (Oxford: Blackwell, 1996).

37. Both Anthony Edwards and James Crow shared this view of Haldane as a mathematician: Kohn, *A Reason for Everything*, 145.

38. To be fair, Haldane didn't follow dialectical materialism blindly. Fisher argued that selection could produce modifier genes that would "negate" the effect of dominant deleterious mutations, a "beautifully dialectical theory" that Haldane rejected.

39. For a critical rendition of the story of the *betularia* see Judith Hopper, *Of Moths and Men: The Untold Story of Science and the Peppered Moth* (New York: W. W. Norton, 2002).

40. Scholars argue about the extent and manner in which dialectical materialism played a role in Haldane's science. Arthur M. Shapiro shows that Haldane's evolutionary papers in the 1920s became increasingly dialectical but puts it down to Hegel, via Uffer, rather than Marx. See his "Haldane, Marxism and the Conduct of Research," *Quarterly Review of Biology* 68 (1993), 69–77. Sahotra Sarkar argues that it was via his mechanistic and reductionist scientific work that Haldane became a Marxist, rather than his politics, or any form of philosophy, influencing his science. See his "Science, Philosophy, and Politics in the Work of J.B.S. Haldane, 1922–1937," *Biology and Philosophy* 7 (1992), 385–409.

41. J. B. S. Haldane, "The Effect of Variation on Fitness," *American Naturalist* 71 (1937), 337–49, and "The Cost of Natural Selection," *Journal of Genetics* 55 (1957), 511–24. This line of thought later led to Motoo Kimura's "neutral theory" of evolution, whereby natural selection is "blind" to much of the genetic mutation in a population, which therefore has little or no effect on population fitness.

42. On Lysenko see David Joravsky, *The Lysenko Affair* (Cambridge, MA: Harvard University Press, 1970), Zhores Medvedev, *The Rise and Fall of T. D. Lysenko*, trans. Michael Lerner (New York: Columbia University Press, 1969), and Nils Roll-Hansen, *The Lysenko Affect: The Politics of Science* (New York: Humanity Books, 2004). A large literature is discussed in Oren Solomon Harman, "C. D. Darlington and the British and American Reaction to Lysenko and the Soviet Conception of Science," *Journal of the History of Biology* 36 (2003), 309–52. On Vavilov see Peter Pringle, *The Murder of Nikolai Vavilov: The Story of Stalin's Persecution of One of the Great Scientists of the Twentieth Century* (New York: Simon & Schuster, 2008).

43. Haldane, *The Inequality of Man*, 136; J. B. S. Haldane, *Heredity and Politics* (New York: W. W. Norton, 1938), 14.

44. Boris Ephrussi said this of Haldane, quoted in Clark, *The Life and Work of J. B. S. Haldane*, 109.

45. Ruse, *Monad*, 367.

46. Sewall Wright, "The Genetical Theory of Natural Selection: A Review," *Journal of Heredity* 21 (1930), 340–56.

47. See Sewall Wright, "Evolution in Mendelian Populations," *Genetics* 16 (1931), 97–159. Also, "The Roles of Mutation, Inbreeding, Crossbreeding and Selection in Evolution," *Proceedings of the Sixth International Congress of Genetics* 1 (1932), 356–66, and "Adaptation and Selection" in Glenn L. Jepson, Ernst Mayr, and George Gaylord Simpson, *Genetics, Paleontology and Evolution* (Princeton: Princeton University Press, 1949), 365–89.

48. Sewall Wright finally produced a book-size exposition of his complete evolutionary worldview late in life in *Evolution: Selected Papers*, ed. William B. Provine (Chicago: University of Chicago Press, 1986). There is also Wright, *Evolution and the Genetics of Populations*, 4 vols. (Chicago: University of Chicago Press, 1968).

49. R. A. Fisher, "The Measurement of Selective Intensity," *Proceedings of the Royal Society of London B* 121 (1936), 58–62.

50. It was T. H. Huxley's grandson and Haldane's old Eton friend Julian Huxley who gave the enterprise its name. See his *Evolution: The Modern Synthesis* (London: Allen and Unwin, 1942).

51. The treatment of this problem came in the chapter "The Evolution of Distastefulness" in Fisher, *The Genetical Theory of Natural Selection*.

52. Ibid., 159.
53. Sewall Wright, "Coefficients of Inbreeding and Relationship," *American Naturalist* 56 (1922), 330–38.
54. This exposition exists in Dugatkin, *The Altruism Equation*, 81–82.
55. Sewall Wright, "Tempo and Mode in Evolution: A Critical Review," *Ecology* 26 (1945), 415–19.
56. On group selection see Elliott Sober and David Sloan Wilson, *Unto Others: The Evolution and Psychology of Unselfish Behavior* (Cambridge, MA: Harvard University Press, 1998).
57. J. B. S. Haldane, "Darwinism Today," in *Possible Worlds*, 35.
58. Haldane, *Causes of Evolution*, 71; quoted in Dugatkin, *The Altruism Equation,* 72.
59. Fisher, *The Genetical Theory of Natural Selection*, 163.
60. J. B. S. Haldane, "Population Genetics," *New Biology* 18 (1955), 34–51, quote on 44.
61. Ibid.
62. Precisely why Fisher, Haldane, and Wright did not work out mathematical models of the evolution of genes for altruism is a matter of speculation. Some possible answers are discussed in Dugatkin, *The Altruism Equation*, 82–85.
63. Legends are as legends go. Before it ballooned into the drunken stupor tale, it was John Maynard Smith who told the pub story, situating it at the now-defunct Orange Tree off the Euston Road in London, and with Haldane calculating alertly on the back of an envelope rather than inebriated. Bill Hamilton was angered by what he took to be Maynard Smith's misplaced—and trivializing—memory. He had worked out the math painstakingly over two years, and it was his own comment, he thought, that Maynard Smith mistakenly attributed to Haldane.
64. Robert N. Proctor, *Racial Hygiene: Medicine Under the Nazis* (Cambridge, MA: Harvard University Press, 1988), 307.
65. Diane Paul, "A War on Two Fronts: J. B. S. Haldane and the Response to Lysenkoism in Britain," *Journal of the History of Biology* 16 (1983), 1–37.
66. Clark, *J. B. S.*, 209. On Haldane in India see Krishna Dronamraju, *Haldane: The Life and Work of J.B.S. with Special Reference to India* (Aberdeen: Aberdeen University Press, 1985).

CHAPTER 4: ROAMING

1. "Class of '44, 1050 Strong, Registers Today at Memorial Hall; Dean Chase Sounds Welcome," *Harvard Crimson*, September 20, 1940.
2. Roger Rosenblatt, *Coming Apart: A Memoir of the Harvard Wars of 1969* (Boston: Little Brown and Company, 1997), 94; *Harvard Crimson*, September 20, 1940; *Harvard University Directory of Students*, published by the university (Cambridge, MA: October 1940), 79.
3. George Price transcript, Harvard College, 1940–41, HUA.
4. Richard D. Edwards, "Harvard Views the War," Class of 1941 Album, Harvard University Archive (HUA), 252–61, quotes on 258, 252, 255.
5. Ibid., 259.
6. Ibid, 260–61.
7. Quoted in Rosenblatt, *Coming Apart*, 98.
8. George Price letter to Gage Avery, August 1941, EPFA; George Price letter to Abraham B. Albert, August 9, 1941, EPFA.

9. David Freeman Hawke, *John D.: The Founding Father of the Rockefellers* (New York: Harper & Row, 1980); Charles R. Morris, *The Tycoons: How Andrew Carnegie, John D. Rockefeller, Jay Gould, and J. P. Morgan Invented the American Supereconomy* (New York: Owl Books, 2006); William J. Barber, "Political Economy in an Atmosphere of Academic Entrepreneurship: The University of Chicago," in *Breaking the Academic Mould: Economics and American Learning in the Nineteenth Century*, ed. William J. Barber (Middletown, CT: Wesleyan University Press, 1988), 241. There had been an original founding of the University of Chicago in 1857, incidentally, under the auspices of Senator Stephen A. Douglas, but a mortgage foreclosure had shut its doors in 1886.

10. Milton Mayer, *Young Man in a Hurry: The Story of William Rainey Harper* (Chicago: University of Chicago Alumni Association, 1957), 22; Edward Shils, ed., *Remembering the University of Chicago: Teachers, Scientists and Scholars* (Chicago: University of Chicago Press, 1991), xii; Alfred North Whitehead, *Dialogues of Alfred North Whitehead*, recorded by Lucien Price (Boston: Little, Brown, 1954), 137.

11. William Michael Murphy and D. J. R. Bruckner, eds., *The Idea of the University of Chicago: Selections from the Papers of the First Eight Executives of the University of Chicago from 1891 to 1975* (Chicago: University of Chicago Press, 1976), 2; Mayer, *Young Man in a Hurry*, 61; Mary Ann Dzuback, *Robert M. Hutchins: Portrait of an Educator* (Chicago: Chicago University Press, 1991), 74; J. L. Laughlin, "Academic Liberty," *Journal of Political Economy* January 1906), 41–43, quote on 43.

12. "When War Came," in *News of the Quadrangles, University of Chicago Magazine* 34 (December 1941), 10–11.

13. Robert M. Hutchins, "The State of the University, November 1, 1943: A Report to the Friends of the University of Chicago," 3–4, University of Chicago Archives (UCA); "Meet the Army Meteorologists," *Daily Maroon*, September 30, 1942, 4.

14. "Theory, Mud, Maneuvers," *University of Chicago Magazine* 34 (December 1941), 11.

15. Don Morris, News of the Quadrangles, *University of Chicago Magazine* 35 (May 1943), 18.

16. Martin Gardner, "All Out for War," *University of Chicago Magazine* 34 (January 1942), 15.

17. Robert M. Hutchins, "The State of the University, September 10th, 1942," 22, UCA; ibid., 7–11.

18. "Prep For Mustache Race," *Daily Maroon*, February 20, 1942, 1; Ellen Baum, "The Traveling Bazaar," *Daily Maroon*, February 22, 1946, 2.

19. Hutchins, "The State of the University," 13–14; George Price transcript, 1941–43, University of Chicago, Office of the Registrar.

20. Communication from Al Somit, May 15, 2003. George had been renting a room in the home of Professor Thorfin Hogness of the Chemistry Department until then: George Price letter to Don Fergusson, Oct. 28, 1960, GPP.

21. "Discrimination Clauses in Club Constitutions," *Daily Maroon*, March 6, 1942, 2; Harold W. Flitcraft, ed., *History of the 57th Street Meeting of Friends* (Chicago: 57th Street Meeting of Friends), 32.

22. Interview with Al Somit, December 6, 2007.

23. Al Somit letter to George Price, January 4, 1943, GPP.

24. Interview with Al Somit, April 16, 2008; Ellis Student Cooperative Handbook, September 1942, GPP.

25. George Price letter to Bob Sheffield, February 5, 1945, GPP.

26. "Coffee Shop Gone out of Business," *Daily Maroon*, April 30, 1943, 2.
27. "The University and the War," official publications of the University of Chicago, 1944, 34, CUA.
28. Richard Rhodes, *The Making of the American Bomb* (New York: Simon & Schuster, 1986).
29. Daniel J. Kevles, *The Physicists: The History of a Scientific Community in Modern America* (New York: Alfred A. Knopf, 1977).
30. George Price transcript; Allie Shah, "Medical Researcher Samuel Schwartz Dies," *Star Tribune*, December 9, 1997; George Price letter to Dr. Herman H. Goldstine, October 28, 1964, GPP.
31. George Price, "Fluorescence of Uranium, Plutonium, Neptunium, and Americium, A Dissertation Submitted to the Faculty of the Division of the Physical Sciences in Candidacy for the Degree of Doctor of Philosophy, the University of Chicago Department of Chemistry, August 6, 1946"; Samuel Schwartz Papers (SSP), University of Minnesota Medical School, Box 12, Schwartz's notes.
32. Ibid., 6–7.
33. Interviews with Kathleen Price, April 12, 2008, and April 13, 2008; interview with Al Somit, April 16, 2008.
34. Chet Opal, "Victory!" *University of Chicago Magazine* 37 (May 1945), 13.
35. Alice Avery Price letter to George Price, August 14, 1945, GPP.
36. George Price letter to Alice Avery Price, August 20, 1945, GPP.
37. "Mecca of the Caffeine Addicts Soon to Reopen," *Daily Maroon*, February 22, 1946, 5.
38. George Price letter to Fred, August 19, 1946, GPP.
39. George Price letter to Dr. Erwin Haas, August 17, 1946, GPP; George Price letter to Fred, August 19, 1946, GPP; Robert M. Hutchins, "The State of the University, September 25, 1945," 28, and "November 25, 1946," 18, UCA.
40. Interview, Kathleen Price, April 13, 2008; Directory of University Officers and Students, November 25, 1946, 118, and June 1947, 138, HUA.
41. Communication from Professor Gilbert Stork, January 11, 2008; communication from Professor Leonard K. Nash, May 5, 2008.
42. Lloyd Shapley letter to George Price, February 3, 1947, GPP; Lloyd A. Wood letter to George Price, October 24, 1947, GPP; "Chess Club Takes Lead in Crucial Match," *Harvard Crimson*, March 1, 1947.
43. Archival materials and historical sources are on the official Argonne National Laboratory Web site, www.anl.gov.
44. Claude E. Shannon, "A Mathematical Theory of Communication," *Bell System Technical Journal* 27 (July 1948), 379–423 (October 1948), 623–66.
45. University of Minnesota, Office of the President, "Request for Information," May 9, 1950, University of Minnesota Archives (UMA).
46. John W. Rae and John W. Rae, Jr., *Morristown's Forgotten Past: "The Gilded Age"* (Morristown, NJ: John W. Rae, 1980).
47. George Price, "Transistor Work Already Started That Would Be Simple and Valuable to Finish," draft, June 27, 1949, GPP.
48. Interviews with Annamarie Price, April 15 and 17, 2008.
49. George Price letter to Al Somit, February 1, 1953, GPP.
50. Stanford Lehmberg and Ann M. Pflaum, *The University of Minnesota 1945–2000* (Minneapolis: University of Minnesota Press, 2000), 51–53; Jay Edgerton, "U. of M. Medical College Had Become One of the World's Greatest," *Minneapolis Star*, January 28, 1955.
51. "Conference Agenda," box 4; "Memo," box 10, Samuel Schwartz Papers (SSP).

The eventual paper was titled, "Some Relationships of Porphyrins, Tumors, and Ionizing Radiations," *University of Minnesota Medical Bulletin* 27 (1955), 7–13.

52. Box 10, SSP.

53. Samuel Schwartz letter to Dr. Avram Goldstein, January 25, 1978, box 33, SSP.

54. Box 1, SSP, undated.

55. George Price letter to Bob and Marjorie Sheffield, December 3, 1953, GPP.

56. George Price letter to Al Somit, February 1, 1953, GPP.

57. John Lewis Gaddis, *The Cold War* (London: Penguin, 2005).

58. Jean and Fairfield Hoban telegram to George Price, September 30, 1953, GPP; Julia Price letter to Alice Avery Price, December 10, 1953, GPP; Alice Avery Price letter to George Price, September 22, 1953, GPP.

CHAPTER 5: FRIENDLY STARFISH, SELFISH GAMES

1. Clay Blair, Jr., "Passing of a Great Mind," *Life*, February 25, 1957, 89–90, quote on 89.

2. William Poundstone, *Prisoner's Dilemma: John von Neumann, Game Theory, and the Puzzle of the Bomb* (New York: Anchor Books, 1992), 5; Blair, "Passing of a Great Mind."

3. Adam Smith, *An Inquiry into the Nature and Causes of the Wealth of Nations* (1776; reprint edited by Edwin Cannan, Chicago: University of Chicago Press, 1976).

4. Veblen first used the term "neoclassical" in "Preconceptions of Economic Science," *Quarterly Journal of Economics* 13 (January 1899).

5. Lucy Sprague Mitchell, *Two Lives: The Story of Wesley Clair Mitchell and Myself* (New York: Simon & Schuster, 1953), 86; Thorstein Veblen, *The Theory of the Leisure Class: An Economic Study of Institutions* (1899; reprint, New York: Macmillan, 1953) and "The Limitations of Marginal Utility," *Journal of Political Economy* 17 (November 1909), 622, 624. On Veblen see Rick Tilman, *Thorstein Veblen and His Critics, 1891–1963* (Princeton: Princeton University Press, 1992).

6. C. Wright Mills of Columbia University wrote this in his introduction to Veblen's own *The Theory of the Leisure Class*, ix.

7. Frank Knight, "The Newer Economics and the Control of Economic Activity," *Journal of Political Economy* (August 1932), 453; *Chicago Tribune*, May 28, 1972; Frank Knight, *Risk, Uncertainty, and Profit* (Boston: Houghton Mifflin, 1921).

8. Frank Knight, *The Economic Organization* (Chicago: University of Chicago Press, 1933).

9. W. C. Allee, "Evolution of a Mechanist," circa 1915, 9–11, quoted in Gregg Mitman, *The State of Nature: Ecology, Community, and American Social Thought, 1900–1950* (Chicago: Chicago University Press, 1992), 52. Mitman's book is the best study of Allee and Chicago ecology.

10. Karl Patterson Schmidt, "Warder Clyde Allee," *Biographical Memoirs, National Academy of Sciences USA* 30 (1957), 3–40; Alfred E. Emerson and Thomas Park, "Warder Clyde Allee: Ecologist and Ethologist," *Science* 121 (May 13, 1955), 686–87; David W. Blight, *Passages to Freedom: The Underground Railroad in History and Memory* (Washington, DC: Smithsonian Books, 2002); on Earlham see Thomas Hamm, "A Brief History of Earlham College," at the school's Web site: www.earlham.edu/EC_history.html.

11. Warder Clyde Allee, "An Experimental Analysis of the Relation Between Physiological States and Rheotaxis in Isopoda," *Journal of Experimental Zoology* 13 (1912), 270–344; Mitman, *The State of Nature*, 53.

12. *Springfield News-Record*, February 9, 1917, quoted in Mitman, *The State of Nature*, 56.

13. See ibid., 58–62.

14. Peter Weikart, *From Darwin to Hitler: Evolutionary Ethics, Eugenics, and Racism in Germany* (New York: Palgrave Macmillan, 2004), 10; Oren Harman, "On the Power of Ideas," *Minerva* 45 (2007), 175–189. Many American biologists connected German militarism to Darwinism. See Mitman, *The State of Nature*, 220, n36.

15. Quotes from Mitman, *The State of Nature*, 60.

16. On Spencer see Robert J. Richards, *Darwin and the Emergence of Evolutionary Theories of Mind and Behavior* (Chicago: University of Chicago Press, 1987), 234–330.

17. Blair, "Passing of a Great Mind."

18. Poundstone, *Prisoner's Dilemma*, 22, 21, 24. See also Stanislaw Ulam, "John von Neumann, 1903–1957," *Bulletin of the American Mathematical Society* 64, no. 3 (May 1958), 1–49.

19. John von Neumann, "Zur Theorie der Gesellschaftsspiele," *Mathematische Annalen* 100 (1928), 295–320. See Robert J. Leonard, "From Parlor Games to Social Science: Von Neumann, Morgenstern and the Creation of Game Theory, 1928–1944," *Journal of Economic Literature* 33 (1995), 730–61.

20. John Maynard Keynes, *General Theory of Employment, Interest and Money* (London: Macmillan 2007;1936), 3.

21. Hyman Minsky, *John Maynard Keynes* (New York: Columbia University Press, 1975); Geoff Tily, *Keynes's General Theory, the Rate of Interest and "Keynesian" Economics* (London: Palgrave Macmillan, 2007).

22. Don Patinkin, *Essay On and In the Chicago Tradition* (Durham, NC: Duke University Press, 1981), 299; Leonard Silk, *The Economists* (New York: Basic Books, 1976), 46; Henry Simons, *A Positive Program for Laissez-Faire: Some Proposals for a Liberal Economic Policy* (Chicago: University of Chicago Press, 1934); Craufurd D. Goodwin, "Martin Bronfenbrenner, 1914–1997," *Economic Journal* (November 1998), 1776.

23. Jacob Viner, "Mr. Keynes and the Causes of Unemployment," *Quarterly Journal of Economics*, November 1936; Johan Van Overtveldt, *The Chicago School: How the University of Chicago Assembled the Thinkers Who Revolutionized Economics and Business* (Canada: B2 Books, 2007), 81. Van Overtveldt provides a thorough overview of the rise of the Chicago School.

24. Edward O. Wilson and Charles D. Michener, "Alfred Edward Emerson, 1896–1976," *Biographical Memoirs, National Academy of Sciences USA* 53 (1982), 159–77, 162. Emerson's collection was donated to the American Museum of Natural History in New York City.

25. Sewall Wright, "Genetics of Abnormal Growth in the Guinea Pig," *Cold Spring Harbor Symposia on Quantitative Biology* 2 (1934), 137–47, quote on 139. On Wright see William Provine, *Sewall Wright and Evolutionary Biology* (Chicago: University of Chicago Press, 1986).

26. For an up-to-date, modern appreciation of the idea of the superorganism see Bert Hölldobler and E. O. Wilson, *The Superorganism: The Beauty, Elegance, and Strangeness of Insect Societies* (New York: W. W. Norton, 2009).

27. See William Morton Wheeler, *The Social Insects* (New York: Harcourt, Brace & Co., 1928), and "Animal Societies," *Scientific Monthly* 39 (1934), 289–301.

28. Alfred E. Emerson, "Biological Basis of Social Cooperation," *Illinois Academy of Science Transactions* 39 (1946), 12, quoted in Mitman, *The State of Nature*, 158.

29. Frank Lillie, "The Department of Biology in Relation to the New Organization," *Daily Maroon*, December 11, 1930; Warder Clyde Allee, "Science Confirms an Old Faith," *American Friend* 35 (1928), 780. Quoted in *Mitman* 52.

30. See Warder Clyde Allee, "Animal Aggregations," *Quarterly Review of Biology* 2 (1927), 367–98, and *Animal Aggregations: A Study in General Sociology* (Chicago: Chicago University Press, 1927).

31. A good description of Allee's aggregation work appears in Dugatkin, *The Altruism Equation*, 41–50.

32. W. C. Allee, "Concerning Biology and Biologists," quoted in ibid., 57.

33. W. C. Allee, "Reexamination of One Fundamental Doctrine in the Light of Modern Knowledge," and "Where Angels Fear to Tread," quoted in ibid., 48, 56.

34. Oskar Morgenstern, Diary, April–May 1942, quoted in Leonard, "From Parlor Games to Social Science," 730.

35. John von Neumann and Oskar Morgenstern, *Theory of Games and Economic Behavior* (Princeton: Princeton University Press, 1944), 2.

36. Van Overveldt, *The Chicago School*, 215. Many of the professors at the Chicago Department of Economics did not endorse the research methods used at the Cowles Commission. All, however, considered von Neumann a genius.

37. Actually, game theory as Neumann and Morgenstern had defined it was probably least applicable to economics compared to other fields: Zero-sum games between two people are negligible in economic situations, whereas cooperative, many-people games were not strictly solved in the book. That would have to wait for John Nash.

38. Mitman, *The State of Nature*, 159.

39. Alfred E. Emerson, "The Biological Basis of Social Cooperation," *Illinois Academy of Science Transactions* 39 (1946) 12, quoted in Mitman, *The State of Nature*, 158.

40. George Gaylord Simpson, "The Role of the Individual in Evolution," *Journal of the Washington Academy of Sciences* 31 (1941), 16, quoted in Mitman, *The State of Nature*, 165. Simpson's accusation was directed against the neurobiologist Ralph Gerard. See Ralph W. Gerard, "Organism, Society and Science," *Scientific Monthly* 50 (1940), 340–50, 530–35.

41. Not that Emerson wasn't a democrat. His problem was how to hold on to a biological view of the democratic state while combating the organicism of fascism. Emerson chose to save organicism by arguing that fascism had misused it by stamping out individual variation—a crucial component of progressive integration of the group.

42. Emerson, "The Biological Basis of Social Cooperation," 17–18 quoted in Mitman, *A State of Nature*, 160.

43. For a recent biography see Lanny Ebenstein, *Milton Friedman* (New York: Palgrave Macmillan, 2007).

44. Milton Friedman, *Capitalism and Freedom* (Chicago: University of Chicago Press, 1962; reprint, 2002), xi.

45. This was the title of a book by the Austrian-born economist Friedrich Hayek, a friend of Friedman's, and, beginning in 1950, a colleague at Chicago. See Friedrich Hayek, *The Road to Serfdom. Fiftieth Anniversary Edition* (Chicago: University of Chicago Press, 1994).

46. Quoted in Friedman, *Capitalism and Freedom*, 26.

47. Ibid., 2.

48. Mitman, *A State of Nature*, 109.

49. Allee worked on the endocrinology of the pecking order. See W. C. Allee, N. E. Collias, and Catherine Z. Lutherman, "Modification of the Social Order in Flocks

of Hens by the Injection of Testosterone Proprionate," *Physiological Zoology* 12 (1939), 412–40; W. C. Allee and N. E. Collias, "The Effect of Estradiol on the Social Organization of Flocks of Hens," *Endocrinology* 27 (1940), 87–94; W. C. Allee, N. E. Collias, and Elizabeth Beeman, "The Effect of Thyroxin on the Social Order in Flocks of Hens," *Endocrinology* 27 (1940), 827–35; Benson Ginsburg and W. C. Allee, "Some Effects of Conditioning on Social Dominance and Subordination in Inbred Strains of Mice," *Physiological Zoology* 15 (1942), 485–506.

50. A. M. Guhl and W. C. Allee, "Some Measurable Effects of Social Organization in Flocks of Hens," *Physiological Zoology* 17 (1944), 320–47.

51. Warder Clyde Allee, "Biology and International Relations," *New Republic* 112 (1945), 816.

52. Mitman, *A State of Nature*, 184.

53. Warder Clyde Allee, *The Social Life of Animals* (New York: W. W. Norton, 1938); "Where Angels Fear to Tread: A Contribution from General Sociology to Human Ethics," *Science* 97 (1943), 517–25; Mitman, *A State of Nature*, 178–201.

54. Poundstone, *Prisoner's Dilemma*, 83–99.

55. Sylvia Nasar, *A Beautiful Mind* (New York: Simon & Schuster, 1998), 104–14, quote on 104

56. In fact, the earliest use of game theory in World War II by the United States involved designing antisubmarine tactics against the Germans. Mathematicians at the Anti-Submarine Warfare Operation Evaluation Group were said to carry around copies of Neumann's 1928 paper on poker.

57. Poundstone, *Prisoner's Dilemma*, 68.

58. Ibid., 118.

59. John Nash's "Equilibrium Points in N-Person Games," *Proceedings of the National Academy of Sciences USA* 36 (1950), 48–49, had been the original spur to the prisoner's dilemma, since Nash's optimal solutions were not satisfied in it. Flood and Dresher themselves now believe that the prisoner's dilemma will never be "solved," and nearly all game theorists agree. On Nash's incredible story see Nasar, *A Beautiful Mind*.

60. Dugatkin, *The Altruism Equation*, 59. Allee was, however, a highly respected scientist and was elected to the National Academy of Sciences in 1951.

61. Ibid., 60.

62. Poundstone, *Prisoner's Dilemma*, 191.

63. Even former RANDers Luce and Raiffa concluded in 1957 that most social scientists were disillusioned with game theory. R. Duncan Luce and Howard Raiffa, *Games and Decisions* (New York: John Wiley & Sons, 1957).

64. Poundstone, 193–94; George Orwell, "As I Please," *Tribune*, 30 June, 1944, in George Orwell, *The Collected Essays, Journalism and Letters: Volume 3* (Harmondsworth: Penguin, 1970), 208.

CHAPTER 6: HUSTLING

1. "Results of Psychical Research," *New York Times*, September 4, 1892.

2. Seymour Mauskopf, ed., *The Reception of Unconventional Science by the Scientific Community* (Boulder, CO: Westview Press, 1979); Joseph Rhine letters to George Price, April 13, 1971, and May 18, 1972, GPP.

3. Seymour Mauskopf and Michael R. McVaugh, *The Elusive Science: Origins of Experimental Psychical Research, 1915–1940* (Baltimore: Johns Hopkins Press, 1980).

4. Samuel George Soal, "Fresh Light on Card Guessing," *Proceedings of the Society for Psychical Research* 46 (1940), 152; Joseph B. Rhine, *New Frontiers of the Mind* (New York: Farrar and Rinehart, 1937).

5. Alice Avery Price letters to herself from W. E. Price, April–May 1946, June 25 1950, EPFA.

6. George Price letter to Bob and Marjorie Sheffield, December 1, 1953, GPP; George Price letter to Henry Noel, February 1, 1953. GPP.

7. George Price, "Science and the Supernatural." *Science* 122 (1955), 359–67, quote on 360; George Price correspondence with Joseph Rhine February 25 and March 15, 1941, GPP.

8. Ibid., 367, 363.

9. Aldous Huxley, "Facts and Fetishes," *Esquire*, September 1955, 43–44, 115–16, quote on 43; George Price, "Science and the Supernatural." *Science* 122 (August 26, 1955), 359–67.

10. Anthony Arthur, *Radical Innocent: Upton Sinclair* (New York: Random House, 2006), 43; Upton Sinclair letter to George Price, August 5, 1956, GPP. Einstein had become friendly with Sinclair via correspondence over social issues, and even attended a séance in his home in California in 1930. When Mrs. Sinclair challenged his views about telepathy, Einsteins wife, Elsa, chided her for her presumption, saying, "You know my husband has the greatest mind in the world." Mrs. Sinclair responded: "Yes, I know, but surely he doesn't know everything." Einstein remained amused and wrote the preface out of friendship rather than belief, a fact that did not disturb Sinclair's overwhelming pride. See Walter Isaacson, *Einstein: His Life and Universe* (New York: Simon & Schuster, 2007), 373–74.

11. *New York Times*, August 27, 1955. A further article in the *Times*, "Scientists Debate 6th Sense of Man," appeared on January 8, 1956; Reverend Norman Boswell letter to George Price, January 15, 1956, GPP C. H. Chalmers letter to George Price, October 28, 1955, and George's reply, February 2, 1956, GPP; Fern Irene Clarke letter to George Price, September 5, 1955, and George's reply, October 27, 1955, GPP; Lin Cutler letter to George Price, January 13, 1955, GPP.

12. *Science* 123 (1956), 7–19, quotes on 8, 9, 15, 13, 11.

13. Ibid., 16.

14. George Price, "Where Is the Definitive Argument?" *Science* 123 (1956), 17–18.

15. Joan Cook, "John Scarne, Gambling Expert," *New York Times*, July 9, 1985; George Price correspondence with Dr. George K. Bennett and Dr. J. Ricardo Musso, July–December 1956, GPP; George Price letter to Dr Keith S. Ditman, November 14, 1964, GPP.

16. Claude Shannon letter to George Price, January 31, 1955, GPP. Scarne later became the technical adviser on the motion picture *The Sting* and his hands doubled for Paul Newman's during the deck-switching scenes. See John Scarne, *The Odds Against Me* (New York: Simon & Schuster, 1966)

17. George Price letter to Al Somit, October 23, 1955, GPP.

18. Gaddis, *The Cold War*, 163, 123.

19. George Price, "Does Poverty Really Lead to Communism: Outline." circa 1953, GPP.

20. George R. Price, "Altman's Theory of Economic Cycles," *Science* 117 (1953), 335–36, quotes on 336; George T. Altman, "Cycles in Economics in Nature," *Science* 115 (1952), 51.

21. George R. Price, "How to Speed Up Invention," *Fortune*, November 1956, 150–53 and 218–28, quote on 223.

22. Ibid., 228; George Price letter to Dr. Gilfillan, July 1, 1972, GPP.

23. George R. Price, "Arguing the Case for Being Panicky," *Life*, November 18, 1957, 125–28, quotes on 125.
24. Ibid., 127, 128.
25. Hubert Humphrey letter to George Price, January 29, 1957, GPP; "Scientist's Answer to Lagging Research: A Machine That Will Help Invent Machines" and "Researcher Devotes His Life to Science," *Minneapolis Sunday Tribune*, January 20, 1957; George Price letters to Martin Kessler, October 2 and 29, 1955, GPP; George Price letter to Al Somit, March 8, 1956, GPP.
26. "Emanuel Piore, 91, Leader and Researcher at IBM," *New York Times*, May 12, 2000.
27. George Price letter to E. R. Piore, August 12, 1957, GPP; George Price letter to Dr. Gilfillan, July 1, 1972, GPP.
28. Interviews with Kathleen Price, April 12 and 13, 2008.
29. Correspondence between George Price and the Law Offices of William R. Lieberman, July 25, 1957, and September 28, 1957, the British Library George Price Collection (BLGPC), KPX1_2.3 and KPX1_7.5.

CHAPTER 7: SOLUTIONS

1. Kohn, *A Reason for Everything*, 227.
2. John Maynard Smith, "In Haldane's Footsteps," in *Leaders in the Study of Animal Behavior*, ed. Donald A. Dewsbury (Lewisburg, PA: Bucknell University Press), 347–54, quote on 348.
3. Quoted in Kohn, *A Reason for Everything*, 212.
4. Ibid., 211.
5. Vero C. Wynne-Edwards, "Backstage and Upstage with 'Animal Dispersion,'" in *Leaders in the Study of Animal Behavior*, 487–512, quotes on 488, 489.
6. Charles Elton, "Periodic Fluctuations in the Numbers of Animals: Their Causes and Effects," *British Journal of Experimental Biology* 2 (1924), 119–63.
7. Alexander M. Carr-Saunders, *The Population Problem: A Study in Human Evolution* (Oxford: Clarendon Press, 1922).
8. Wynne-Edwards's descriptions of nature in the Arctic are very similar to those of Kropotkin in *Mutual Aid*: "Except perhaps among carnivorous predators, competition between individuals for space and nourishment seems commonly to be reduced to a low level among members of the Arctic flora and fauna; they live somewhat like weeds, the secret of whose success lies in their ability to exploit transient conditions while they last, in the absence of serious competition. In the Arctic the struggle for existence is overwhelmingly against the physical world, now sufficiently benign, now below the threshold for successful reproduction, and now so violent that life is swept away, after which recolonization alone can restore it." V. C. Wynne-Edwards, "Zoology of the Baird Expedition," *The Auk* 69, no. 4 (1952), 353–91, quote on 384.
9. V. C. Wynne-Edwards, "Intermittent Breeding of the Fulmar (*Fulmarus glacialis*) with Some General Observations on Non-Breeding in Sea-Birds," *Proceedings of the Zoological Society of London* A109 (1939), 127–32.
10. There are a number of known density-dependent reproduction-control mechanisms, such as reducing the rate of ovulation via a change in hormone output, or resorption of the embryos in the uterus as a result of stress. Wynne-Edwards, however, was talking about voluntary restraint accomplished via ritualized convention, as in territoriality displays, migration, and abstention.

11. This had been the original insight of Georgii Gause, which he called the "principle of competitive exclusion." See Sharon E. Kingsland, *Modeling Nature: Episodes in the History of Population Ecology* (Chicago: University of Chicago Press, 1985), 146-62.

12. David L. Lack, *The Natural Regulation of Animal Numbers* (Oxford: Clarendon Press, 1954). See also Lack, "My Life as an Amateur Ornithologist," *Ibis* 115 (1973), 421-31, *The Life of the Robin* (London: Witherby, 1943), *Swifts in a Tower* (London, Methuen, 1956), and *Darwin's Finches: An Essay on the General Biological Theory of Evolution* (London: Methuen, 1973).

13. V. C. Wynne-Edwards, *Animal Dispersion in Relation to Social Behavior* (Edinburgh: Oliver and Boyd, 1962), 20. The fishing analogy first appeared in V. C. Wynne-Edwards, "The Control of Population Density Through Social Behaviour: A Hypothesis," *Ibis* 101 (1959), 436-41, quote on 437.

14. Wynne-Edwards called the kind of behavior necessary for such convention "epideictic." His main study of such behavior was on the red grouse, *Lagopus lagopus scotica*, in the vicinity of Aberdeen. Olavi Kalela, too, developed this idea. See his "Über die Funktion der Mandibeln bei den Soldaten von *Neocapritermes opacus* (Hagen)," *Zoologischer Anzeiger* 152 (1954), 228-34.

15. Charles Darwin, *The Origin of Species*, 139, emphasis added.

16. Quoted in Kohn, *A Reason for Everything*, 227. For a study of Wynne-Edwards' place in the history of the debate on group selections see Mark Borrello, *Evolutionary Restraints: The Contentious History of Group Selection Theory* (Chicago: University of Chicago Press, 2010).

17. Ibid., 224.

18. David Lack, *Population Studies of Birds* (Oxford: Oxford University Press, 1966); Charles Elton, "Self-Regulation of Animal Populations," *Nature* 197 (1963), 634.

19. V. C. Wynne-Edwards, "Population Control in Animals," *Scientific American* 211 (1964), 68-74.

20. "The Nature of Social Life," *Times Literary Supplement*, December, 14, 1962, 967. More reviews are discussed in Borrello, *Evolutionary Restraints*. I thank Mark Borrello kindly for sharing with me a prepublication version of his manuscript.

21. David Cort, "The Glossy Rats: A Review of *Animal Dispersion in Relation to Social Behavior*," *The Nation*, November 16, 1963, 326-29, quote on 327. Wynne-Edwards himself often made the comparison to man.

22. August Weismann, *Ueber die Dauer des Lebens* [On the Duration of Life] (Jena, 1882); *Ueber Leben und Tod* [On Life and Death] (Jena, 1884). Both appear in English in Weismann, *On Heredity* (Oxford: Clarendon Press, 1891).

23. See John Maynard Smith, "The Causes of Ageing," in Review Lectures on Senescence, *Proceedings of the Royal Society of London B* 157 (1962), 115-27. Peter Medawar and George C. Williams had already been thinking along these "pleiotropic" lines. See Medawar, *An Unsolved Problem of Biology* (London: H.K. Lewis, 1952), and Williams, "Pleiotropy, Natural Selection, and the Evolution of Senescence," *Evolution* 11 (1957), 398-411.

24. This formulation belongs to Richard Dawkins, *The Selfish Gene* (Oxford: Oxford University Press, 1976; reprint, 1989), 175.

25. The haplodiploid hypothesis turned out to be flawed: many haplodiploid species have not evolved eusociality, whereas other, diplodiploid species have. Still, the "spirit" of the hypothesis survived in the notion that there is genetic conflict between sisters' interest in producing more female reproductives than males versus the queen's interest in an equal investment in the sexes. See R. L. Trivers and

H. Hare, "Haplodiploidy and the Evolution of the Social Insects," *Science* 91 (1976), 249–63, and more recently, N. J. Mehdiabadi, H. K. Reeve, and U. G. Mueller, "Queens Versus Workers: Sex Ration Conflict in Eusocial Hymenoptera," *Trends in Ecology and Evolution* 18, no. 2 (2003), 88–93. For a definitive presentation of current knowledge on social insects see Bert Hölldobler and E. O. Wilson, *The Superorganism: The Beauty, Elegance, and Strangeness of Insect Societies* (New York: W. W. Norton, 2009).

26. For a full-length biography of Hamilton see Ullica Segerstråle, *Nature's Oracle: A Life of W. D. Hamilton* (Oxford: Oxford University Press, 2010). There are many shorter appreciations of Hamilton's life and science. See Alan Grafen, "William Donald Hamilton, 1936–2000," *Biographical Memoirs of Fellows of the Royal Society* 50 (2004), 109–32.

27. This was the writer Alisdair Gray's remark, quoted in Kohn, *A Reason for Everything*, 259; 260; interview with Janet Hamilton (Bill's sister), October 24, 2007.

28. "I realized that I had little talent in mathematics and even less training for it," he later wrote, "so that my efforts to teach myself what was necessary to understand even the merely standard theoretical population genetics of the day were tedious in the extreme."

29. On the jacket of the 1999 edition of Fisher's book.

30. W. D. Hamilton, *Narrow Roads of Gene Land, Volume 1: Evolution of Social Behavior* (Oxford: Spektrum, 1996), 21; the quote on 22 is from Vincent Wigglesworth, *The Life of Insects* (London: Weidenfeld and Nicolson, 1966).

31. Letters to Mary quoted in Kohn, *A Reason for Everything*, 265, 266.

32. Charles Darwin, *On the Origin of Species*, facsimile of the first edition with an introduction by Ernst Mayr (Cambridge, MA: Harvard University Press), 189.

33. Hamilton, *Narrow Roads*, 22.

34. On the decline of eugenics including Penrose's role see Daniel J. Kevles, *In the Name of Eugenics: Genetics and the Uses of Human Heredity* (Cambridge, MA: Harvard University Press, 1995), especially 148–222.

35. Hamilton, *Narrow Roads*, 14.

36. Ibid., 4.

37. Ibid., 12–13.

38. Ibid., 15, 5.

39. In fact Hamilton's supervisor Cedric Smith had once introduced the two, but Maynard Smith couldn't remember the encounter. Hamilton was shy and very unassuming, and Maynard Smith, like everyone else at the time, wasn't interested in the genetics of altruism.

40. Ullica Segerstråle, *Defenders of the Truth: The Sociobiology Debate* (Oxford: Oxford University Press, 2000), 64.

41. The other two Oxford men at the meeting were the ethologist Niko Tinbergen and the geneticist Arthur Cain. Maynard Smith recalled that it was Cain who came up with the term "kin selection" though Cain himself had no such recollection.

42. John Maynard Smith, "Kin Selection and Group Selection," *Nature* 201 (1964), 1145–47. Maynard Smith's model became known as the "haystack model" and has excited a steady amount of discussion and criticism ever since. See Sober and Wilson, *Unto Others*, 67–71. For Wynne-Edwards's own reply see *Nature* 201 (1964), 1147.

43. William Hamilton, "The Evolution of Altruistic Behavior," *American Naturalist* 97 (1963), 354–56. Here Hamilton collapsed cost and benefit into one variable, k, and so the equation was even simpler: $k > 1/r$. This was the only math that appeared in the article.

44. Hamilton, *Narrow Roads*, 29–30.
45. Sober and Wilson, *Unto Others*, 35–36.
46. David Lack, *Population Studies of Birds* (Oxford: Clarendon Press, 1966).
47. G. C. Williams and D. C. Williams, "Natural Selection of Individually Harmful Social Adaptations among Sibs with Special Reference to Social Insects," *Evolution* 11 (1957), 32–39.
48. Fisher, *The Genetical Theory of Natural Selection*, 158.
49. This formulation is taken from Sober and Wilson, *Unto Others*, 38–39.
50. George C. Williams, *Adaptation and Natural Selection* (Princeton: Princeton University Press, 1966), 273.
51. W. D. Hamilton, "Extraordinary Sex Ratios," *Science* (156) 1967, 477–88.
52. Hamilton, *Narrow Roads*, 131–42.
53. Hamilton, "Extraordinary Sex Ratios," 477.
54. The first to point this out was Robert Colwell, "Group Selection Is Implicated in the Evolution of Female-Biased Sex Ratios," *Nature* 190 (1981), 401–4.
55. Hamilton, "Extraordinary Sex Ratios"; see the section on 'sex ratios and polygyny." Hamilton did, however, make a mathematically imprecise allusion to the effect that certain optimal strategies may reflect a compromise between conflicting individual and group-based selection pressures. He buried it, however, in footnote 43.
56. Hamilton, *Narrow Roads*, 186. I thank Peter Henderson for a wonderful description of his time with Hamilton in the Amazon.

CHAPTER 8: NO EASY WAY

1. George Price letter to Hubert Humphrey, May 2, 1957 GPP.
2. George Price correspondence with Senator Hubert Humphrey, February 1956–December 1959, GPP; Donald S. Harrington, "To Deal with China Crisis; Adjudication by International Court of Justice Proposed," *New York Times*, September 19, 1958.
3. George R. Price articles in *Popular Science Monthly* and *THINK*, March 1959–November 1960; George Price letters to Claude Shannon, October 11 and December 8, 1960; George Price–H. J. Muller correspondence, January–February 1960, GPP.
4. On Skinner see Daniel W. Bjork, *B. F. Skinner: A Life* (Washington, DC: American Psychological Association, 1997); William T. O'Donohue and Kyle E. Ferguson, *The Psychology of B.F. Skinner* (Thousand Oaks, CA: Sage, 2001); Daniel N. Weiner, *B. F. Skinner: Benign Anarchist* (Boston: Allyn and Bacon, 1996).
5. "Skinner's Utopia: Panacea, or Path to Hell?" *Time*, September 20, 1971, 47–53, found in GPP. See also Philip J. Pauly, *Controlling Life: Jacques Loeb and the Engineering Ideal in Biology* (New York: Oxford University Press, 1987).
6. George R. Price, "The Teaching Machine," *THINK*, March 1959, 10–14, quote on 10.
7. George Price letter to B. F. Skinner, October 27, 1957, GPP.
8. Alexis de Tocqueville, *Democracy in America* (1840), quoted in drafts of *No Easy Way*, chapter 6, GPP. Also George Price, "The Warning from the Distant Past," draft, probably of chapter 10, GPP.
9. Cass Canfield letter to George Price, March 3, 1958, GPP; A. Boardman letter to George Price, August 11, 1966, GPP.
10. Ross Wetzsteon, *Republic of Dreams: Greenwich Village, the American Bohemia, 1910–1960* (New York, Simon & Schuster, 2002); George Price passport, GPP.

11. George Price letter to Hubert Humphrey, December 8, 1959, GPP; George Price–Richard Winslow Correspondence, 1959–1961, GPP; *No Easy Way* draft, circa 1960, GPP.
12. George Price letter to Dr. Nathan S. Kline, October 28, 1958, GPP.
13. George Price letter to Richard Winslow, February 29, 1960, GPP.
14. George Price letter to Robert Latimer, February 13, 1961, GPP; Joan letter to George Price, November 12, 1961, GPP; George Price letter to Henry Noel, February 10, 1961, GPP; George Price letter to Al Somit, October 12, 1961, GPP; Fred Schneider letter to George Price, April 10, 1960, BLGPC, KPX1_1.5.
15. Donald Ferguson letter to George Price, September 21, 1960, GPP.
16. George R. Price, drafts of "Fallacies of Random Neural Networks and Self-Organization" and "A Theory of the Function of the Hymen," GPP.
17. *IBM Annual Report* (1961), 7.
18. Ibid., 6; George Price letter to Donald Ferguson, October 28, 1960, GPP; George Price letter to Richard Winslow, October 15, 1962, GPP; George Price letter to Al Somit, June 19, 1962, GPP.
19. Graham DuShane (editor) letter to George Price, July 19, 1962, GPP; George Price letter to Ralph Graves, September 5, 1970, GPP.
20. Alice Avery Price letters to George Price, November 1962–November 1963, GPP.
21. B. F. Skinner, "The Science of Learning and the Art of Teaching," *Harvard Educational Review* 24 (1954), 86–97.
22. George R. Price, "Some Suggestions About Programmed Instruction," Market Requirements Memorandum, IBM, January 31, 1963, GPP.
23. George Price letter to Richard Winslow, December 17, 1963, GPP.
24. George Price letter to Emanuel Piore, March 17, 1969, GPP; George Price letter to John Isaacson, October 28, 1964, GPP; George Price letter to Stephen (last name not known), January 30, 1972, GPP.
25. George R. Price, "The Climate of Invention," *THINK*, June 1959, 24.
26. George Price–Paul Samuelson correspondence, December 17, 1965–September 14, 1966, GPP; George R. Price, "Report on Marginal Optimization," draft, IBM-DSD Poughkeepsie, January 1964, GPP.
27. George Price letter to Paul A. Samuelson, December 17, 1965, GPP. Bob Solow and Andreas Papandreou both became prominent economists.
28. Paul A. Samuelson letter to George Price, January 13, 1966, GPP.
29. George Price letters to Fred Brooks, October 12 and 21, 1964, GPP; George Price letter to Tatiana, May 31, 1964, GPP; George Price letter to Fairfield, August 6, 1964, GPP.
30. John C. Eccles letter to George Price, January 20, 1965, BL:KPX1_3.2, BLGPC.
31. Donald Ferguson letter to George Price, September 21, 1960, GPP; George Price letter to Howard Klevens, October 31, 1967, GPP.
32. George Price letter to Donald Ferguson, March 12, 1966, GPP.
33. George Price letter to Donald Ferguson, September 4, 1966, GPP; Donald Ferguson letter to George Price, January 2, 1967, GPP.
34. St. Hilda's and St. Hugh's "Reservation Agreement," September 6, 1966, GPP; George Price letter to Annamarie Price, October 12, 1966, GPP; George Price–A. Boardman correspondence, August 26, 1966-June 12, 1967, GPP; Edison Price letter to George Price, September 11, 1970, GPP.
35. George Price letter to W. A. Brocker, July 11, 1967, GPP; George Price letter to Heinrich Kluver, July 8, 1967, BL:KPX1_1.4, BLGPC; George Price letter to

Annamarie and Kathleen Price, November 13, 1967, GPP; George Price letter to Howard Klevens, October 31, 1967, GPP.

CHAPTER 9: LONDON

1. It wouldn't last long, though. The boutique closed eight months after opening, a victim of local business interests and shoplifting. The Fool's mural was removed by civic order, but not before Paul McCartney graffitied the windows with the name of the Beatles' upcoming new hit, "Hey Jude."
2. Andrew Loog Oldham, *Stoned: A Memoir of London in the 1960s* (London: St. Martin's Press, 2001).
3. George Price letters to Tatiana, November 11 and December 12, 1967, GPP; George Price letter to Thomas Meyer, December 18, 1967, GPP.
4. Ibid.; George Price to Howard Klevens, October 31, 1967, GPP.
5. George Price letter to A. Boardman, December 25, 1967, GPP.
6. Ibid.
7. On the New Left in postwar England see Michael Kenny, *The First New Left: British Intellectuals After Stalin* (London: Lawrence and Wishart, 1995). Descriptions of the clashes at Grosvenor Square can be found in Ian Bone, *Bash the Rich: True-Life Confessions of an Anarchist in the UK* (London: Naked Guide Ltd., 2006) and in "Mick Farren—the Battle of Grosvenor Square London 1968" on YouTube.
8. Tatiana letter to George Price, December 30, 1967, GPP; George Price letter to Thomas Meyer, December 18, 1967, GPP; Edison Price letter to George Price, August 1, 1946, GPP; George Price letter to Alive Avery Price, February 27, 1968, GPP.
9. George Price letters to UCL librarian, December 12, 1967 and May 10, 1968, GPP; George Price letter to UCL Medical School Librarian, January 15, 1968, GPP; George Price letter to Thomas Meyer, December 18, 1967, GPP.
10. George Price letter to Bill Hamilton, March 5, 1968, GPP/BLGPC.
11. George Price letter to Thomas Meyer, December 18, 1967, and January 29, 1968, GPP; George Price letter to Kathleen Price, April 11, 1968, GPP.
12. Bill Hamilton letter to George Price, March 26, 1968, BL:EPX1_4 5.5.
13. Hamilton describes his own perception of George's depressing revelation in *Narrow Roads*, 320.
14. Draft of an unsent letter from George Price to Bill Hamilton, March 26, 1968, BL:KPX1_4.5.8.
15. The letter is addressed to "W.D. Hamilton, ? ?, Brazil."
16. George Price letter to Kathleen, April 11, 1968, GPP.
17. Charles Darwin, *The Descent of Man*, chapter 17. Antlers are effective shields against other branching antlers, Darwin wrote, but would not protect against unbranched antlers projecting forward.
18. G. Stonehouse, "Thermoregulatory Function of Growing Antlers," *Nature* 218 (1968), 870–72.
19. Draft of George Price, "Antlers, Intraspecific Combat, and Altruism," 1–32, BLGPC.
20. Ibid., 16.
21. The precise description was this: "Basically, *get even* behaviour requires that animal A in conflict with B should hold in memory a measure, S_B, of the seriousness of recent Category II (all-out combat) acts by B, and a measure, S_A, of its own recent category II acts against B. Then the tendency, R_{AB}, for A to retaliate against

B is given by $R_{AB} = S_B - S_A$. The strategy for A against a roughly equal opponent, B, can be stated as: 'Fight as hard as possible at Category I level (deescalated). If $S_B > S_A$, retaliate at level II. If $S_B \leq S_A$, fight at level I; except that when combat has been at level I for a long time, try a probe'" Ibid., 17.

22. Ibid., 19.
23. George did not consider that punishing exacts a price, and that free riders who fail to punish would therefore hold an advantage over those who did. This kind of free-riding game-theoretic thinking would occupy modelers in later years.
24. George Price letter to Bill Hamilton, August 3, 1968, GPP.
25. George Price letter to Thomas Meyer, October 14, 1968, GPP; John Orr declaration on Flat 3, 1a, Little Titchfield, July 30, 1968, GPP.
26. Bill Hamilton letter to George Price, August 29, 1968, GPP.
27. The condition was that there should be statistical association between the *genotypes* of donor and receiver; this is what is meant by "altruists finding one another": organisms that share those genes that contribute to their behaving altruistically coming together in groups.
28. Strictly speaking, George had not shown how the covariance equation could translate into Hamilton's kin-selection equation, replacing relatedness with association, but he understood immediately that this was now possible. For a formal derivation that does this, see Appendix 1: Covariance and Kin Selection.
29. George Price letter to Alice Avery Price, September 21, 1968, GPP.
30. George Price letter to Alice Avery Price, September 24, 1968, GPP.
31. George Price letter to Annamarie Price, October 2, 1968, GPP; George Price letter to Howard Klevens, October 9, 1968, GPP.
32. Kohn, *A Reason for Everything*, 221.
33. Steve Jones, "View from the Lab: A Mastermind for a Number of Reasons," *UK Telegraph*, January 23, 2002.
34. Ibid.; Newton E. Morton, "Professor Cedric B. Smith, pioneer in statistical genetics: Died January 10, 2002, at the age of 84," *Genetic Epidemiology* 22, no. 4 (2002), 283–84.
35. George Price letter to Annamarie Price, October 2, 1968; interview with Annamarie Price, April 13, 2008.
36. George Price, "Supplementary Details of Intended Research," draft proposal to the Science Research Council, BLGPC, KPX_5.4.
37. A corollary of this system would be for males to take advantage of any mating opportunity outside of marriage, but not to provide for any children born. Thus, two distinct categories of female sexual partners would become recognized, with dichotomous treatment of each.
38. George Price letter to Ludwig Luft, January 29, 1969, GPP. George asked Luft, a German-born Jewish émigré to America, for the lyrics from Schumann's *Dichterliebe* for a paper he was preparing on the evolution of love.
39. George Price letter to Edison Price, February 3, 1969, GPP; Edison Price letters to George, February 22 and March 2, 1969, GPP; George Price letter to Kathleen Price, March 5, 1969, GPP.
40. George Price letter to Kathleen Price, May 1, 1969, GPP.
41. Interview with Kathleen Price, April 13, 2008; George Price letter to Ludwig Luft, July 22, 1969, GPP.
42. George Price letter to Tatiana, May 11, 1969, GPP; George Price letter to Kathleen Price, March 5, 1969, GPP. In truth Edison had paid for George's airfare from London and expenses in New York, but George was sour over an unpaid loan and little help with the apartment.

43. George Price letter to Ludwig Luft, July 22, 1969, GPP; George Price letter to Tatiana, May 11, 1969, GPP.
44. Garrett Hardin, "The Tragedy of the Commons," *Science* 162 (1968), 1243–48.
45. The amateur mathematician was William Forster Lloyd (1794–1852) and the source, W. F. Lloyd, *Two Lectures on the Checks to Population* (Oxford: Oxford University Press, 1833), reprinted (in part) in *Population, Evolution, and Birth Control*, ed. Garret Hardin (San Francisco: Freeman, 1964), 37.
46. Hardin, "The Tragedy of the Commons," 1244.
47. Sewall Wright letter to George Price, May 22, 1968, GPP
48. George Price, "Supplementary Details of Intended Research," draft proposal to the Science Research Council, BLGPC, KFX_5.4. These observations would later be challenged in the ethological literature.
49. SRC grant proposal, written by Cedric Smith, March 31, 1969, and signed by the head of the department, Harry Harris, BLGPC, KPX1_10.1.
50. George Price letter to Tatiana, May 11, 1969, GPP.
51. John S. Price, "The Ritualization of Agonistic Behaviour as a Determinant of Variation along the Neuroticism/Stability Dimension of Personality," *Proceedings of the Royal Society of Medicine* 62 (1969), 37–40.
52. Ibid., 39.
53. See S. A. Frank and M. Slatkin, "The Distribution of Allelic Effects under Mutation and Selection," *Genetics Research* 55 (1990), 111–17.
54. See Appendix 2: The Full Price Equation and Levels of Selection for a derivation of the full Price equation and an explanation of how it can be partitioned and expanded.
55. Hamilton, *Narrow Roads*, 173.
56. W. D. Hamilton, "Selection of Selfish and Altruistic Behaviour in Some Extreme Models," in *Man and Beast: Comparative Social Behavior*, ed. J. F. Eisenberg and W. S. Dillon (Washington, DC: Smithsonian Press, 1971), 57–91.
57. For Hamilton's entertaining description of the conference see *Narrow Roads*, 185–97; quotes on 187, 189. Hamilton's reply, following his hero Fisher before him, was eugenics. In the long run the best humans would have to be consciously selected.
58. Ibid., 173.
59. George Price letter to Bill Hamilton, July 28, 1969, BLGPC, EL:KPX1_4.4. The precise correction George alluded to was that in his 1964 paper Hamilton had failed to state that his kin-selection model only held for an indefinitely large population where the average relatedness between individuals was zero, whereas in realistic finite populations that wasn't quite the case, and it was there that spite could arise.
60. Hamilton, *Narrow Roads*, 172–73. Hamilton wrote to a friend at the time that he hadn't yet thought deeply about Price's covariance equation, "apart from seeing how it provides just the key I was looking for in about 1962 when trying to get the ideas of social selection into mathematical form as simply as possible." But Hamilton sensed nonetheless that the covariance equation "may be something like the introduction of Fisher's idea of analysis of variance into statistics." Hamilton letter to Colin Hudson, July 26, 1970, GPP.
61. W. D. Hamilton, "Geometry for the Selfish Herd," *Journal of Theoretical Biology* 31 (1971), 295–311; George Price letter to A. Somit, July 20, 1969, GPP; George Price letter to Aunt Ethel, July 30, 1969, GPP; George Price letter to Ludwig Luft, July 22, 1969, GPP.
62. George Price letter to Annamarie Price, November 11, 1969, GPP.

63. John Maddox letter to George Price, February 11, 1970, BLWHC, Z1X102.1.1.2.3; George Price letter to Edison, April 19, 1970, GPP.

64. George Price letter to Ludwig Luft, June 8, 1970, BLGPC, KPX1_1.6; The paper was revised and sent to *Science* as "The Nature of Selection" in December 1970 and rejected again. It was published posthumously, through the efforts of Steven Frank, in the *Journal of Theoretical Biology* 175 (1995), 389–96; George even wrote to Shannon on October 16, 1969 (GPP), to ask whether he thought the covariance equation could be interpreted "in terms of information channel capacity."

65. George Price correspondence with Al Somit, March 5, 1969–May 20, 1970, GPP; interview with Al Somit, April 16, 2008.

CHAPTER 10: "COINCIDENCE" CONVERSION

1. Bill Hamilton letters to George Price, December 5, 1969, January 11, 1970, BLGPC, BL:KPX1_4.5, and BL:KPX1_4.13.

2. J. H. Morris, assistant editor at *Nature*, letter to Bill Hamilton, April 10, 1970, KPX1_10.6.2; W. D. Hamilton, "Selfishness and Spiteful Behaviour in an Evolutionary Model," *Nature* 228 (1970), 1218–20.

3. George Price letter to J. B. Rhine, September 10, 1971, GPP; George Price letter to Anne Sheffield, December 10, 1970, GPP.

4. George Price letter to Edison Price, October 23, 1970, GPP.

5. George Price letter to Anne Sheffield, December 10, 1970, GPP. In fact Anne returned to London with her friend Rickie, though George later claimed that she had written the letter in such a way so as to hint that she'd be alone, a fact that had gotten him excited, even though she would claim that he had misinterpreted her.

6. George Price letter to Dr. Gilfillan, July 1, 1972, GPP; George Price letter to Henry Noel, August 11, 1970, GPP.

7. George Price letter to Anne Sheffield, May 27, 1970, GPP.

8. George Price letter to Henry Noel, July 13, 1970, GPP.

9. George Price letter to Dr. Gilfillan, July 1, 1972, GPP; George Price letter to Anne Sheffield, December 10, 1970.

10. "A History of All Souls Church at Langham Place," booklet distributed at the church.

11. George Price letter to Anne Sheffield, December 10, 1970, GPP.

12. George Price diary 1970, GPP.

13. Timothy Dudley-Smith, *John Stott: The Making of a Leader* (Leicester: Inter-Varsity Press, 1999). In a November 30, 2004, *New York Times* editorial, "Who Is John Stott?," columnist David Brooks cited Michael Cromartie of the Ethics and Public Policy Center as saying that "if evangelicals could elect a pope, Stott is the person they would likely choose."

14. When Jim Schwartz spoke with the elderly John Stott in December 1999, he didn't remember George but said that it was quite possible that he'd forgotten, and that All Souls "got a lot of kooks." I thank Jim Schwartz for this information.

15. Class of 1944, *Triennial Report*, Harvard University Archive (HUA), 273–74; Class of 1944, *Ten Year Report*, HUA, 209–10; "Harvard Alumnus Renounces U.S.," *Boston Traveler*, February 17, 1948; "Citizen of the World Detained by French," *Plainfield Courier-News*, November 17, 1948; "News Briefs," *Dallas Times-Herald*, November 18, 1948.

16. Henry Noel letter to George Price, July 7, 1970, GPP.
17. Henry Noel letter to George Price, July 16. 1970, GPP.
18. George Price letter to Henry Noel, July 27. 1970, GPP.
19. Henry Noel letter to George Price, July 16. 1970, GPP.
20. George Price letters to Henry Noel, July 27. August 9, and August 12, 1970, GPP. Henry didn't particularly appreciate being lectured by George. "Please be so good as to lay off for a while," he wrote to him in November, and the correspondence broke off temporarily.
21. George Price letter to Anne Sheffield, December 10, 1970, GPP.
22. Ibid.; George Price letter to Julia Price. October 13, 1970, GPP.
23. Ibid.; George Price letter to Henry Noel, August 11, 1970, GPP; George Price letter to Howard Klevens, December 18, 1970, GPP.
24. Bill Hamilton letter to Edison Price. February 15, 1975, BLWHC, Z1X102_1.1.20; Bill Hamilton letter to Colin Hudson, July 26, 1970. I thank Janet Hamilton for sending me this letter.
25. Interview with Janet Hamilton, October 24, 2007.
26. Bill Hamilton letter to George Price, July 2, 1970, BLGPC, BL:KPX1_4.8.
27. George Price letter to Louis J. Vorhous, December 18, 1970, GPP.
28. George Price letter to Caroline Doherety, August 9, 1970, GPP; George Price letter to Marie Lynch, October 14, 1970, GPP; George Price letter to Al Somit, October 19, 1970, GPP. Somit was actually not all that surprised by either development. "I had been expecting the former given the devotedness of your atheism. . . . As for the latter development, I had perhaps been prepared for it by the strength of your expressed anti-Semitism." Al Somit letter to George Price, December 2, 1970, GPP.
29. George Price letter to Caroline Doherety, August 9, 1970, GPP.
30. Communication with Kathleen Price, March 8, 2009; "Guest List: Kathy's 21st Birthday," GPP.
31. George Price letter to Rosemarie Hudson, August 29, 1970, GPP.
32. George Price letter to Bill Hamilton, September 21, 1970, BLGPC, BL:KPX1_4.9.
33. Edison Price letter to George Price, September 11, 1970, GPP.
34. Rosemarie Hudson letter to George Price, November 2 , 1968, GPP; George Price letter to Julia Price, October 13, 1970. GPP; George Price letter to Al Somit, October 19, 1970, GPP; George Price letter to Ludwig Luft, November 4, 1970, GPP.
35. George Price letter to Rosemarie Hudson, August 29, 1970, GPP; Rosemarie Hudson letter to George, undated (circa end of August 1970), GPP; George Price letters to Rosemarie Hudson, October 23 and 25, 1970, GPP.
36. George Price letter to Annamarie and Kathy Price, November 7, 1970. GPP.
37. George Price letter to Kathleen Price, November 30, 1970, GPP.
38. George Price letter to Rosemarie Hudson, October 25, 1970, GPP.
39. George Price, "The Twelve Days of Easter," manuscript dated March 7, 1971, GPP; George Price letter to Dr. Gilfillan, April 4, 1971, GPP.
40. George Price letter to Dr. Lukas Vischer, April 22, 1971, GPP; George Price letter to the Reverend Alan N. Stibbs, January 18, 1971, GPP; George Price correspondence with Jack Finegan, January–February, October 1971, GPP; George Price letter to F. F. Bruce, July 5, 1971. GPP. A reply from the World Council of Churches arrived in May, promising, somewhat unconvincingly, that the "paschal problem" was "on the agenda."
41. Bill Hamilton letter to George Price, April 15, 1971, BLGPC, BL:KPX1_4.10.
42. George Price letter to Bill Hamilton, May 3, 1971, BLGPC, BL:KPX1_5.9.

43. Bill Hamilton postcard to George Price, May 5, 1971, GPP.
44. George Price letter to Bill Hamilton, November 15, 1971, BLGPC, BL:KPX1_5.3.
45. George Price letter to Bill Hamilton, July 29, 1970, BLGPC, BLKPX1_2.7.
46. George Price letter to Annamarie Price, July 15, 1971, GPP; ibid.; Revelation 13:18 reads: "this calls for wisdom: let the one who has understanding calculate the number of the beast, for it is the number of a man, and his number is 666." George mentions a number of times in his letters that he had cracked the true meaning, but not once what it actually was.
47. George Price letter to Bill Hamilton, op. cit.
48. Fisher first presented the theorem in his classic 1930 book, *The Genetical Theory of Natural Selection*. But he repeated it in "Average Excess and Average Effect of a Gene Substitution," *Annals of Eugenics* 11 (1941), 53–63, and in the second edition of *The Genetical Theory*, published in 1958.
49. J. F. Crow and M. Kimura, *An Introduction to Population Genetics Theory* (New York: Harper & Row, 1970); J. R. Turner, "Changes in Mean Fitness Under Natural Selection," in *Mathematical Topics in Population Genetics*, ed. K. Kujima (Berlin: Springer-Verlag, 1970); O. Kempthorne, *An Introduction to Genetical Statistics* (London: Champan and Hall, 1957); A. W. F. Edwards, "Fundamental Theorem of Natural Selection," *Nature* 215 (1967), 537–38.
50. George R. Price, "Fisher's 'Fundamental Theorem' Made Clear," *Annals of Human Genetics* 36 (1972), 129–40, quote on 140. To see how George's covariance mathematics prepared him to understand Fisher's theorem, see Appendix 3.
51. George Price correspondence with Henry Morris, June–October 1971, GPP; John C. Whitcomb and Henry Morris, *The Genesis Flood: The Biblical Record and Its Scientific Implications* (Philadelphia: Presbyterian & Reformed Publishing, 1961); Matt Schudel, "Obituary: Henry M. Morris, Father of 'Creation Science,'" *Seattle Times*, March 5, 2006.
52. George Price letter to Henry Morris, September 10, 1971, GPP.
53. George Price letter to Henry Morris, October 10, 1971, GPP.
54. George Price letter to Rosemarie Hudson, undated, circa February 1972, GPP
55. George Price letter to Ludwig Luft, August 8, 1971, BLGPC, KPX1_1.6.
56. This includes gene interaction, for even two respectively "fit" genes (here defined as the "environment" one of the other) may interact so as to produce fitness degrading effects.
57. George R. Price, "Fisher's 'Fundamental Theorem' Made Clear."
58. Warren J. Ewens later published a similar conclusion to George's: "An Interpretation and Proof of the Fundamental Theorem of Natural Selection," *Theoretical Population Biology* 36 (1989), 167–80. I thank Warren for extended communications over Fisher and the true meaning of his fundamental theorem.
59. Bill Hamilton letter to George Price, November 28, 1971, BLGPC, BL:KPX1_4.15.
60. Ibid.
61. Ibid. The article Hamilton had read was Richard Levins, "Extinction," in *Some Mathematical Problems in Biology*, ed. M. Gesternhaber (Providence, RI: American Mathematical Society, 1970), 77–107.
62. George Price letter to Bill Hamilton, March 14, 1972, BLGPC, BL:KPX1_4.11.
63. George Price letters to Joseph Rhine, December 19, 1970, April 4 and September 10, 1971, GPP; George Price correspondence with Hugh Ross, July, 1971–March 1972, GPP.
64. Joseph Rhine letters to George Price, December 29, 1970, and April 13, 1971,

GPP. For a history of the psychic and parapsychological movements see Mauskopf and McVaugh, *The Elusive Science.*

65. Joseph Rhine letter to George Price, November 22, 1971 GPP; George Price letter to Joseph Rhine, December 19 1971, GPP.

66. George R. Price, "Apology to Rhine and Soal," *Science* 75 (1972), 359; Price-Rhine correspondence, January–October 1972; last quote from letter dated October 19, 1972, GPP.

CHAPTER 11: "LOVE" CONVERSION

1. Randall E. King, "When Worlds Collide: Politics, Religion, and Media at the 1970 East Tennessee Billy Graham Crusade (Appearance by President Richard M. Nixon)," *Journal of Church and State* 39, no. 2 (1997), 273–96; George Price letter to Billy Graham, September 1970; reply from assistant to Billy Graham, September 24, 1970, GPP.

2. George Price letter to Annamarie and Kathleen, August 2, 1970, GPP; George Price letter to Al Somit, August 13, 1970, GPP; Michael Simpson letter to George Price, July 25, 1970, GPP.

3. Al Somit letter to George Price, December 2, 1970, GPP; George Price letter to Al Somit, December 12, 1970, GPP.

4. Al Somit letter to George Price, February 8, GPP.

5. George Price letter to Al Somit, September 11, 1971, GPP.

6. Al Somit letter to George Price, October 12, 1971, GPP.

7. Lorenz shared the 1973 Nobel Prize in Physiology or Medicine with Niko Tinbergen and Karl von Frisch. On Lorenz, his co-Nobelists, and the history of ethology, see Richard W. Burkhardt, Jr., *Patterns of Behavior: Konrad Lorenz, Niko Tinbergen, and the Founding of Ethology* (Chicago: University of Chicago Press, 2005).

8. Lorenz eventually published his theories on animal fighting in book form in *On Aggression* (London: Methuen, 1963). The assumption of the wide prevalence of nonviolent ritualized combat has since been softened—animals, it seems, escalate conflict more than was previously suspected.

9. John Maynard Smith, "Equations of Life" in *It Must Be Beautiful: Great Equations of Modern Science*, ed. Graham Farmelo (London: Granta, 2003), 161–80, 166; John Maynard Smith, "In Haldane's Footsteps," 351. Alongside Lorenz, T. H. Huxley's grandson Julian Huxley also espoused a group selectionist argument to explain ritualized combat: See J. S. Huxley, "Ritualization of Behaviour in Animals and Man," *Philosophical Transactions of the Royal Society B* 151 (1966), 249–71.

10. R. Duncan Luce and Howard Raiffa, *Games and Decisions: Introduction and Critical Survey* (New York: John Wiley, 1953).

11. R. C. Lewontin, "Evolution and the Theory of Games," *Journal of Theoretical Biology* 1 (1961), 382–403; interview with Richard Lewontin, December 31, 2007; on the history of the divide Lewontin alludes to, see James Schwartz, "Population Genetics and Sociobiology: Conflicting Views of Evolution," *Perspectives in Biology and Medicine* 45, no. 2 (2002), 224–40.

12. Maynard Smith eventually published *The Evolution of Sex* (Cambridge: Cambridge University Press, 1978).

13. John Maynard Smith, "Evolution and the Theory of Games," *American Scientist* 64 (1976), 41–45.

14. Maynard Smith later explained that he was aware of the "unbeatable strategy"

term in Hamilton's 1967 paper, but hadn't registered this as influencing his ideas when formalizing the ESS. "In Haldane's Footsteps," 352.

15. Ibid.; George Price letter to Richard Lewontin, September 15, 1970. I thank Dick Lewontin for providing me with their correspondence.

16. George Price letter to John Maynard Smith, August 9, 1971, GPP.

17. George R. Price, "Extension of Covariance Selection Mathematics," *Annals of Human Genetics* 35 (1972), 485–90. George had recently learned (and acknowledged in this paper) that Alan Robertson had published a covariance selection equation prior and similar to his own in a somewhat obscurely placed article, "A Mathematical Model of the Culling Process in Dairy Cattle," *Animal Production* 8 (1966), 95–108. Robertson, however, a researcher more concerned with breeding practice than evolution, had not added the second, transmission term to his equation, meaning that it was not expansible to multiple levels of selection, which was the whole beauty of George's approach. A few months after George's own derivation of the equation, the mathematical biologist Joel E. Cohen, then at the Society of Fellows at Harvard University, independently derived the covariance relationship (again, without its expansible component) in an attempt to determine whether the differential use of legal abortions by people of varying socioeconomic or educational status had a positive or negative selective effect on measured IQ. It was published nine months after George's *Nature* paper as "Legal Abortions, Socioeconomic Status, and Measured Intelligence in the United States," *Social Biology* 18 (1971), 55–63. Cohen had derived the covariance from what he took to be "an easy consequence" of the formula for the selection differential published by the eminent Edinburgh quantitative geneticist Douglas S. Falconer, and was therefore not all impressed by either his or, when he learned of it, George's, originality. See, D. S. Falconer, "Genetic Consequences of Selection Pressure," in *Genetic and Environmental Factors in Human Ability*, ed. J. E. Meade and A. S. Parkes (Edinburgh: Oliver and Boyd, 1966), 219–32. There was a further antecedent to the covariance equation that George was unaware of, once again without the second term: C. C. Li, "Fundamental Theorem of Natural Selection," *Nature* 214 (1967), 505–6. I thank Joel Cohen, now a professor at the Laboratory of Populations at Rockefeller and Columbia Universities, for correspondence about this matter throughout May 2008.

18. George Price letter to Frieda (last name not known), May 29, 1971, GPP.

19. Lewontin would summarize his finding in the classic book *The Genetic Basis of Evolutionary Change* (New York: Columbia University Press, 1974).

20. Charles Darwin, *The Origin of Species*, 84; Motoo Kimura original paper was "Evolutionary Rate at the Molecular Level," *Nature* 217 (1968), 624–26. His major findings were later summarized in his book *The Neutral Theory of Molecular Evolution* (Cambridge: Cambridge University Press, 1983). There is much literature on the history of the neutralism debate, see in particular Michael Dietrich, "The Origins of the Neutral Theory of Molecular Evolution," *Journal of the History of Biology* 27 (1994), 21–59, and James F. Crow, "Motoo Kimura and the Rise of Neutralism," in *Rebels, Mavericks, and Heretics in Biology*, ed. Oren Harman and Michael Dietrich (New Haven: Yale University Press, 2008), 265–81.

21. George viewed the genetic polymorphism work as a "digression" from the "Sex and Rapid Evolution" paper he was working on at the time but undertook it nevertheless since he felt it would help him better understand a point involved in the sex paper: George Price letter to John Maynard Smith, April 20, 1972; "Antlers" File, John Maynard Smith Papers, British Library (BLJMSC).

22. Harry Harris, "Polymorphism and Protein Evolution: The Neural Mutation–

Random Drift Hypothesis," *Journal of Medical Genetics* 8, no. 4 (December 1971) 444–52; "A Theoretical Investigation into Genetic Polymorphism," MRC application for a project grant, signed by Cedric A. B. Smith, January 31, 1973, BL:KPX1_10.2.

23. George Price letter to Paul Samuelson, August 19, 1972, GPP. Samuelson had become interested in population genetics and written a short article on "The Hardy-Who Law of Genetics?" in March 1971, which he now sent to George.

24. George Price letter to John Maynard Smith, April 20, 1972.

25. The model used the following probabilities: Probability of serious injury from a single D play = 0.10. Probability that a "Prober-Retaliator" will probe on the opening move or after opponent has played C = 0.05. Probability that "Retaliator" or "Prober-Retaliator" will retaliate against a probe (if not injured) by opponent = 1.0. Payoffs were calculated as follows: Payoff for winning = +60. Payoff for receiving serious injury = -100. Payoff for each D received that does not cause serious injury (a "scratch") = -2. Payoff for saving time and energy (awarded to each contestant not seriously injured) varied from 0 for a contest of maximum length to +20 for a very short contest.

26. George Price letter to Kathleen Price, August 5, 1972, GPP.

27. George Price letter to John Maynard Smith, October 19, 1972, BLJMSC.

28. Ibid.

29. John Maynard Smith letter to George Price, October 24, 1972, BLJMSC; Bill Hamilton letter to George Price, November 22, 1972, BL:KPX1_4.14.

30. George Price letter to Freda (last name not known), May 29, 1971, GPP; Interview with Sam Berry, May 5, 2008.

31. George Price letter to Dr. Webb, December 21, 1972, GPP; George Price letter to the Home Office, Immigration and Nationality Department, December 28, 1972, GPP.

32. George Price letter to Annamarie Price, January 23, 1973, GPP.

33. The Latin is: "Amicabilia quae sunt ad alterum vererunt amicabilibus quae sunt ad seipsum." Aristotle, *Nichomachean Ethics*, book IX, chapter 4.

34. Thomas Aquinas, *The Summa Theologica of St. Thoma Aquinas* 5 vols., trans. Fathers of the English Dominican Province (Westminster, MD: Christian Classics, 1981); David Hume, *A Treatise of Human Nature* (1739), ed. L. A. Selby-Bigge and P. H. Nidditch (Oxford: Clarendon Press, 1978), 521.

35. Adam Smith, *An Inquiry into the Nature and Causes of the Wealth of Nations*, 2 vols., ed. R. H. Campbell and A. S. Skinner (Indianapolis: Liberty Fund, 1981), I.ii.2, 26–27; Bernard Mandeville, *The Fable of the Bees: or, Private Vices, Publick Benefits*, 4th ed. (London: J. Tonson, 1725), vol. 1, 9.

36. Andrew Brown, "The Kindness of Strangers," *The Guardian*, August 27, 2005; Roger Bingham, "Trivers in America," *Science* 80 (March–April 1930), 56–67; communication with Robert Trivers, May 27, 2008.

37. Robert Trivers, "The Evolution of Reciprocal Altruism," *Quarterly Review of Biology* 46 (1971), 35–57. For a description of writing this paper see Trivers, *Natural Selection and Social Theory* (Oxford: Oxford University Press, 2002), 3–18.

38. "A Theoretical Investigation," MRC application for a project grant.

39. Ibid.

40. George Price letter to Bill Hamilton, May 3, 1971, BLGPC, BL:KPX1_5.9.

41. George Price letter to Bill Hamilton, May 30, 1973, BLGPC, KPX1_2.4.

42. George Price letter to Annamarie Price, March 3, 1973, GPP.

43. George Price letter to Rosemarie Hudson, February 23, 1973, GPP; Henry Noel letter to George Price, January 30 1973, GPP.

44. George Price letter to Kathleen Price, March 24, 1973, GPP.
45. George Price letter to Julia Price, March 10, 1973, GPP.
46. George Price letter to Julia Price, March 24, 1973, GPP.
47. Jack London, *The People of the Abyss* (London: Macmillan, 1903).
48. See Exploring 20th Century London—Homelessness, Museum of London, www.20thcenturylondon.org.uk/server.php?show=conInformationRecord.77, for a well-researched, colorful exhibition. For a classic study of homelessness see Christopher Jencks, *The Homeless* (Cambridge, MA: Harvard University Press, 1994).
49. George Price letter to Jack (surname unknown—a functionary at Saint Mark's Church), August 18, 1973, GPP.
50. John Price letter to George Price, April 5, 1973, GPP; Henry Noel letter to George Price, April 7, 1973, GPP; Tatiana letter to George Price, August 17, 1973, GPP; Edison Price letter to George Price, April 16, 1973, GPP.
51. George Price to Jack, August 18, 1973.
52. The Reverend R. F. H. Howarth letter to George Price, July 16, 1973, GPP; George Price letter to Henry Noel, July 22, 1973, GPP.
53. George Price letter to Kathleen Price, March 19, 1973, GPP.

CHAPTER 12: RECKONINGS

1. UCL letter to George Price (signed by the assistant secretary of personnel), April 26, 1973, GPP; George Price letter to Julia Price, May 6, 1973, GPP.
2. George Price letter to Mr. Norman Ingram-Smith, May 27, 1973, GPP.
3. Ibid.
4. George Price letter to Jack, August 18, 1973.
5. Communication from Ursula Mittwoch, May 4, 2008; George Price letter to Dr. Gilfillan, July 3, 1973, GPP.
6. Trevor Russell (Smoky) to George Price, July 3, 1973, GPP.
7. Trevor Russell letter to George Price, July 9, 1973, GPP.
8. Muriel Challenger letter to George Price, September 3, 1973, GPP.
9. Julia Price letter to George Price, July 30, 1973, GPP.
10. George Price letter to Smoky, September 14, 1973, GPP.
11. Ibid.
12. George Price letter to John Maynard Smith, October 19, 1973, BLJMSC, "Conflict Draft" folder.
13. George Price letter to John Maynard Smith, February 12, 1973, BLJMSC, "Conflict Draft" folder.
14. John Maynard Smith letter to George Price, October 24, 1972, BLJMSC, "Conflict Draft" folder. In fact Maynard Smith had been more than scrupulous and entirely generous both to have initially contacted George to get his permission to cite his unpublished antlers paper, and then in offering joint authorship. As for the 1964 Hamilton affair, Maynard Smith always had a conscience about it and did his best to set the record straight. He often remarked that he had a knack, blessed or cursed, for getting ideas from papers he reviewed.
15. J. Maynard Smith and G. R. Price, "The Logic of Animal Conflict," *Nature* 246 (1973), 15–18, quote on p. 15. Half a year later, in "On Fighting Strategies in Animal Combat," *Nature* 250 (1974), 354, Valerius Geist of the University of Calgary accused John and George of not acknowledging his prior explication of the retaliation principle in V. Geist, "On the Evolution of Hornlike Organs,"

Behaviour 27 (1966), 175. It had been an oversight that George and John were sorry about, and George wrote a letter of apology. See George Price letter to V. Geist, March 24, 1974, BLJMSC, "Antlers File."

16. Al Somit letter to George Price, September 27, 1973, GPP.

17. George Price letter to Edison Price, September 26, 1973, GPP.

18. Cedric Smith note to George Price, November 7, 1973, GPP; Paul Garvey letter to George Price, November 11, 1973, GPP.

19. George Price letter to Joan Jenkins, November 15, 1973. I thank Jim Schwartz for providing me with copies of the correspondence between Joan and George.

20. "Monthly Message No. 202," June 1974, London Healing Mission, GPP.

21. George Price letter to Kathleen Price, December 6, 1973, GPP.

22. George Price letter to Morris Goodman, June 7, 1974, GPP; phone interview with Professor Sam Berry, May 5, 2008.

23. Interviews with Al Somit, December 6, 2007 and April 10, 2008, and communication on May 6, 2009.

24. Al Somit letter to George Price, November 27, 1973, GPP.

25. On the speculating bonanzas in London in the sixties see Oliver Marriot, *The Property Boom* (London: Hamish Hamilton, 1967); Nick Wates, *The Battle for Tolmers Square* (London: Routledge and Kegan Paul, 1976), 41.

26. Wates, *The Battle for Tolmers Square*, 23, 24.

27. Only 4 percent of the properties in the neighborhood were occupied by their owners, whereas 80 percent, three-quarters of which unfurnished, were being rented from a private landlord. Wates; ibid., 8, 9.

28. *Private Eye*, February 9, 1973; Wates, *The Battle for Tolmer Square*, 43–73, quotes on 48, 68, 71, 43.

29. Communication from Paul Nicholson, February 4, 2008; Wates, *The Battle for Tolmers Square*, 120.

30. Wates, *The Battle for Tolmers Square*, 129; communication with Paul Nicholson, May 10, 2009.

31. Communications with Ches Chesney, January 14 and 15, 2008.

32. Paul Nicholson, "Room At the Top," unpublished short story.

33. Wates, *The Battle for Tolmers Square*, 162. Also see Nick Wates and Christian Wolmer, eds., *Squatting: The Real Story* (London: Bay Leaf Books, 1980).

34. Communication with Paul Nicholson, February 4, 2008.

35. Paul Nicholson, "Room At the Top."

36. Ibid.

37. George Price letter to Dr. P. J. D. Heaf, December 16, 1973.

38. George Price letter to the official solicitor, Royal Courts of Justice, January 30, 1974, GPP.

39. James Schwartz, "Death of an Altruist," *Lingua Franca* 10, no. 5 (July/August 2000), 51–61, quote on 59,

40. Beginning in February, George had embarked on a project involving polymorphism analysis of macaque data from Morris Goodman, a professor of anatomy at Wayne State University, via Harris's collaborator Dr. Nigel Barnicot, an anthropologist at UCL.

41. Bill Hamilton letter to Dr. Kelly, undated (circa December 10, 1974), BLWHC, Z1X102_1.1.21; Hamilton, *Narrow Roads* 174.

42. George Price letter to Bill Hamilton, March 4, 1974, BLGPC, KPX1_2.5; George Price letter to Annamarie Price, April 27, 1974, GPP.

43. George Price letter to Bill Hamilton, March 4, 1974, *op. cit.*

44. Bill Hamilton letter to George Price, undated (circa March 6, 1974) BLGPC,

BL:KPX1_4.12; Bill Hamilton letter To Whom It May Concern, on Imperial College letterhead, March 14, 1974, BLGPC KPX1_6 arrow KPX1_10.7.2.

45. George Price letter to Bill Hamilton, May 18, 1974, BLGPC, KPX1_2.5.

46. I thank Jim Schwartz for kindly sharing with me his notes from his meeting and conversations with Joan Jenkins from 2000. Joan Jenkins died shortly after. For her work on estrogen replacement therapy she was granted an OBE.

47. George Price letter to Kathleen Price, January 25, 1974, GPP; George Price letter to Joan Jenkins, May 19, 1974; Schwartz notes.

48. George Price letters to Joan Jenkins, May 20, June 6, June 9, 1974. Often the liaison between the two was Ruth Lang, CABS's secretary at the Galton.

49. George Price letter to Kathleen Price, January 25, 1974, GPP; interview with Kathleen Price, April 12, 2008.

50. George Price letter to Kathleen Price, May 31, 1974, GPP.

51. Bill Hamilton letter to George Price, June 11, 1974, BLGPC, BL:KPX1_4.6.; George Price letter to Bill Hamilton, August 21, 1974, BLGPC, KPX1_2.5.

CHAPTER 13: ALTRUISM

1. Immanuel Kant, *Critique of Practical Reason and Other Works on the Theory of Ethics*, trans. Thomas Kingsmill Abbott, 4th rev. ed. (London: Kongmans, Green and Co., 1889), 260.

2. Williams, *Adaptation and Natural Selection*, 93.

3. Richard Dawkins, *The Selfish Gene* (Oxford: Oxford University Press, 1996), 201, 254.

4. Gene-centrism became wedded to "adaptationism," and huddled together within a worldview called sociobiology, both were furiously attacked from different quarters. In particular see the anthropologist Marshall Sahlins, *The Use and Abuse of Biology: An Anthropological Critique of Sociobiology* (Ann Arbor: University of Michigan Press, 1976); the philosopher Mary Midgley, "Gene-Juggling," *Philosophy* 54 (1979), 439–58; and the biologists Steven J. Gould and Richard R. Lewontin, "The Spandrels of San Marco and the Panglossian Paradigm: A Critique of the Adaptationist Programme," *Proceedings of the Royal Society of London B* 205 (1979), 581–98.

5. The notion of altruism as a mistake (due to imperfect design) is strengthened by the observation that some species of host birds have developed anticuckoo behavior. See A. Moksnes, E. Roskaft, A. T. Braa, L. Korsnes, H. M. Lampe, and H. C. Pedersen, "Behavioral Responses of Potential Hosts Towards Artificial Cuckoo Eggs and Dummies," *Behaviour* 116 (1990), 64–89.

6. P. W. Sherman, J. Jarvis, and R. Alexander, eds., *The Biology of the Naked Mole-Rat* (Princeton: Princeton University Press, 1991); D. W. Pfennig, H. K. Reeve, and P. W. Sherman, "Kin Recognition and Cannibalism in Spadefoot Toads," *Animal Behavior* 46 (1993), 87–94; Robert B. Payne, Michael D. Sorenson, and Karen Klitz, *The Cuckoos: Cuculidae* (Oxford: Oxford University Press, 2005).

7. John Tyler Bonner, *The Social Amoebae: The Biology of Cellular Slime Molds* (Princeton: Princeton University Press, 2009).

8. For a nontechnical description of this research program, see "Altruism among Amoebas," Joan E. Strassman and David C. Queller, *Natural History* 116 (September 2007), 24–29.

9. Martin Daly and Margo Wilson, *The Truth about Cinderella: A Darwinian View of Parental Love* (New Haven: Yale University Press, 1999). There has been, of

course, much criticism of Daly and Wilson's claims. For a study presenting contradictory data see H. Temrin, S. Buchmayer, and M. Enquist, "Step-Parents and Infanticide: New Data Contradict Evolutionary Predictions," *Proceedings of the Royal Society of London* 267 (2000), 943–45.

10. J. S. Gale and L. J. Evans showed that Retaliator was not, in fact, an ESS. See their "Logic of Animal Conflict," *Nature* 254 (1975), 463–64.

11. The name "Mouse" appeared only once in the literature—in George and Maynard Smith's 1973 paper. From then on it was "Dove."

12. G. A. Parker and E. A. Thompson, "Dung Fly Struggles: A Test of the War of Attrition," *Behavioral Ecology and Sociobiology* 7 (1980), 37–44; John Maynard Smith laid down the logic of games in evolution in his classic book *Evolution and the Theory of Games* (Cambridge: Cambridge University Press, 1982); for a good explanation of the limits and advantages of game theory for understanding natural evolution see G. A. Parker and J. Maynard Smith, "Optimality Theory in Evolutionary Biology," *Nature* 348 (1990), 27–33.

13. Amos Twersky, Daniel Kahneman, and Paul Slovic, eds., *Judgment under Uncertainty: Heuristics and Biases* (Cambridge: Cambridge University Press, 1980); Larry Samuelson, *Evolutionary Games and Equilibrium Selection* (Cambridge, MA: MIT Press, 1997), provides a technical account of how theoretical economists have adopted evolutionary ideas; Ran Shpigler, "The Invisible Hand and Natural Selection," *Odyssey* 2 (2009), 45–51 (in Hebrew)

14. Trivers, "The Evolution of Reciprocal Altruism," and Trivers, *Natural Selection and Social Theory*, 3–18, 51–55; Robert Axelrod and William Hamilton, "The Evolution of Cooperation," *Science* 211 (1981), 1390–96

15. C. Packer, "Reciprocal Altruism in *Papio anubis*," *Nature* 265 (1977), 441–43; G. S. Wilkinson, "Reciprocal Food Sharing in Vampire Bats," *Nature* 309 (1984), 181–84; M. Melinski, "Tit for Tat in Sticklebacks and the Evolution of Cooperation," *Nature* 325 (1987), 433–35; H. Godfray, "The Evolution of Forgiveness," *Nature* 355 (1992), 206–7; A. H. Harcourt and F. B. M. de Waal, eds., *Coalitions and Alliances in Humans and Other Animals* (New York: Oxford University Press, 1992); S. Frank, "Mutual Policing and Repression of Competition in the Evolution of Cooperative Groups," *Nature* 377 (1995), 520–22; T. H. Clotton-Brock and G. Parker, "Punishment in Animal Societies," *Nature* 373 (1995), 209–16; Lee Alan Dugatkin, *Cooperation Among Animals: An Evolutionary Perspective* (New York: Oxford University Press, 1997); M. Nowak and K. Sigmund, "Evolution of Indirect Reciprocity by Image Scoring," *Nature* 393 (1998), 573–77 and "The Dynamics of Indirect Reciprocity," *Journal of Theoretical Biology* 194 (1998), 561–74; G. Roberts and T. Sherratt, "Development of Cooperative Relationships Through Increasing Investment," *Nature* 394 (1999), 175–79; J. L. Sachs, "The Evolution of Cooperation," *Quarterly Review of Biology* 79 (2004), 135–60; Martin Nowak, "Five Rules for the Evolution of Cooperation," *Science* 314 (2006), 1560–63; S. A. West, A. S. Griffin, and A. Gardner, "Social Semantics: Altruism, Cooperation, Mutualism, Strong Reciprocity and Group Selection," *Journal of Evolutionary Biology* 20 (2007), 415–32. The above are choice citations.

16. Amotz Zehavi, "Mate Selection—a Selection for a Handicap," *Journal of Theoretical Biology*, 53, 1975, 205–14, and Amotz Zahavi and Avishag Zahavi, *The Handicap Principle: A Missing Piece of Darwin's Theory* (Oxford: Oxford University Press, 1997). See also Joseph Laporte, 'Selection for Handicaps," *Biology and Philosophy*, 16, 2001, 239–49.

17. Michael Ghiselin, *The Economy of Nature and the Evolution of Sex* (Berkeley: University of California Press, 1974), 274.

18. Richard Dawkins, "Twelve Misunderstandings of Kin Selection," *Zeitschrift für Tierpsychologie* 51 (1979), 184–200, quote on 190.
19. Helena Cronin, *The Ant and the Peacock* (Cambridge: Cambridge University Press, 1991), 265.
20. For a counterargument based on the rejection of sexual selection in favor of social selection, see Joan Roughgarden, *The Genial Gene: Deconstructing Darwinian Selfishness* (Los Angeles: University of California Press, 2009).
21. Dawkins made a point of arguing in the 1989 edition of *The Selfish Gene* that his language was not merely metaphorical. In his next book, *The Extended Phenotype* (New York: Oxford University Press, 1982), he even went so far as to argue for the boundlessness of the individual organism as due to the all-importance of the gene. For a counterargument see Eva Jablonka, "From Replicators to Heritably Varying Phenotypic Traits: The Extended Phenotype Revisited," *Biology and Philosophy* 19 (2004), 353–75.
22. Sober and Wilson, *Unto Others*; Samir Okasha, *Evolution and the Levels of Selection* (Oxford: Oxford University Press, 2006); David Sloan Wilson and Kevin M. Kniffin, "Altruism from an Evolutionary Perspective," in *Research on Altruism and Love: An Annotated Bibliography of Major Studies in Psychology, Sociology, Evolutionary Biology, and Theology*, ed. Stephen G. Post, Byron Johnson, Michael E. McCullough, and Jeffrey P. Schloss (Philadelphia: Templeton Foundation Press, 2003), 117–36. It's important to add that the exclusive role of genetic inheritance in evolution has been seriously challenged in the last decade. See in particular Eytan Avital and Eva Jablonka, *Animal Traditions: Behavioural Inheritance in Evolution* (Cambridge: Cambridge University Press, 2000), Eva Jablonka and Marion Lamb, *Evolution in Four Dimensions: Genetic, Epigenetic, Behavioral, and Symbolic Variation in the History of Life* (Cambridge, MA: MIT Press, 2005), and Scott F. Gilbert and David Epel, *Ecological Developmental Biology: Integrating Epigenetics, Medicine, and Evolution* (Sunderland: Sinauer Associates, 2008).
23. Sober and Wilson, *Unto Others*, 332.
24. Actually, the preferred term is "interactors," since "vehicles" implies being under the control of the "replicators." For a clear explanation of the interactor concept see David Hull, "Individuality and Selection," *Annual Review of Ecology and Systematics* 11 (1980), 311–32.
25. For a recent illustration of this principle see Jeffrey A. Fletcher and Michael Doebeli, "A Simple and General Explanation for the Evolution of Altruism," *Proceedings of the Royal Society of London B* 276 (2009), 13–19.
26. For a clear explanation of the errors in Maynard Smith's "Haystack Model," see Sober and Wilson, *Unto Others*, 67–71.
27. W. D. Hamilton, "Innate Social Aptitudes in Man, An Approach from Evolutionary Genetics," in *Biological Anthropology*, ed. Robin Fox (London: Malaby Press, 1975), 133–53.
28. This is a game in which a pair constitutes a group, but N-person evolutionary game theory can handle groups comprised of any number. The advantage of modeling an N-person game, as opposed to a two-person game, is that it more closely resembles what happens in nature: Whereas in a two-person game altruism always incurs a cost to the altruist, in a many-player situation where many altruists are acting, altruism can benefit the group, including the altruists, which in turn means that it can more easily evolve.
29. For a firsthand description of the vagaries of group selection, see a series of online, ongoing articles at the *Huffington Post* by David Sloan Wilson titled, "Truth and Reconciliation for Group Selection," beginning December 27, 2008.

30. This is the case whether or not one sees it as a general assault on "methodological individualism," both from within biology and from other disciplines like psychology and sociology.

31. R. K. Colwell, "Group Selection Is Implicated in the Evolution of Female-Biased Sex Ratio," *Nature* 190 (1981), 401–4.

32. R. C. Lewontin, "The Units of Selection," *Annual Review of Ecology and Systematics* 1 (1970), 1–18, quote on 14–15. Lewontin, however, stopped short of implicating group selection directly.

33. G. C. Williams and R. M. Nesse, "The Dawn of Darwinian Medicine," *Quarterly Review of Biology* 66 (1991), 1–22; Randolph M. Nesse and George C. Williams, *Why We Get Sick: The New Science of Darwinian Medicine* (New York: Times Books, 1994). For more recent appreciations see Wenda R. Trevathan, E. O. Smith, and James McKenna, *Evolutionary Medicine and Health: New Perspectives* (New York: Oxford University Press, 2007), and Stephen C. Stearns and Jacob C. Koella, *Evolution in Health and Disease*, 2nd ed. (New York: Oxford University Press, 2008).

34. J. L. Brown, "Types of Group Selection," *Nature* 211 (1966), 870–71; M. J. Wade, "Group Selection Among Laboratory Populations of *Tribolium*," *Proceedings of the National Academy of Science* 73 (1976), 4604–7, and "An Experimental Study of Group Selection," *Evolution* 31 (1977), 134–53; D. S. Wilson, "The Group Selection Controversy: History and Current Status," *Annual Review of Ecology and Systematics* 14 (1983), 159–87; S. W. Rissing, G. B. Pollock, M. R. Higgins, R. H. Hagen, and D. R. Smith, "Foraging Specialization Without Relatedness or Dominance Among Co-Founding Ant Queens," *Nature* 338 (1989), 420–22; D. C. Queller, "Quantitative Genetics, Kin Selection, and Group Selection," *American Naturalist* 139 (1992), 540–58; Leticia Avilé, "Interdemic Selection and the Sex-Ration: A Social Spider Perspective," *American Naturalist* 142 (1993), 320–45; W. M. Muir, "Group Selection for Adaptation to Multiple-Hen Cages: Selection Program and Direct Responses," *Poultry Science* 75 (1995), 447–58; C. J. Goodnight and L. Stevens, "Experimental Studies of Group Selection: What Do They Tell Us About Group Selection in Nature?" *American Naturalist* 150 (1997), 59–79; P. B. Rainey and K. Rainey, "Evolution of Cooperation and Conflict in Experimental Bacterial Populations," *Nature* 425 (2003), 72–74; P. J. Werfel and Y. Bar-Yam, "The Evolution of Reproductive Restraint Through Social Communication," *Proceedings of the National Academy of Sciences of the United States of America* 101 (2004), 11019–20; Benjamin Kerr, Claudia Neuhauser, Brendan Bohannan, and Antony Dean, "Local Migration Promotes Competitive Restraint in a Host–Pathogen 'Tragedy of the Commons,'" *Nature* 442 (2006), 75–78.

35. Samir Okasha, "Why Won't the Group Selection Controversy Go Away?" *British Journal of the Philosophy of Science* 52 (2001), 25–50; Aye et Shavit, "Shifting Values Partly Explain the Debate Over Group Selection," *Studies in History and Philosophy of Biological and Biomedical Sciences* 35 (2004), 697–720, and *One for All? Facts and Values in the Debates Over Group Selection* (Jerusalem: Magnes Press, 2008) (in Hebrew).

36. David Sloan Wilson and Edward O. Wilson, "Rethinking the Theoretical Foundations of Sociobiology," *Quarterly Review of Biology* 82 (2007), 327–48. See also Bert Hölldobler and E.O. Wilson, *The Superorganism: The Beauty, Elegance, and Strangeness of Insect Societies* (New York: W. W. Norton, 2009). Wilson, in fact, was never as opposed to group selection as either Dawkins or Williams—see Segerstråle, *Defenders of the Truth*, 37–38. Hölldobler, on the other hand, is not as

convinced as Wilson of the importance of group selection. Dawkins, too, has yet to be won over, and the debate remains acrimonious. See his "The Group Delusion," from January 10, 2009, at RichardDawkins.net. Also, see Ayelet Shavit and Roberta L. Millstein, "Group Selection is Dead! Long Live Group Selection?" *BioScience* 58 (2008), 574–75. For an interesting argument for group selection based on the plasticity of circadian clocks in bees, see Guy Bloch, "Plasticity in the Circadian Clock and the Temporal Organization of Insect Societies," in *Organization of Insect Societies*, ed. Jürgen Gadau and Jennifer Fewell (Cambridge, MA: Harvard University Press, 2009), 402–31.

37. Each differs in perspective. See S. A. Frank, "George Price's Contributions to Evolutionary Genetics," *Journal of Theoretical Biology* 175 (1995), 373–88, and "The Price Equation, Fisher's Fundamental Theorem, Kin Selection, and Causal Analysis," *Evolution* 51 (1997), 1712–29; Okasha, "Why Won't the Group Selection Controversy Go Away?"; A. Grafen, "Developments of the Price Equation and Natural Selection Under Uncertainty." *Proceedings of the Royal Society London B*, 267 (2000), 1223–27, and "The First Formal Link Between the Price Equation and an Optimization Program," *Journal of Theoretical Biology* 217 (2002), 75–91; but also see I. L. Heisler and J. Damuth, "A Method for Analyzing Selection in Hierarchically Structured Populations," *American Naturalist* 130 (1987), 582–602, which provides a "contextualized" alternative to the Price equation.

38. Benjamin Kerr and Peter Godfrey-Smith, "Generalization of the Price Equation for Evolutionary Change," *Evolution* 63, no. 2 (2009), 531–36.

39. The Price equation assumes fixed heritability, and is entirely devoid of mechanism. Though Frank decomposes it neatly, others are concerned that this precludes the interactivity inherent in nature, rendering it good exclusively for border cases. See Okasha, "Why Won't the Group Selection Controversy Go Away?" for a thorough discussion of the relative merits of the equation. Also Massimo Pigliucci and Jonathan Kaplan, *Making Sense of Evolution: The Conceptual Foundations of Evolutionary Biology* (Chicago: University of Chicago Press, 2004), especially chapter 4, in which they discuss the relationship between statistical formalism and causal analysis. I thank Jim Griesemer for discussions on this point.

40. William B. Langdon, "Evolution of GP Populations: Price's Selection and Covariance Theorem," in *Genetic Programming and Data Structures*, (Norwell, MA: Kluwer Academic Press, 1998) 167–208; M. Van Veelen, "On the Use of the Price Equation," *Journal of Theoretical Biology* 237 (2005), 412–26; T. Day, "Insights from Price's Equation into Evolutionary Epidemiology," *DIMACS Series in Discrete Mathematics and Theoretical Computer Science* 71 (2006), 23–43; J. W. Fox, "Using the Price Equation to Partition the Effects of Biodiversity Loss on Ecosystem Function," *Ecology* 87 (2006), 2687–96; Stephen C. Stearns, "Are We Stalled Part Way Through a Major Evolutionary Transition from Individual to Group?" *Evolution* 61 (2007), 2275–80.

41. Darwin, *The Origin of Species*, 395.

42. Robert Trivers, *Social Evolution* (Menlo Park, CA: Benjamin/Cummings, 1985). For a defining volume on what became known as "evolutionary psychology" see Jerome H. Barkow, Leda Cosmides, and John Tooby, eds., *The Adapted Mind: Evolutionary Psychology and the Generation of Culture* (New York: Oxford University Press, 1992); Sarah F. Brosnan and Frans B. M. de Waal, "Monkeys Reject Unequal Pay," *Nature* 425 (2003), 297–99. For a recent wide treatment of animal morality see Mark Bekoff and Jessica Pierce, *Wild Justice: The Moral Lives of Animals* (Chicago: University of Chicago Press, 2009).

43. R. Boyd and P. J. Richerson, "Punishment Allows the Evolution of Cooperation (or Anything Else) in Sizable Groups," *Ethology and Sociobiology* 13 (1992), 171–95; E. Fehr and S. Gachter, "Altruistic Punishment in Humans," *Nature* 415 (2002), 137–40; R. Boyd, H. Gintis, S. Bowles, and P. J. Richerson, "The Evolution of Altruistic Punishment," *Proceedings of the National Academy of Sciences of the United States of America* 100 (2003), 3531–35; E. Fehr and U. Fischbacher, "The Nature of Human Altruism," *Nature* 425 (2003), 785–91, are challenged by A. Dreber, D. G. Rand, D. Fudenberg, and M. A. Nowak, "Winners Don't Punish," *Nature* 452 (2008), 348–51.

44. See George Loewenstein, Scott Rick, and Jonathan D. Cohen, "Neuroeconomics," *Annual Review of Psychology* 59 (2007), 647–72 fo examples. Also, for a window into "positive evolutionary psychology," see Dacher Keltner's *Born to Be Good: The Science of a Meaningful Life* (New York: W. W. Norton, 2009), as well as new findings on cooperative infant behavior in Michael Tomasello, *Why We Cooperate* (Cambridge, MA: MIT Press, 2009).

45. Robert H. Frank, *Passions Without Reason: The Strategic Role of the Emotions* (New York: W. W. Norton, 1988). For an update of this argument from the field, see Frans B. M. de Waal, "Putting the Altruism Back into Altruism: The Evolution of Empathy," *Annual Review of Psychology* 59 (2008), 279–300. For an update from neurophysiology see the work of Joshua D. Green et al., "An fMRI Investigation of Emotional Engagement in Moral Judgment," *Science* 293 (2001), 2105–8, and "The Neural Bases of Cognitive Conflict and Control in Moral Judgment," *Neuron* 44 (2004), 389–400, showing how emotion and moral reasoning relate in the brain. For a recent study on the importance of imitation as an empathy-building mechanism see A. Paukner, S. J. Suomi, E. Visalberghi, and P. F. Ferrari, "Capuchin Monkeys Display Affiliation Toward Humans Who Imitate Them," *Science* 325 (2009), 880–83.

46. Robert Trivers, "The Elements of a Scientific Theory of Self-Deception." *Proceedings of the New York Academy of Science* 187 (1999), 111–26.

47. C. Daniel Batson and Laura L. Shaw, "Evidence for Altruism: Toward a Pluralism of Prosocial Motives," *Psychology Inquiry* 2 (1991), 107–22; Lise Wallach and Michael A. Wallach, "Why Altruism, Even Though It Exists, Cannot Be Demonstrated by Social Psychological Experiments," *Psychological Inquiry* 2 (1991), 153–55; Sober and Wilson, *Unto Others*, presents the case for pluralism.

48. Charles Darwin, *The Descent of Man, and Selection in Relation to Sex*, selections and commentary by Carl Zimmer (New York: Plume, 2007), 157.

49. Peter J. Richerson and Robert Boyd, *Not By Genes Alone: How Culture Transformed Human Evolution* (Chicago: University of Chicago Press, 2005).

50. Trivers did not mean this as a criticism, but rather as a reflection of how others had viewed it. See Hamilton, *Narrow Roads*, 316; for an update of Hamilton's argument, see Samuel Bowles, "Group Competition, Reproductive Leveling, and the Evolution of Human Altruism," *Science* 314 (2006), 569. Also J. Heirich, "Cultural Group Selection, Co-evolutionary Processes and Large Scale Cooperation," *Journal of Economic Behavior and Organization* 53 (2004), 3–35.

51. Claudia Rutte and Michael Taborsky, "General Reciprocity in Rats," *PloS Biology* 5 (2007), 196.

52. However, see David Sloan Wilson, *Darwin's Cathedral: Evolution, Religion, and the Nature of Society* (Chicago: University of Chicago Press, 2002), for an argument showing religion functioning as a positive unifying system. From a neuropsychological angle, see also Jonathan Haidt, *The Happiness Hypothesis: Finding the Truth in Ancient Wisdom* (New York Basic Books, 2005).

53. Peter Singer, *The Expanding Circle: Ethics and Sociobiology* (New York: Farrar, Straus & Giroux, 1981); Christopher Boehm, *Hierarchy in the Forest: The Evolution of Egalitarian Behavior* (Cambridge, MA: Harvard University Press, 1999), argues that we are laden with two conflicting evolutionary substrates: the foundational animal substrate reflecting hierarchy and status, and the overlaying cultural substrate allowing for genuine altruism. He finds this deeply embedded ambiguity in human nature both worrisome and hopeful. See also Stearns, "Are We Stalled Part Way Through a Major Evolutionary Transition from Individual to Group?" for an argument that humanity is stalled between individual and group selection.

54. J. Moll, F. Krueger, R. Zahn, M. Pardini, R. de Oliveira-Souza, and J. Grafman, "Human Fronto-mesolimbic Networks Guide Decisions About Charitable Donation," *Proceedings of the National Academy of Sciences* 103 (2006), 15623–28.

55. Marc Hauser, *Moral Minds: How Nature Designed Our Universal Sense of Right and Wrong* (New York: Ecco, 2006); S. Anderson, H, Damasio, D. Tranel, and A. R. Damasio, "Impairment of Social and Moral Behavior Related to Early Damage in Human Prefrontal Cortex," *Nature Neuroscience* 2 (1999), 1032–37. See also Marco Iacoboni, "Imitation, Empathy and Mirror Neurons," *Annual Review of Psychology* 60 (2009), 653–70 for a presentation of the ways in which recent discoveries on the action of "mirror-neurons" complement cognitive models of imitation and social psychology studies on empathy.

56. Zoe R. Donaldson and Larry J. Young, "Oxytocin, Vasopressin, and the Neurogenetics of Sociality," *Science* 332 (2008), 900–904; S. Israel, E. Lerer, I. Shalev, F. Uzefovsky, M. Riebold, E. Laiba, R. Bachner-Melman, A. Maril, G. Bornstein, A. Knafo, and R. Ebstein, "The Oxytocin Receptor (OXTR) Contributes to Prosocial Fund Allocations in the Dictator Game and the Social Value Orientations Task," *PloS ONE* 4 (2009), e5535.

57. "Academic Appeals to Oxbridge Students for an Egg Donor," *Daily Telegraph*, December 4, 2008.

CHAPTER 14: LAST DAYS

1. Wates, *The Battle for Tolmers Square*, 9, 64; G. A. Hobbins, *St. Pancras Chronicle*, March 24, 1972, quoted in ibid., 65.

2. George Price letter to Kathleen Price, August 28, 1974, GPP.

3. Communications with Sylvia Stevens, February 15 and March 18, 2008, May 11, 2009.

4. On Edison Price see Stanley Abercrombie, "Edison Price: His Name Is No Accident," *Architecture Plus* 1 (August 1973); Jonathan Glancey, "The Century's Leading Light," *The Guardian*, October 25, 1997; Robert M. Thomas, Jr., "Edison A. Price, Lighting Designer, Dies at 79," *New York Times*, October 17, 1997.

5. Communication with Sylvia Stevens, March 18, 2008.

6. Henry Noel letter to George Price, August 7, 1974, GPP.

7. George Price letter to Bill Hamilton, August 21, 1974, BLGPC, KPX1_2.5.

8. George Price letter to Miss Sara Smith, August 27, 1974, BLGPC, BL:KPX1_1.8.

9. Ibid.

10. George Price letter to Kathleen Price, August 28, 1974, GPP.

11. George Price letter to Richard Lewontin, May 17, 1969 and Lewontin's response, June 16, 1969. I thank Richard Lewontin for kindly sending me their correspondence.

12. Richard Lewontin letter to George Price, March 13, 1974
13. James Crow letter to C. A. B. Smith, October 24, 1974. I thank Jim Crow for kindly sending me this correspondence.
14. C. A. B. Smith letter to James Crow, November 1, 1974.
15. Ibid.
16. George Price letter to Richard Lewontin, June 13, 1974.
17. Richard Lewontin letter to George Price, August 27, 197-.
18. George Price letter to Joan Jenkins, GPP, June 28, 1974.
19. Communication with Sylvia Stevens, March 18, 2008.
20. George Price letter to Joan Jenkins, October 7, 1974; Jim Schwartz notes.
21. Joan Jenkins letter to George Price, GPP, October 16, 1974; Lewis Florman letter to George Price, October 11, 1974, GPP.
22. George Price letter to Trevor Russell, November 5, 1974, GPP.
23. Interview with Shmulik Atia, May 4, 2008; interview with Asher Dahan, May 7, 2008; George Price letter to Bill Hamilton, November 20, 1974, GPP.
24. George was referring to Dr. Ronald Ng, whom he had met some months back in the computer room at UCL: George Price letter to Dr. Ronald Ng, November 26, 1974, GPP.
25. George Price letter to Edison Price, November 25, 1974, BLWHC, Z1X102_1.1.18; drafts of the letters, GPP.
26. Interviews with Atia and Dahan.
27. George Price letter to Kathleen Price, November 5, 1974, GPP.
28. For two rather interesting appreciations of Thompson see Caoimghghin S. Breathnach, "Francis Thompson (1859–1907): A Medical Truant and His Troubled Heart," *Journal of Medical Biography* 16, no. 1 (2008), 57–62, and "Francis Thompson—Student, Addict, Poet," *Journal of the Irish Medical Association* 45 (1959), 98–103.
29. J. R. R. Tolkien, *The Book of Lost Tales*, part 1, ed. Christopher Tolkien (Boston: Houghton Mifflin Company, 1984), 29.
30. Bernardo and Chrissy letter and note to George Price, November 30 and 31, 1974, GPP.
31. Dr. O. W. Hill letter to George Price, December 2, 1974, GPP. George Price letter to Bill and Chris Hamilton, December 10, 1974, BLWHC, Z1X102_1.1.19.
32. Communication with Chris Hamilton, November 18, 2007; Bill Hamilton letter to Edison Price, February 15, 1975, BLWHC, Z1X102_1.1.20; Bill Hamilton letter to Al Somit, October 16, 1996, BLWHC, Z1X102_1.2.2.2.
33. The paper was eventually published as "Innate Social Aptitude in Man: An Approach from Evolutionary Genetics," in *Biosocial Anthropology*, ed. Robin Fox (London: Malaby Press, 1975), and is discussed and printed, respectively, in Hamilton, *Narrow Roads*, 315–28 and 329–51. I discuss the paper in chapter 13.
34. Bill Hamilton letter to Dr. Kelly, undated (circa December 19, 1974), BLWHC, Z1X102_1.1.18.

EPILOGUE

1. Interview with Shmulik Atia, May 4, 2008.
2. Ibid.; interview with Asher Dahan, May 7, 2008.
3. Handwritten notes taken by Hamilton at the inquest, BLWHC, Z1X102_1.2.3.2.
4. Ibid.; draft of a letter to Sylvia found in George's papers, GPP.

5. "Jesus Hot Line," *Sennet*, January 15, 1975.

6. Bill Hamilton letter to Edison Price, February 15, 1975, BLWHC, Z1X102_1.1.20. The American Consulate sent what it collected to Kathleen; what Hamilton found he sent to George's brother, Edison.

7. Hamilton, *Narrow Roads*, 174; George had been working on methods of programming and using a significance test suggested by Warren Ewens in 1972, and applying them to data on enzyme polymorphisms collected at the Galton. "Test of the Neutrality Hypothesis," written on behalf of George R. Price, came out in the *Annals of Human Genetics* (care of C. A. B. Smith) 39 (1976), 471–73. Years later, Steven Frank was responsible for bringing "The Nature of Selection" to print in the *Journal of Theoretical Biology* 175 (1995), 389–96.

8. Bill Hamilton letter to Edison Price, February 15, 1975; Hamilton, *Narrow Roads*, 321.

9. The posters and pamphlets appear in Wates, *The Battle for Tolmers Square*, 178–79.

10. A full description of the court proceedings and subsequent events can be found in ibid., 180–88.

11. Ibid., 183; "A Square Deal for the People," *Hampstead and Highgate Express*, June 6, 1975.

12. Ludwig Wittgenstein, *Tractatus Logico-Philosophicus*, trans. D. F. Pears and B. F. McGuinness (London: Routledge, 1974), 6.52, 7.

13. Frans de Waal, *Primates and Philosophers: How Morality Evolved* (Princeton: Princeton University Press, 2006), 163. See also his new book, *The Age of Empathy: Nature's Lessons for a Kinder Society* (New York: Harmony, 2009).

14. In fact it is very likely that the new, post-neo-Darwinian umbrella paradigm, being called the "Epigenetic Turn," which includes all forms of genetic, epigenetic, behavioral, and symbolic inheritance and their interaction with the broader environment, will necessitate a meaningful reworking of the foundations of our understanding of the evolution of human behavior.

15. Adam Smith, *The Theory of Moral Sentiments* (1759; reprint, New York: Modern Library, 1937), 9.

16. Hamilton, too, might have benefited from this insight. An uncompromising rationalist, he came to believe that there is nothing interesting that man cannot hope to understand, including his own morality. This led him, among other things, to support eugenic measures that seemed to many outrageous.

17. See Jonathan Haidt's essay "Moral Psychology and the Misunderstanding of Religion," posted at *The Edge*, September 22, 2007, for an interesting argument about religious/conservative people's usage of an expanded moral-salience plate (compared to liberals) to direct their behavior.

18. Since the Price equation is a statistical, not a causal, decomposition of selection, there is an interesting reflection of this problem in the very bones of the math itself. In the end the Price equation can tell you that a trait evolved, but why it evolved remains elusive.

19. Bill Hamilton letter to Kathleen Price, June 25, 1997, BLWHC, BL:Z1X101_1.2.2.3.

20. "Love is the Greatest!" found in George's scattered papers, GPP.

APPENDIX 1: COVARIANCE AND KIN SELECTION

1. Here I follow S. A. Frank, "George Price's Contributions to Evolutionary Genetics," *Journal of Theoretical Biology* 175 (1995), 374–75.

2. D. C. Queller, "Kinship, Reciprocity and Synergism in the Evolution of Social Behaviour," *Nature* 318 (1985), 366–67. See also his "A General Model for Kin Selection," *Evolution* 46 (1992), 376–80.
3. W. D. Hamilton, "Selfish and Spiteful Behaviour in an Evolutionary Model," *Nature* 228 (1970), 1218–20.

APPENDIX 2: THE FULL PRICE EQUATION AND LEVELS OF SELECTION

1. Once again, I follow Frank, "George Price's Contributions to Evolutionary Genetics," 375–79. Samir Okasha, *Evolution and the Levels of Selection* (Oxford: Oxford University Press, 2006) provides a full, fleshed-out interpretation for those interested.
2. A. Robertson, "A Mathematical Model for the Culling Process in Dairy Cattle," *Animal Production* 8 (1966), 95–108; C C. Li, "Fundamental Theorem of Natural Selection," *Nature* 214 (1967), 505–5.

Index

Page numbers in *italics* refer to illustrations.